Recent Titles in This Series

132 **Hajime Urakawa,** Calculus of variations and harmonic maps, 1993
131 **V. V. Sharko,** Functions on manifolds: Algebraic and topological aspects, 1993
130 **V. V. Vershinin,** Cobordisms and spectral sequences, 1993
129 **Mitsuo Morimoto,** An introduction to Sato's hyperfunctions, 1993
128 **V. P. Orevkov,** Complexity of proofs and their transformations in axiomatic theories, 1993
127 **F. L. Zak,** Tangents and secants of algebraic varieties, 1993
126 **M. L. Agranovskiĭ,** Invariant function spaces on homogeneous manifolds of Lie groups and applications, 1993
125 **Masayoshi Nagata,** Theory of commutative fields, 1993
124 **Masahisa Adachi,** Embeddings and immersions, 1993
123 **M. A. Akivis and B. A. Rosenfeld,** Élie Cartan (1869–1951), 1993
122 **Zhang Guan-Hou,** Theory of entire and meromorphic functions: Deficient and asymptotic values and singular directions, 1993
121 **I. B. Fesenko and S. V. Vostokov,** Local fields and their extensions: A constructive approach, 1993
120 **Takeyuki Hida and Masuyuki Hitsuda,** Gaussian processes, 1993
119 **M. V. Karasev and V. P. Maslov,** Nonlinear Poisson brackets. Geometry and quantization, 1993
118 **Kenkichi Iwasawa,** Algebraic functions, 1993
117 **Boris Zilber,** Uncountably categorical theories, 1993
116 **G. M. Fel'dman,** Arithmetic of probability distributions, and characterization problems on abelian groups, 1993
115 **Nikolai V. Ivanov,** Subgroups of Teichmüller modular groups, 1992
114 **Seizô Itô,** Diffusion equations, 1992
113 **Michail Zhitomirskiĭ,** Typical singularities of differential 1-forms and Pfaffian equations, 1992
112 **S. A. Lomov,** Introduction to the general theory of singular perturbations, 1992
111 **Simon Gindikin,** Tube domains and the Cauchy problem, 1992
110 **B. V. Shabat,** Introduction to complex analysis Part II. Functions of several variables, 1992
109 **Isao Miyadera,** Nonlinear semigroups, 1992
108 **Takeo Yokonuma,** Tensor spaces and exterior algebra, 1992
107 **B. M. Makarov, M. G. Goluzina, A. A. Lodkin, and A. N. Podkorytov,** Selected problems in real analysis, 1992
106 **G.-C. Wen,** Conformal mappings and boundary value problems, 1992
105 **D. R. Yafaev,** Mathematical scattering theory: General theory, 1992
104 **R. L. Dobrushin, R. Kotecký, and S. Shlosman,** Wulff construction: A global shape from local interaction, 1992
103 **A. K. Tsikh,** Multidimensional residues and their applications, 1992
102 **A. M. Il'in,** Matching of asymptotic expansions of solutions of boundary value problems, 1992
101 **Zhang Zhi-fen, Ding Tong-ren, Huang Wen-zao, and Dong Zhen-xi,** Qualitative theory of differential equations, 1992
100 **V. L. Popov,** Groups, generators, syzygies, and orbits in invariant theory, 1992
99 **Norio Shimakura,** Partial differential operators of elliptic type, 1992
98 **V. A. Vassiliev,** Complements of discriminants of smooth maps: Topology and applications, 1992

(*Continued in the back of this publication*)

Translations of

MATHEMATICAL MONOGRAPHS

Volume 100

Groups, Generators, Syzygies, and Orbits in Invariant Theory

V. L. Popov

American Mathematical Society
Providence, Rhode Island

В. Л. ПОПОВ

ГРУППЫ, ОБРАЗУЮЩИЕ,
СИЗИГИИ И ОРБИТЫ
В ТЕОРИИ ИНВАРИАНТОВ

Translated from the Russian by A. Martsinkovsky
Translated edited by Sergei Gelfand

2000 *Mathematics Subject Classification.* Primary 14–XX, 14L30;
Secondary 14J10, 20G20.

ABSTRACT. The book contains a detailed and complete account of recent development in invariant theory for the action of reductive groups on algebraic varieties. The author presents general results related to the 14th Hilbert problem and its generalizations. Among these results are the solution of the fundamental problem of the constructive invariant theory for connected semisimple Lie groups, the proof of the finiteness theorem for finite-dimensional rational modules of the given homological dimension, and classification of G-modules with the free algebra of invariants.

Based on general results, certain classification theorems are proved for homogeneous and quasi-homogeneous varieties of small dimension (curves, surfaces, threefolds).

Library of Congress Cataloging-in-Publication Data

Popov, V. L. (Vladimir Leonidovich)
[Gruppy, obrazuiushchie, sizigii i orbity v teorii invariantov. English]
Groups, generators, syzygies, and orbits in invariant theory/V. L. Popov.
p. cm.—(Translations of mathematical monographs, ISSN 0065-9282; v. 100)
Translation of: Gruppy, obrazuiushchie, sizigii i orbity v teorii invariantov.
Includes bibliographical references and index.
ISBN 0-8218-4557-8
1. Invariants. I. Title. II. Series.
QA201.P6613 1992 92-10604
512′.944–dc20 CIP

AMS softcover ISBN 978-0-8218-5335-1

This volume was typeset using $\mathcal{A}\mathcal{M}\mathcal{S}$-TEX,
the American Mathematical Society's TEX macro system.
Visit the AMS home page at `http://www.ams.org/`

10 9 8 7 6 5 4 3 2 1 16 15 14 13 12 11

Contents

Introduction

1. Invariant theory has a history that is almost one-and-a-half centuries long. It owes its existence to certain problems from number theory, algebra, and geometry, which appeared in the works of Gauss, Jacobi, Eisenstein, and Hermite. By the middle of the 19th century it had evolved into an algebraic discipline in its own right through the efforts of mathematicians of the following three schools: British (Boole, Cayley, Sylvester, and Salmon), German (Hesse, Aronhold, Clebsch, and Gordan), and Italian (Brioschi, Cremona, and Beltrami).

The point of view on the fundamental problems of invariant theory and the methods of their solution changed several times. There is no doubt, however, that invariants came into existence as a tool to distinguish (and, ultimately, to classify) nonequivalent objects in the algebraic problems, where the equivalence relation is given by the orbits of a group G acting on a set X. On the other hand, invariant theory noticeably influenced the theories that were connected (in one context or another) to classification of orbits of a group action (e.g., [92, 93, 114, 103, 117]).

Although the connection between invariants and orbits was essentially discovered in the works of Aronhold [54] and Boole [59] (despite the fact that the term "orbit" did not exist at that time), a clear understanding of this connection had not been achieved until recently, when invariant theory was in fact subsumed by a general theory of algebraic groups of transformations.

In the classical setup, the set X is a finite-dimensional vector space V over a field k (of real or complex numbers), and G is a finite group of linear transformations of V. The fundamental problem is to describe the structure of the subalgebra of invariants $k[V]^G$ of the algebra $k[V]$ of all polynomials on V. In the 19th century mostly invariants of forms were considered (i.e., V was the space of degree d forms in n variables and G was the group of all unimodular linear transformations of these variables), with the case of binary forms (i.e., $n = 2$) given special attention.

In the 1930s H. Weyl [11], prompted by developments in the theory of Lie groups and their representations, went further by letting G be any group and defining the action via an arbitrary linear representation $G \to \mathrm{GL}(V)$.

In particular, the invariants of an arbitrary system of tensors, and not only invariants and covariants of systems of forms, drew attention. In this setup it suffices to consider only algebraic groups G (see [120]). Finally, the developments in algebraic geometry and the theory of algebraic groups and their representations in the second half of this century brought about the problem of studying the algebra $k[X]^G$ of invariant regular functions on an affine algebraic variety X on which a linear algebraic group G acts regularly. This point of view was first clearly formulated by Mumford in his book [114].

2. From the point of view of classification of orbits, the algebra $k[V]^G$ (compared, say, with the algebra of all functions constant on orbits) has the drawback that it does not necessarily separate any two orbits. However, it has the substantial advantage that in many important cases it admits a "finite" description in the sense that it is finitely generated (in this case one says that "the first fundamental theorem of invariant theory" is true). It should be remarked that (as was shown by Luna [109]) by naturally enlarging $k[X]^G$ (say, to the class of invariant holomorphic functions) we do not obtain any new information about orbits.

Over the years the role of "the fundamental theorem" in invariant theory changed several times. From the classical point of view its proof was an important part of the general problem. It was considered as a direct, or possible in principle, determination of a minimal system of generators of $k[X]^G$ ("a complete system of invariants"). Nowadays this theorem for reductive groups (i.e., groups with a trivial unipotent radical) has become one of the instruments in classification of closed orbits of G on X: affine variety $X/G = \operatorname{Specm} k[X]^G$ parametrizes, via the natural morphism $\pi_{X,G}\colon X \to X/G$, all such orbits. This result is the basis for applications of invariant theory to problems of moduli spaces, i.e., the subtle questions of construction of quotient structures for families of algebro-geometric objects like curves, surfaces, abelian varieties, and such (see [114, 117, 115, 116, 73]).

The first fundamental theorem was proved by Gordan in 1868 for invariants of binary forms [74] (and in 1870 for systems of binary forms [75]). In 1890 Hilbert gave a general nonconstructive proof for any system of forms in any number of variables [80]. Although in his version he deals with special representations of GL_n, the idea of the proof can be carried over directly to the general case. H. Weyl [11] accomplished this for any representation of a complex semisimple group, and Mumford [114] extended this result to the general case of a reductive algebraic group acting regularly on an affine algebraic variety over a field k of characteristic zero. Finally, through the joint efforts of Nagata [119], Haboush [78], and Mumford [114], the restriction on the characteristic of k was removed (in the case $\operatorname{char} k \neq 0$ Hilbert's approach does not work). In this final form the theorem is now called Hilbert's theorem on invariants.

The aforementioned results (along with the wealth of concrete information accumulated during the last century and a half) pushed reductive groups to the forefront of invariant theory. It is this class of groups for which we now have a rich geometric invariant theory, whereas for no other class of groups does such a theory exist. The question about the role of group G in the fundamental problem of invariant theory already appeared in the 19th century and is known as Hilbert's 14th problem (see [82]): for which subgroups G of $\mathrm{GL}(V)$ is the algebra $k[V]^G$ finitely generated? In 1958 Nagata (see [118]) constructed an example in which $k[V]^G$ was not finitely generated and also considered a more general question called "the generalized 14th problem" (see [120]): when is the algebra of invariants of a linear algebraic group G acting by automorphisms of an affine algebraic variety X finitely generated? Hilbert's theorem on invariants asserts that this is the case when G is reductive.

3. Being nonconstructive, Hilbert's proof of the first fundamental theorem does not provide us with an algorithm for finding a complete system of invariants. Nowadays, however, the geometrization of invariant theory not only has not retreated from the classical point of view on invariant theory as a constructive theory, but it has enriched it even further because an explicit description of the algebra of invariants is usually followed by a meaningful geometric interpretation either in the form of a direct application to geometry (in the spirit of Klein's Erlangen program [48, 100]), as a description of the corresponding moduli variety [114, 117, 48, 143, 141], or as examples of varieties with remarkable properties (e.g., X/G, or the fibers of $\pi_{X,G}$, often have interesting singularities [145, 105, 63, 62, 137]). In certain situations the algebra of invariants helps to clarify the structure of other algebras (e.g., the cohomology algebras of homogeneous spaces [60, 83] and the algebras of modular forms [86]). Finally, an explicit description of the algebra of invariants is important for applications in coding theory and combinatorics [150, 149, 144, 83] and some special problems in differential equations [47].

The determination of a complete system of invariants is usually a rather complicated problem (for degree d binary forms the classical papers explicitly described the solutions only for $d \leq 6$ (see [48])). The classical results [48, 76, 143, 132, 11, 21, 64, 134], along with those obtained by powerful modern techniques [13, 95, 108, 110, 111], allow us to single out whole classes of G-modules of reductive groups G for which the problem is solved (see [2, 19, 64, 66, 94, 135, 149, 156]). However, the success in each particular case was conditioned by some special properties of the modules under consideration.

In its general form the problem of constructive invariant theory for a given reductive group $G \subset \mathrm{GL}(V)$ is to find an algorithm that would allow us to determine in principle a minimal system of generators of $k[V]^G$ by means of a finite number of explicitly performable operations. Because the spaces of homogeneous elements in $k[V]^G$ of any given degree can be explicitly described (via the "symbolic method" developed in classical papers; in the general case the answer can be obtained by methods of representation theory (see Chapter 2)), the problem of the existence of (and finding) such an algorithm reduces to finding an a priori upper bound M on the degrees of the elements of a complete system of invariants.

For a finite group G such a bound was obtained in 1915 by E. Noether who showed that one can take $M = |G|$ (see [121]). As H. Weyl [11] said "one cannot ask for anything more specific." The case of a connected group (over an algebraically closed field k with $\operatorname{char} k = 0$) is developed to a much lesser degree. It was treated (aside from Gordan's papers [74, 75] on binary forms) only by Hilbert [81] in 1893 and (following his ideas) by Dieudonné and Carrell [21] (in both works G is the image of SL_m under a polynomial representation).

Hilbert wrote his paper as a response to the criticism (from Gordan, Poincaré, and others) of the nonconstructive nature of his proof of the first fundamental theorem. He considers two similar approaches to the problem. The first one reduces the problem to the "solution of an elementary problem from arithmetic theory of algebraic functions" [81], or, more precisely, to Kronecker's method for construction of the integral closure of a finitely-generated integral domain in its field of fractions. The second method is a modification of the first and, by using some special properties of the algebras and fields under consideration, it allows us to avoid the reference to Kronecker's result. Hilbert was trying to avoid Kronecker's method because this method only gives a general strategy (by reducing the problem in each particular case to finding an estimate on the degrees of certain equations, calculation of determinants, etc.) and leaves the problem of the explicit description of M unsolved. In fact, his second approach has the same drawback (in modern language, the result of [81] was that for the groups and representations of the type in question only an explicit bound on the degrees of the system of homogeneous parameters of the algebra $k[V]^G$ was found, but not the constant M). Apparently these considerations were the reason that the authors of [21], trying to save Hilbert's general idea, changed the final part of his approach by replacing the description of integral elements (in one form or another) with a reference to the Nullstellensatz. However, they did not find the constant M. In fact, as shown in Chapter 2 of this book, the approach to the solution of the problem of constructive invariant theory proposed by Dieudonné and Carrell in [21] is flawed and does not lead to the goal even theoretically.

4. A description of a complete system of invariants for a G-module V is only one part of a method developed in the previous century for describing the algebra $k[V]^G$. It also consists of finding basis relations between basis invariants (i.e., first syzygies), then relations between those relations (second syzygies), etc. In modern language, this means constructing the corresponding free resolution for $k[V]^G$; the kernels of the homomorphisms in this resolution are precisely the syzygy modules, and the chain of these kernels is the chain of syzygies. The classics called the assertion about the finiteness of the first syzygies (interpreted as their explicit description) "the second fundamental theorem of invariant theory." Hilbert proved in [81] that the first fundamental theorem implies the finiteness of all syzygies (whence the second theorem) as well as the finiteness of the syzygy chain itself.

The problem of describing all the syzygies for a fixed G and various G-modules V is extremely difficult. It has been solved only in certain special cases. In the second half of the 19th century it was completely solved for invariants of binary forms of degree ≤ 6 (and partially for degree ≤ 8), covariants of binary forms of degree ≤ 4 (partially for degree ≤ 8), and certain very simple cases of invariants and covariants of a system of forms [76, 48]. Only recently have the corresponding resolutions been constructed, by Shioda [143] for invariants of a binary form of degree 8 and by Lascoux [107] for invariants of a system of vectors in n-space. Nowadays, because of the developments in the theory of Schur functors [157], the problem of constructing free resolutions for algebras of invariants draws much more attention. In particular, the results of Lascoux were extended to orthogonal and symplectic groups [90] (a constructive proof of the second fundamental theorem for classical groups and representations was obtained much earlier (see [11])). These and some other very special results [144, 149], in spite of their importance, are of a sporadic nature. Only one class of linear groups, the groups with no syzygies (i.e., $k[V]^G$ is free), was subjected to systematic investigation. At the present time there are substantial gains in this area (in the case of an algebraically closed field k of characteristic zero). The results of Chevalley [66] and Shepard and Todd [142] give a classification of the finite groups of this type. In recent years there has been significant progress for connected groups: Kač, Popov, and Vinberg [94] found all such irreducible simple groups. Then, using the same method as in [94], Schwarz [135], and independently, Adamovich and Golovina [2] also found all such reducible simple groups. A posteriori these results imply that each finite or connected simple group G has only a finite number of G-modules V without nonzero fixed vectors with free algebra of invariants $k[V]^G$.

The reason that for a fixed G one can explicitly describe the syzygy chains only for few G-modules V is that in the "general case" the algebra $k[V]^G$ is very complicated. It was already realized by the classical researchers; however, until recently this assertion was based on empirical evidence and its precise meaning was not clarified.

A natural measure of the complexity of the algebra $k[V]^G$ is the length of its syzygy chain. In modern terminology this number is known as the homological dimension $\operatorname{hd} k[V]^G$ of $k[V]^G$. In the overwhelming majority of cases, when the resolution can be explicitly constructed, $\operatorname{hd} k[V]^G$ happens to be small (≤ 3 for invariants of binary forms, ≤ 1 for covariants, and 0 for a free algebra $k[V]^G$), which explains the success. The study of $\operatorname{hd} k[V]^G$ is the basis for a systematic approach to the syzygy problem. It allows us to single out classes of G-modules with algebras of invariants of the "same complexity." One can hope that for some values of $\operatorname{hd} k[V]^G$ (say, not very large) these classes can be completely described (as they were for finite or connected simple groups with $\operatorname{hd} k[V]^G = 0$). This approach is in the spirit of the invariant theory of the last decade, which resulted in realization of the fact that there exist classes of G-modules with some "good" properties and these classes can be completely (or almost completely) and explicitly described (see [94, 1, 37, 135, 136, 2, 92, 36, 23, 91]). The modules outside these classes seem to be "wild" in some sense. No effort has been made, however, to study systematically homological dimension of algebras of invariants. Such an effort would solidify a basis for the aforementioned program.

5. Besides the study of syzygies there is another approach to the solution of the fundamental problem of invariant theory: describing the algebra $k[V]^G$. It is given (in the case when G is reductive and $\operatorname{char} k = 0$) by the recently proved Hochster–Roberts theorem [85]. According to this theorem, for any system of homogeneous parameters $a_1, \dots, a_m$ of the algebra $k[V]^G$, this algebra is a free $k[a_1, \dots, a_m]$-module, and, therefore, $k[V]^G = k[a_1, \dots, a_m]b_1 \oplus \cdots \oplus k[a_1, \dots, a_m]b_l$ for some homogeneous elements $b_i \in k[V]^G$. Because the structure of $k[a_1, \dots, a_m]$ is very simple (it is just the algebra of polynomials), the structure of $k[V]^G$ is completely determined by the "multiplication table" of the elements $b_1, \dots, b_l$. Moreover, this decomposition yields a unique representation of each invariant as a linear combination of $b_1, \dots, b_l$ with coefficients from $k[a_1, \dots, a_m]$. In other words, it gives a "canonical form." In the case of a finite G, Dade gave an algorithm for constructing a system of homogeneous parameters and found an upper bound for their degrees (see [149]). For connected groups no such bounds are currently known (one of Hilbert's key results in [81] was a bound for homogeneous polynomial representations of GL_s; Kempf [97] characterizes this problem in the general case as "most difficult").

6. Invariants is not the only gadget for separating orbits produced by the classical invariant theory. Let G be reductive and $\operatorname{char} k = 0$. Any invariant $f \in k[V]^G$ can be considered an equivariant morphism $V \to k$ under the assumption that G acts on k trivially. Replacing k with arbitrary G-module W we obtain a covariant, a generalization of the notion of an invariant. The existence of a covariant is equivalent to the existence of a G-module homomorphism $W^* \to k[V]$. If W is irreducible, then this homomorphism is an embedding of W^* in the (finite-dimensional) G-module $k[V]$, which is a direct sum of its finite-dimensional irreducible submodules. Thus the question about the structure of covariants reduces to the question about the G-module structure of $k[V]$ or, in other words, about the $k[V]^G$-module structure of $k[V]$. This module is called the module of covariants of the G-module V (in classical papers these structures were described using the so-called algebra of covariants, i.e., the algebra of highest vectors in $k[V]$). In the especially simple and important case when the module of covariants is free, we have $k[V] = H \otimes_k k[V]^G$ where H is the space of "harmonic polynomials" [101]; moreover, the multiplicity of any finite-dimensional irreducible G-module in H can be easily computed via the stabilizer of the generic closed orbit (see [101, 136]). The class of G-modules with free module of covariants is not only interesting from the point of view of representation theory. It also has remarkable geometric properties: e.g., Slodowy showed in [145] that if G is connected and $V^G = \{0\}$, then the canonical morphism $\pi_{V,G} : V \to V/G$ is the semiuniversal deformation of the null-variety $\mathfrak{N} = \pi_{V,G}^{-1}(\pi_{V,G}(0))$, which sometimes exhibits very interesting singularities (see [145, 63, 105, 137, 151]) (for the adjoint modules one recovers in this way the famous connection between simple surface singularities and corresponding simple groups, discovered by Brieskorn [63]).

In the case of a finite group G the module of covariants is free if and only if the algebra of invariants is free [9], hence the results of Chevalley [66] and Shepard–Todd [142] give the solution to the classical problem. For connected groups this statement is no longer true. Kostant posed a classification problem for such modules in [101], where he showed that the module of covariants of the adjoint G-module is free. Then in [102], Kostant and Rallis showed that the same is true for the module defined by the isotropy representation of a symmetric space.

7. Let k be an algebraically closed field of characteristic zero, and let G a reductive group acting regularly on an affine algebraic variety X. As was mentioned before, invariants do not solve the problem of separating the orbits in the strict sense; they only separate the closed orbits. The study of this phenomenon in its "pure form" leads to the study of varieties X without nonconstant invariants: $k[X]^G = k$. As an example one can consider a fiber of the orbit morphism $\pi_{V,G}$. On such an X, however, there may exist nonconstant (rational) invariants, and they, according to Rosenlicht (see [131]), separate almost all orbits. There are no such invariants exactly when X contains a dense orbit. Thus one is naturally led to the study of new objects: quasihomogeneous varieties (i.e., varieties with a dense orbit). A great number of the classical varieties happen to fall into this category. Examples include affine cones over projective quadrics, Grassmannians, flag varieties, Schubert varieties, varieties of nilpotent elements in semisimple Lie algebras, and others, not to mention homogeneous spaces.

It was only recently that various classes of (not necessarily affine) quasihomogeneous varieties were subject to systematic study, see [18, 16, 99, 67,

6, 55, 123, 70, 7, 133, 122, 14, 140] and the literature quoted therein (we should also remark that quasihomogeneous varieties in the form of orbit closures played an important role already in Hilbert's paper [81] and, even earlier, in a theorem of Roberts (see [48, 134])). Moreover, invariant theory proper was not always the motivation and purpose of those investigations (e.g., they naturally appear in the study of degenerations of flat families of finite-dimensional modules over finitely-generated associative k-algebras (see [103])). The classification problem for quasihomogeneous varieties makes sense only under certain restrictions (e.g., on the dimension, smoothness, stabilizer of a generic point, complement to the dense orbit, algebro-geometric properties like being affine, projective, or complete, etc.). Mumford and his colleagues developed a theory of toroidal embeddings, which is just a theory of quasihomogeneous varieties of algebraic tori (see [99, 122] for the reference to some of the earlier ideas in this theory). Its affine variant is a particular case of the theory of affine quasihomogeneous varieties with the stabilizer of a generic point containing a maximal unipotent subgroup of G (known as S-varieties), which was developed by Popov and Vinberg in [16]. Recently this theory was generalized by Pauer [123] to not necessarily affine varieties, which yields a generalization of the theory of toroidal embeddings. In [6] Akhiezer classified quasihomogeneous normal projective varieties with two ends in the sense of Borel, and in [7] with homogeneous divisorial boundary (i.e., the complement of the dense orbit). Gizatullin [18] determined all smooth affine surfaces that are quasihomogeneous with respect to an algebraic group and have at most zero-dimensional boundary. Another area of research is the classification of quasihomogeneous G-modules (or, in other words, locally transitive representations) and the study of their connections with problems in number theory and analysis; see papers of Sato and Kimura [133], Elashvili [52, 53], Kač [23, 91], and Servedio [140]. Kimelfeld and Vinberg [14] studied quasihomogeneous varieties with the simple spectrum of the G-module of regular functions. For more problems related in one way or another to quasihomogeneous varieties, see [67, 77, 55, 70, 83, 61, 145, 155, 92, 104, 105, 69, 68] (as well as the literature quoted therein).

Having evolved naturally in the framework of invariant theory, the problem of classification of quasihomogeneous varieties in the utmost generality gave rise to a separate area of research. For quasihomogeneous affine varieties, however, it is directly related to the classification problem for orbits, the fundamental geometric problem of the traditional invariant theory; such varieties are exactly the closures of orbits of finite-dimensional rational G-modules (see Chapter 6). A purely algebraic reformulation of this problem reveals connections with no less traditional aspects of invariant theory: this problem turns out to be equivalent to the classification problem of finitely-generated left-invariant subalgebras of the algebra of regular functions $k[G]$ on group G (see Chapter 8). In this language the original Hilbert's 14th problem can be naturally reformulated (according to Grosshans [77]) as the problem of construction of quasihomogeneous affine varieties with boundary of codimension ≥ 2.

It is indicative that in the majority of the mentioned papers the tools for classification of quasihomogeneous varieties (and their orbits, in particular) are of a discrete nature. They include rational polyhedral cones, lattices, semigroups of integral points, fans (simple and colored), standard monomials, Young diagrams, highest weights, etc. This is one of the ways in which the modern invariant theory (and algebraic geometry) is connected with combinatorics.

Next we shall describe the main results discussed in this book.

8. Chapter 1 deals with the role group G plays in the solution of fundamental problems of invariant theory. The main result here is a converse to Hilbert's theorem on invariants. Together these two results completely characterize the groups for which the generalized Hilbert's 14th problem has an affirmative answer (the field k has an arbitrary characteristic):

THEOREM. *The following are equivalent*:

a) *G is reductive.*

b) *for any affine k-algebra A and any rational action of G on A by k-algebra automorphisms, the algebra of invariants A^G is also affine.*

This theorem explains why a good invariant theory is possible only for reductive groups.

In the same chapter we consider the connections between the generalized and original (in Grosshans' form [77]) Hilbert's problem and obtain a bound on the dimension of the boundary of a quasihomogeneous variety. From this we deduce a description of quasihomogeneous affine varieties X with zero-dimensional boundary; they turn out to be S-varieties in the sense of [16] (which gives, besides the description itself, some meaningful information about their algebro-geometric properties). Such varieties have important applications (for example, in algebraic geometry, see Bogomolov's paper [8]). As a consequence, one obtains a description of minimal nonzero-dimensional orbits in the closure of a nonsemistable point of a rational module of a reductive group. They turn out to be the orbits of the corresponding sums of highest vectors.

9. Chapter 2 deals with the solution of the fundamental problem of constructive invariant theory for any connected semisimple group G and any finite-dimensional rational module V over G (we assume that char $k = 0$).

First, using a new approach, we show how one can eliminate the ambiguity stemming from Kronecker's method and its modifications (as was mentioned, the methods of [81] and [21] also lead to this ambiguity). Second, we offer a method for the solution of this problem, which works for an arbitrary semisimple group G and any of its representations (not only for homogeneous polynomial representations of SL_m). It also explains the somewhat surprising (and, at first glance, accidental) role that the function det played in Hilbert's theory [81]. This way we obtain an explicit upper bound M on the degrees of a minimal system of homogeneous generators of $k[V]^G$ (we assume that $G \subset \mathrm{GL}(V)$). Namely, if $n = \dim V$, $s = \dim G$, $r = \operatorname{rk} G$, and m is the transcendence degree of $k[V]^G$ over k (which easily can be found either by geometric methods or with the aid of the Poincaré series, (see Chapter 2)) then we show that M can be taken to be

$$m \cdot \mathrm{LCM}\left(a \in \mathbb{Z},\ 0 < a \leq \frac{2^{r+s} n^{s+1} (n-1)^{s-r} t^r (s+1)!}{3^s ((\frac{s-r}{2})!)^2} \right) \tag{1}$$

(the number t here is determined by the choice of a maximal torus of G (see precise definition in Chapter 2); for example, if G is the image of SL_d under the usual representation in the space of degree p forms in d variables, then one can take $t = p$).

In Chapter 2 we also obtain an explicit upper bound on the degrees of a system of homogeneous parameters $a_1, \dots, a_m$ of $k[V]^G$ (we elaborate on that in Chapter 3) as well as on the degrees of elements $b_1, \dots, b_l$ (see §5 above for notation). In the conclusion of that chapter we show, using the degree 6 binary forms as an example, that the null-variety $\mathfrak{N} = \pi_{V,G}^{-1}(\pi_{V,G}(0))$ can be nonreduced. This means that the approach proposed by Dieudonné and Carrell [21] to the solution of the problem of constructive invariant theory will not work (even in the case of polynomial representations of the unimodular group).

10. Chapters 3 and 4 provide a realization of the program suggested in §4 above. We develop a general constructive method for classification of rational G-modules of a connected semisimple algebraic group over a field k of characteristic zero, such that the corresponding algebra of invariants has any predetermined homological dimension.

As previously mentioned, Hilbert's finiteness theorems on generators, relations, and the length of syzygy chain play a key role in the study of algebras of invariants. The main result of those two chapters is another theorem of that sort:

THEOREM (Finiteness Theorem). *For any integer $d \geq 0$ there exist only finitely many, up to isomorphism and addition of trivial direct summands, finite-dimensional rational G-modules V such that* $\operatorname{hd} k[V]^G = d$.

It follows from this theorem that if G is nontrivial then for any d there exists a G-module V such that $\operatorname{hd} k[V]^G > d$ (and moreover the set of G-modules satisfying this condition is infinite, and the set of the modules not satisfying this condition is finite). It is this corollary that makes precise the assertion about the complexity of the algebra of invariants in the general case.

A remarkable consequence of the finiteness theorem is that the problem of "complexity" is not as hopeless as it might have seemed a priori. This theorem naturally leads to a new point of view: classification of G-modules with given homological dimension d of its algebra of invariants. At the same time it offers us a method for the solution of this problem in principle: being constructive, its proof gives an explicit finite list of the G-modules (say, with the bounds on the numerical labels of highest weights), among which are all the G-modules V with $\operatorname{hd} k[V]^G \leq d$. Then methods of Chapter 2 in conjunction with the theorem of Hochster and Roberts [85] allow us to single out in principle all the modules V with $\operatorname{hd} k[V]^G = d$. When $d = 0$, the theorem implies that there are only finitely many G-modules V with a free algebra of invariants (for simple groups G this follows from [94, 135, 2]). We should remark that in Chapter 4 the finiteness theorem is also proved for a finite group G. In this case the proof simplifies significantly because G has only finitely many irreducible modules; for a connected group that is no longer true, and this is the source of substantial difficulties.

The developed method allows us to classify explicitly, not only in principle, all G-modules V with $\operatorname{hd} k[V]^G = d$ for given G and d. Because our method suggests induction, it is natural to start the classification for smaller values of d and then to move on to the higher values. In this way we obtain new "good" classes of modules, and one hopes that they will exhibit interesting geometric properties. One such property is obvious: if for the classes considered above [94, 135, 2, 66, 142, 1, 37, 136, 92, 36, 23, 91] the geometry of V/G is not very interesting (V/G is just an affine space), then for $d > 0$ the variety V/G acquires singularities, with simplest degenerations occurring when d is small. If $d = 1$, then V/G is an affine hypersurface. Some of these hypersurfaces are classical: for example, rational singularities of types A, D, and E in $\mathbb{A}^3$ or the cone over the Grassmannian of lines in $\mathbb{P}^3$ canonically embedded in $\mathbb{P}^5$. In §10 of Chapter 4 we give examples of such explicit classifications. The first one treats the classical case of binary forms. We show that SL_2 has exactly eleven representations whose algebras of invariants are hypersurfaces (see the list in §10 of Chapter 4). A posteriori, it turns out that they were all considered in classical papers (in other contexts). We also determine all l for which the algebra of invariants of degree l binary forms has homological dimension ≤ 10. It turns out that there are no such algebras with $\operatorname{hd} = 2, 4, 5, 6, 7, 8, 9, 10$: there is exactly one with $\operatorname{hd} = 3$ (when $l = 8$), there are two with $\operatorname{hd} = 1$ (when $l = 5, 6$), and there are four with $\operatorname{hd} = 0$ (when $l = 1, 2, 3, 4$). The second example, on the other hand, treats the "most nonclassical" case: we show that the algebra of invariants of any irreducible representation of an exceptional connected simple algebraic group is never a hypersurface; or, more precisely, it has a rather large homological dimension if the representation is different from either the simplest representation, or its dual, or the adjoint representation (for E_7 it exceeds 26,334).

This finiteness theorem allows us to introduce the characteristic function $S_G(t) = \sum_{d\geq 0} R_G(d)t^d$ where $R_G(d)$ is the number of nonequivalent G-modules V with $V^G = \{0\}$ and $\operatorname{hd} k[V]^G = d$. We give bounds on its coefficients, which implies that the series $S_G(t)$ converges in a neighborhood of zero when either G is finite or $G = \mathrm{SL}_2$. In general this function is not rational (for example, if G is the group of order two then $S_G(t)$ is the right-hand side of one of the classical Jacobi identity for the theta-function θ_3).

The case $d = 0$ in the finiteness theorem is of exceptional nature and requires an approach different from that used for $d > 0$. The former is treated in Chapter 3 and the latter is treated in Chapter 4. The proof in the case $d > 0$ is based on: 1) the $d = 0$ case, 2) the Hochster–Roberts theorem, 3) monotonicity theorems (established in §1 of Chapter 4) according to which the homological dimension does not increase when passing to subrepresentations and slice-representations, 4) a technique for constructing the so-called test-slice representations (§§3, 5, 6 of Chapter 4), 5) a detailed investigation of rank 1 groups, and 6) majorizing theorems for multiplicities, generic stabilizers, and stability (§4, Chapter 4). In the simplest (from the point of view of our method) case $d = 0$ this approach contains, as a particular case, the method used in [94] for classification of simple groups with free algebra of invariants (however it does not yield the finiteness theorem for connected groups in the case $d = 0$).

Our proof in the case $d = 0$ is based on quite different ideas, namely, on the analysis of the functional equation for the Poincaré series $F(k[V]^G, x)$ of the algebra of invariants. This analysis, in turn, leads to certain purely geometric problems about inclusions between certain convex polyhedra (in fact, they turn out to be zonohedra) in the space of characters of a maximal torus of G that are solved in Chapter 3. The Poincaré series was an effective tool for mathematicians in the last century [48, 153], and it is no less important in the modern invariant theory [149, 150, 144, 9, 65, 148, 64]. Interpretations for certain numerical characteristics of the rational function $F(k[V]^G, x)$ in terms of the properties of the algebra of invariants (for example, the order of the pole of $F(k[V]^G, x)$ at 1 is equal to the transcendence degree of $k[V]^G$ over k) were known for a long time (see [48]). In fact, the proof of the finiteness theorem when $d = 0$ comes out as a consequence of an interpretation, established in Chapter 4, of yet another such characteristic: we show that for all, except for a finite (explicitly bounded) number, of G-modules V the degree of the rational function $F(k[V]^G, x)$ is equal to $-\dim V$ if G is simple. As a rather special case, we find the degree of the algebra of invariants of any system of binary forms (before only this degree was known, via the explicit form of the Poincaré series, for invariants of a binary form of degree ≤ 16 (see [65, 48]) and covariants of degree ≤ 12 (see [153])). As another application, we investigate the relation between the first and second coefficients, γ and τ, of the Laurent series of the function $F(k[V]^G, x)$ at 1 (the question about the coefficients of this series is also a classical problem: Hilbert found an explicit formula for γ for invariants of a degree l binary form [81]). We show that $2\tau/\gamma = \dim G$ for all, except for finitely many, irreducible V (and that $2\tau/\gamma \in \mathbb{Z}$ for any V). From this we deduce an explicit formula for τ in the case of binary forms of any degree l (we should remark that when G is finite the number $2\tau/\gamma$ also has a remarkable interpretation: it is equal to the number of reflections in G, see Chapter 4).

11. Chapter 5 deals with the classification of G-modules V with free module of covariants.

From the results of Chapters 3 and 4 we deduce another finiteness theorem: for any connected semisimple group G there are only finitely many G-modules V without fixed nonzero vectors with free module of covariants. Any such module is called equidimensional (i.e., all the fibers of $\pi_{V,G}$ are equidimensional). In fact, in Chapter 5 we consider the more general problem of classification of equidimensional G-modules (notice that this problem contains, as a particular case, a classification of G-modules with some other good properties, for example, a classification, discovered by Kač [23, 91], of G-modules with finitely many orbits in the null-variety $\mathfrak{N} = \pi_{V,G}^{-1}(\pi_{V,G}(0))$. We prove that their number is also finite. This assertion is proved with the aid of explicit bounds on its dimensions (e.g., we show that $\dim V \leq 3(\dim G - 1)$). In their turn, these bounds are deduced from explicit bounds on the dimension of the null-variety $\mathfrak{N} = \pi_{V,G}^{-1}(\pi_{V,G}(0))$. This way we obtain an explicit classification of all irreducible equidimensional modules of connected simple groups (see the list in Chapter 5). A case-by-case investigation shows that the module of covariants of each of those modules is free. A direct comparison shows that this list coincides with lists of modules that have a free algebra of invariants (found by Kač, Popov, and Vinberg [94]), with the list of modules with nontrivial generic stabilizer (found by A. M. Popov [26–28] and by Andreev, Vinberg, and Elashvili [3]), as well as with the list of modules that have finitely many orbits in the null-variety and finitely many orbits in any fiber of $\pi_{V,G}$, [23, 91]. A posteriori this means that for irreducible modules of connected simple groups the following properties are equivalent: 1) $k[V]^G$ is a free algebra, 2) $k[V]$ is a free $k[V]^G$-module, 3) $\pi_{V,G}$ is an equidimensional morphism, 4) the null-variety has finitely many orbits, 5) any fiber of $\pi_{V,G}$ has finitely many orbits, and 6) the generic stabilizer in V is not contained in the kernel of the representation. It turns out (see Chapter 5) that these properties are also equivalent to the following: 7) each irreducible component of the null-variety contains an open orbit, 8) each irreducible component of any fiber of $\pi_{V,G}$ contains an open orbit, 9) V has a Cartan subspace, and 10) V has a section (for definitions see Chapter 5).

The method developed in that chapter can also be applied to classification of reducible modules of connected simple groups with free module of covariants (this was done by Schwarz in [136]), as well as to classification of equidimensional reducible modules over connected simple groups (as was done by Adamovich in [1]). (In the reducible case properties (1)–(9) are no longer equivalent; however, the freeness of the module of covariants still follows from equidimensionality. This supports the general conjecture asserting the equivalence of properties 2) and 3) for connected semisimple groups G, which was formulated by the author in [37] and is still open; among the specialists it is now known as the "Russian conjecture.")

In this same chapter we also present a classification of irreducible modules V over connected simple groups G having the following properties: a) $k[V]^G$ is free and b) each fiber of $\pi_{V,G}$ contains an open orbit whose complement is of codimension ≥ 2. This classification problem was posed by Igusa [87] who, while generalizing Siegel formulas [87, 88], discovered that all the modules he had found for this purpose had these remarkable properties (Igusa called such $\pi_{V,G}$ excellent fibrations of V, and we call such modules exceptional in the sense of Igusa).

12. Chapters 6 and 7 deal with quasihomogeneous varieties and some of their applications ($\operatorname{char} k = 0$).

In Chapter 6 we first discuss general questions (connections with classification of left-invariant subalgebras, classification of smooth quasihomogeneous affine varieties, equivariant embeddings, etc.). Here, using the results of Chapter 1, we obtain a classification of quasihomogeneous affine varieties with two orbits under an action of a reductive group. Next we consider the group $G = \mathrm{SL}_2$, a traditional object of study of the classical invariant theory, and obtain the main result of that chapter: a classification of all of its normal quasihomogeneous affine varieties X up to G-isomorphism.

The cases $\dim X = 0, 1$ are trivial. We then show that the class of (not necessarily normal) surfaces of this type consists of two homogeneous ones (namely, $\mathbb{P}^2 - C$ and $\mathbb{P}^1 \times \mathbb{P}^1 - \Delta$ with the natural action of G, where C is a nondegenerate quadric and Δ is the diagonal), and a countable series of nonhomogeneous ones, for which the complement of the open orbit $\mathcal{O}$ is just a point. There is a natural bijection between those nonhomogeneous surfaces and nonzero finitely-generated subsemigroups of the additive semigroup $\mathbb{N}$ of natural numbers with zero (given any such semigroup $\mathfrak{Z}$ one can explicitly construct the corresponding surface $X(\mathfrak{Z})$). The surfaces $X(\mathfrak{Z})$ are affine cones over the projective line $\mathbb{P}^1$ embedded in a product of projective spaces via the product of Veronese embeddings. The surfaces X from our classification exhaust all irreducible affine surfaces with a nontrivial action of SL_2.

Homogeneous three-dimensional X naturally corresponds to binary polyhedral groups, whereas the classification of nonhomogeneous ones is the most difficult. For such an X the stabilizer of the generic point is always a cyclic group (of order n_X). If such a (not necessarily small) X has no fixed points, then the boundary $\Gamma_X = X - \mathscr{O}_X$ is a homogeneous surface. If Γ_X is isomorphic to $\mathbb{P}^1 \times \mathbb{P}^1 - \Delta$ (resp., $\mathbb{P}^2 - C$) then X is a homogeneous bundle over Γ_X, with fiber an irreducible affine curve Y, which is quasihomogeneous but not homogeneous with respect to a (one-dimensional) maximal torus T of G (resp., a reducible affine curve Z, which is a union of two copies of the aforementioned curve Y, intersecting at a T-fixed point). The curve Y is uniquely determined by a nonzero finitely-generated subsemigroup (with zero) $\mathfrak{Z}$ of $\mathbb{N}$. The correspondence $X \mapsto \mathfrak{Z}$ yields a bijection between the set of isomorphism classes of such varieties (with a fixed type of the boundary) and the set of nonzero finitely-generated subsemigroups of $\mathbb{N}$ with zero. If X is normal, then it is automatically smooth, $\Gamma_X = \mathbb{P}^1 \times \mathbb{P}^1 - \Delta$, and the correspondence $X \to n_X$ is a bijection between the set of isomorphism classes of such varieties and the set of positive integers $\mathbb{Z}^+$. If X contains a fixed point, then, besides n_X, there is another numerical invariant, the so-called height h_X. It is a positive rational number smaller than $1/2$. Roughly speaking, h_X is a numerical characteristic of certain curves in SL_2 that move the fixed point to a finite point of the boundary $\Gamma_X = X - \mathscr{O}_X$. Height controls the domination relation between such normal varieties: if X and Y are two varieties of this type and $n_X = n_Y$ then there exists a birational morphism $\varphi : X \to Y$ (commuting with the action of G) if and only if $h_X \geq h_Y$. If also $h_X > h_Y$, then φ has an exceptional subvariety, the boundary Γ_X, which deforms to a point. Variety Γ_X is always irreducible and, if X is normal, it is isomorphic to the surface $X(\mathfrak{Z})$, where $\mathfrak{Z}$ is the semigroup with generator $n_X p/\text{g.c.d.}(n, p-2q)$ (here p/q is an irreducible representation of the rational number h_X). The stabilizer and height are a complete and independent system of invariants: the correspondence $X \to (n_X, h_X)$ is a bijection between the set of isomorphism classes of the normal varieties of this type and the set $\mathbb{Z}^+ \times \{q \in \mathbb{Q} \mid 0 < q < 1/2\}$.

As an application we obtain in this chapter a complete description of the closures of orbits in any finite-dimensional rational SL_2-module (in the simplest case of an irreducible module this result was obtained by Hadzhiev in [49]).

13. In §1 of Chapter 7 we give a complete classification of irreducible quasihomogeneous (not necessarily affine) curves and their automorphism groups when the field k is of characteristic zero. Only the case of singular curves is nontrivial. We show that such curves are classified by some special semigroups or by pairs of such semigroups in the additive group of natural numbers (see Chapter 7 for details). As an application we obtain examples showing that the fundamental theorems of Sumihiro [152] on the existence of an equivariant projective embedding and an invariant affine neighborhood are no longer true for nonnormal varieties.

14. The curves found in §1 of Chapter 7 admit an algebraic group of automorphisms acting transitively on the complement of a finite set of points. It turns out that one can classify varieties of higher dimensions having this property: in §§2 and 3 of Chapter 7 we give classifications of, respectively, irreducible affine surfaces and three-dimensional varieties (solids) of this type (i.e., admitting an algebraic group of automorphism acting transitively on the complement of a finite set of points).

Smooth surfaces of this type were found by Gizatullin in [18]. We determine all the surfaces of this type: ($\operatorname{char} k = 0$)

1) (smooth):

a) $\mathbb{A}^p \times (\mathbb{A}^1_*)^q$, $p+q=2$, where $\mathbb{A}^1_*$ is a punctured affine line,
b) $\mathbb{P}^1 \times \mathbb{P}^1 - \Delta$, where Δ is the diagonal,
c) $\mathbb{P}^2 - C$, where C is a nondegenerate quadric,

2) (singular):
the surfaces $X(\mathfrak{Z})$ (see §12 above) (with $X(\mathfrak{Z})$ isomorphic to $X(\mathfrak{Z}')$ if and only if $\mathfrak{Z} = \mathfrak{Z}'$).

Three-dimensional varieties of this type are as follows:

1) (smooth):

a) the series $\mathbb{X}_n$, where $\mathbb{X}_n$ is the 1-bundle over $\mathbb{P}^1 \times \mathbb{P}^1 - \Delta$ corresponding to $n \in \mathbb{Z} \cong \operatorname{Pic}(\mathbb{P}^1 \times \mathbb{P}^1 - \Delta)$,
b) the series $\mathbb{X}_n^* = \mathbb{X}_n -$ (the zero section),
c) $\mathbb{Y}_0$, where $\mathbb{Y}_n$ is the 1-bundle over $\mathbb{P}^2 - C$ corresponding to $n \in \{0, 1\} = \mathbb{Z}/2\mathbb{Z} \cong \operatorname{Pic}(\mathbb{P}^2 - C)$,
d) $\mathbb{Y}_n^* = \mathbb{Y}_n -$ (the zero section),
e) the series $\mathbb{W}_n$, $n \in \mathbb{Z}^+$, where $\mathbb{W}_n$ is SL_2 modulo the binary dihedral group D_n (equivalently, X_{2n}^* modulo some standard involution),
f) $\mathbb{S}_n$, $n = 3, 4, 5$, which are SL_2 modulo, respectively, the binary tetrahedral, binary octahedral, and binary icosahedral groups,
g) $\mathbb{A}^p \times (\mathbb{A}^1_*)^q$, $p+q=3$;

2) (singular):
certain three-dimensional analogs of the surfaces $X(\mathfrak{Z})$, which are affine cones over $\mathbb{P}^1 \times \mathbb{P}^1$ or $\mathbb{P}^2$ embedded in a product of projective spaces via the Veronese embedding (for details see §3 of Chapter 7).

We also determine when those solids are isomorphic to each other, see §3 of Chapter 7.

The method developed there can also be used for classification of the varieties of this type in higher dimensions (for example, Kaga and Watabe [96] used it for the classification of four-dimensional affine varieties with a reductive algebraic group of automorphisms one of whose orbits has at most zero-dimensional complement).

15. Preceding Chapter 1 the reader will find a list of notation; it is used in the text without reference.

The numbering of sections, formulas, and propositions is uniform within each chapter. The same is true for subsections inside each section. Within each chapter, the chapter number is not included in the references; the same is true for the references to a subsection within a section.

16. The results presented in this book were included in lectures delivered at the International Conference on Schur Functors (Poland, 1980, [124, 125]), the International Conference on Algebraic Groups in Oberwolfach (FRG, 1982, T. Springer's lecture [146]), the XV All-Union Conference on Algebra (Krasnoyarsk, 1979, [41]), a meeting of the Moscow Mathematical Society (Moscow State University, 1982), and various seminars, including the seminar at the Department of Algebra of the Steklov Institute (1981), the Algebra and Topology seminar (Minsk, 1981), the seminars on Lie groups and invariant theory at Moscow State University (1976, 1978, 1981, 1982, 1983). The results of Chapters 1 and 5 are described in the appendix to the last edition of Mumford's *Geometric invariant theory* [117]. The results of Chapter 6 constitute the third part of Chapter 3 of Kraft's *Geometric methods in invariant theory* [103]. These results were used by other authors (see, e.g., [135, 136, 1, 6, 57, 71, 72, 96, 98, 25, 91, 64, 56]).

The results related to those presented here were published in [31, 32, 34–37, 39, 40–44, 124, 125].

Notation and Terminology

The ground field k is algebraically closed. All algebraic varieties, groups, actions, and morphisms are defined over k. Varieties are identified with the sets of their k-points. Homomorphisms and subgroups of algebraic groups are always assumed to be algebraic. An action of an algebraic group G on an algebraic variety X is called regular if the map $G \times X \to X$ defining it is a morphism. Topological terms, unless stated otherwise, are related to the Zariski topology; the overbar denotes the closure.

k^* and k_+ are, respectively, the multiplicative and additive group of k.

Let X be an algebraic variety. Then

$k[X]$ is the algebra of regular functions on X,
$k(X)$ is the field of rational functions on (irreducible) X,
$\operatorname{sing} X$ is the singular locus of X,
$\operatorname{Aut} X$ is the group of biregular automorphisms of X,
$\Theta_{x,X}$ is the tangent space to X at x,
$\mathcal{O}_{x,X}$ is the local ring of point $x \in X$, and
$\mathfrak{m}_{x,X}$ is the maximal ideal of $\mathcal{O}_{x,X}$.

Let G be a transformation group of M. Then

$GY = \{gy \mid g \in G, y \in Y\}$ is the orbit of a set $Y \subset M$,
$G_Y = \{g \in G \mid gY \subset Y\}$ is the stabilizer of a set $Y \subset M$, and
$M^G = \{m \in M \mid gm = m \ \forall g \in G\}$.

The group $\bigcap_{m \in M} G_m$ is called the noneffectivity kernel of the action of G on M; an action is called effective if its noneffectivity kernel is trivial.

We say that G acts transitively on a subset $N \subset M$ if N is contained in the orbit of a point of M.

Let G be an algebraic group and H a subgroup of G. Then

G^0 is the connected component of the identity of G,
$Z_G(H)$ and $N_G(H)$ are, respectively, the centralizer of H in G,
$Z_G = Z_G(G)$ is the center of G,
G' is the commutator of G,

Rad G and $\mathrm{Rad}_u G$ are, respectively, the radical and the unipotent radical of the (linear algebraic) group G, and

$\mathrm{Hom}(G, k^*)$ is the group of rational characters (i.e., homomorphisms into k^*) of G.

When working with $\mathrm{Hom}(G, k^*) \otimes_{\mathbb{Z}} \mathbb{Q}$, we always use the additive notation. Any character $\chi \in \mathrm{Hom}(G, k^*)$ is considered a one-dimensional representation of G (which is also denoted by χ). If $\chi \in \mathrm{Hom}(G, k^*)$ and V is a G-module then we set $V_\chi = \{v \in V \mid gv = \chi(g)v \ \forall g \in G\}$ and call V_χ a G-weighted subspace of weight χ (it may be zero dimensional). Vectors from V_χ are called G-weighted of weight χ (similar terminology is used for the Lie algebra $\mathfrak{G}$ of G and an element $\chi \in \mathfrak{G}^*$). If $\lambda\colon k^* \to G$ is a one-parameter subgroup and V is a G-module then $V_{t^a} = \{v \in V \mid \lambda(s)v = s^a v \ \forall s \in k^*\}$. A G-module V (as well as the action of G on V) is called locally transitive if there is a point $v \in V$ whose orbit is open in V; it is called locally effective if its noneffectivity kernel is finite.

The trivial group $\{e\}$ is considered semisimple.

The irreducible representation or module (of a given nontrivial connected semisimple group or a nonzero Lie algebra, depending on the context) with highest weight $\Lambda = \sum_{i=1}^{l} n_i\omega_i$, where $n_i \in \mathbb{Z}$, $n_i \geq 0$, and $\omega_1, \dots, \omega_l$ is a system of fundamental weights, is denoted $R(\Lambda)$. The numbers $n_1, \dots, n_l$ are called the numerical labels of Λ (or $R(\Lambda)$); writing them over the corresponding vertices of the Dynkin diagram of the group (or algebra) we obtain the Dynkin diagram of the module $R(\Lambda)$. The one-dimensional representation (module) $R(0)$ is denoted by 1; this notation is also used for the trivial group (and the zero Lie algebra). The representation in zero-dimensional space (the trivial representation) is denoted by 0. We set $R(\sum_{i=1}^{l} n_i\omega_i) = 0$ if at least one of the n_i is negative. $S^m R$, $\Lambda^m R$, R^*, and mR are, respectively, the mth symmetric and exterior power of the representation (module) R, the representation (module) contragredient to R, and the m-fold direct sum $R \oplus \cdots \oplus R$ of R. We set $S^m R = \Lambda^m R = 0$ for $m < 0$ and $S^0 R = \Lambda^0 R = 1$, $0R = 0$. $R_1 \cdot \cdots \cdot R_s = \prod_{i=1}^{s} R_i$ is the Cartan product of irreducible representations (modules) $R_1, \dots, R_s$, i.e., the highest irreducible component of $R_1 \otimes \cdots \otimes R_s$. If $R_i = R(\Lambda_i)$, then $R_1 \cdot \cdots \cdot R_s = R(\Lambda_1 + \cdots + \Lambda_s)$. We set $R^d = R \cdot \cdots \cdot R$ (d times) and $R^0 = 1$, Ad is the adjoint representation (module), and $\mathscr{R}(G)$ is the set of all finite-dimensional irreducible representations of an algebraic group G considered up to equivalence.

If $\mathfrak{h}$ is the Lie algebra of automorphisms of a vector space V then we set $V^{\mathfrak{h}} = \{v \in V \mid Xv = 0 \ \forall X \in \mathfrak{h}\}$.

Let X be an algebraic variety. We say that a certain condition holds for a point in general position of X if it holds for all points of a (depending on the condition) nonempty open subset Ω of X. The points of Ω are then called points in general position.

Let an algebraic group G act regularly on X. This action is called stable if Gv is closed in X for a point in general position $v \in X$. The action of G is stable if and only if the corresponding action of G^0 is stable.

Let G be reductive and X affine. The algebra $k[X]^G$ is finitely generated and we set $X/G = \operatorname{Specm} k[X]^G$ and $\pi_{X,G}\colon X \to X/G$, where $\pi^*_{X,G}\colon k[X]^G \hookrightarrow k[X]$ is the natural embedding. Morphism $\pi_{X,G}$ is surjective and if Y is closed in X and $GY = Y$, then $\pi_{X,Y}(Y)$ is closed in X/G, isomorphic to Y/G, and $\pi_{X,G}|_Y$ can be naturally identified with $\pi_{Y,G}$ (see [108, 115]). If the action of G on X is stable then $\dim X/G = \dim X - \max_{v\in X} \dim Gv$ (see [29, 108]).

$k[b_1, \dots, b_d]$ (respectively, $k(b_1, \dots, b_d)$) is the algebra (resp., field) generated over k by $b_1, \dots, b_d$. $\operatorname{ed} A$, $\operatorname{hd} A$, $\dim A$ are, respectively, the embedding dimension, homological dimension, and Krull dimension of a graded k-algebra A.

$\mathbb{N}$ is the additive semigroup of nonnegative integers.

$\mathbb{Z}^+ = \{1, 2, 3, \dots\} = \mathbb{N} - \{0\}$.

$\mathbb{A}^1_*$ is the punctured affine line $\mathbb{A}^1$.

$|S|$ is the number of elements in a set S.

$[a]$ is the integer part of a number a.

$\binom{m}{n} = \frac{m!}{n!(m-n)!}$; we set $\binom{m}{n} = 0$ if $n < 0$ or $n > m$.

$a|b$ means that integer a divides integer b.

$\langle\langle M\rangle\rangle$ is the linear hull of a nonempty subset M of a vector space V.

Let $\mathfrak{Z}$ be an abelian semigroup, written additively (resp., a semigroup with zero). We say that it is finitely generated if there exist elements $\Lambda_1, \dots, \Lambda_n$ of $\mathfrak{Z}$ (called a system of generators of $\mathfrak{Z}$) such that any element Λ of $\mathfrak{Z}$ (resp., $\Lambda \neq 0$) is of the form $\Lambda = \Lambda_{i_1} + \dots + \Lambda_{i_s}$, where $s \geq 1$ and $1 \leq i_1 \leq \dots \leq i_s \leq n$. We set $m\mathfrak{Z} = \{M_1 + \dots + M_n | M_i \in \mathfrak{Z}\ \forall i\}$. If $M_1, \dots, M_n \in \mathfrak{Z}$ then $\langle\langle M_1, \dots, M_n\rangle\rangle$ denotes the subsemigroup of $\mathfrak{Z}$ (resp., subsemigroup with zero) generated by $M_1, \dots, M_n$. We also set $\langle\langle M_1, \dots, M_n\rangle\rangle_0 = \langle\langle M_1, \dots, M_n\rangle\rangle - \{0\}$.

We assume that $\dim \varnothing = -1$.

The symbols $\subset$ and $\supset$ denote not necessarily strict inclusions.

CHAPTER 1

The Role of Reductive Groups in Invariant Theory

Unless stated otherwise, throughout this chapter the characteristic of the field k is arbitrary.

§1. Reductive groups and the generalized Hilbert's 14th problem

1. By an affine k-algebra we will mean a finitely-generated commutative associative k-algebra with identity and without nilpotent elements. Those are precisely the algebras of regular functions on reduced affine algebraic varieties.

Let H be a linear algebraic group and A a (not necessarily finite-dimensional) vector space. An action of H by linear transformations of A will be called rational if for any $a \in A$ the subspace $\langle Ha \rangle$ of A is finite dimensional and the action of H on it is given by a rational representation $H \to \mathrm{GL}(\langle Ha \rangle)$ (see [50]). In this case we will speak of a rational H-module A or a rational representation $H \to \mathrm{GL}(A)$. If A is an affine algebra then rational actions of H by k-automorphisms of A are precisely the actions defined by regular actions of H on $\operatorname{Specm} A$.

2. H will be called reductive if $\mathrm{Rad}_u H = \{e\}$. When $\operatorname{char} k = 0$ reductive groups can be characterized in terms of representation theory (see [120]): those are the groups all of whose rational representations are completely reducible (which is not true if $\operatorname{char} k > 0$, see [120]). If H is reductive, $\operatorname{char} k$ is arbitrary, and A is an affine k-algebra on which H acts rationally by k-automorphisms, then the following two conditions, which form a basis for all applications of reductive groups to invariant theory, hold:

1) (Hilbert's theorem on invariants) A^H is an affine algebra.
2) (separation of closed invariant subsets of $\operatorname{Specm} A$ by $A\uparrow$). If I_1 and I_2 are invariant ideals of A and $A = I_1 + I_2$, then there is an element $a \in I_1^H$ such that $1 - a \in I_2$.

In fact, condition 2) characterizes reductive groups in terms of invariant theory (see [78, 119]): H is reductive if and only if condition 2) is satisfied for any A and any rational action of H on A.

We will show in this section that condition 1) also characterizes reductive groups; in fact, we will prove a theorem converse to Hilbert's theorem. This result shows that, first, reductive groups are exactly the groups for which the generalized Hilbert's 14th problem [120] is always solved positively, and, second, those are exactly the groups for which a good geometric invariant theory exists (meaning that conditions 1) and 2) are always satisfied).

3. THEOREM 1. *The following conditions are equivalent*:

a) *for any affine* k*-algebra* $\widehat{A}$ *and any rational action of* H *on* A *by* k*-algebra automorphisms the* k*-algebra of invariants* A^H *is finitely generated* (*and hence affine*);

b) H *is reductive.*

PROOF. By Hilbert's theorem we only have to prove the implication a) $\Rightarrow$ b). Suppose that H is not reductive. Then $H_u = \mathrm{Rad}_u H \neq \{e\}$. First we will show that there exist an affine k-algebra B and a rational action of H_u on B by k-automorphisms such that k-algebra B^{Hu} is not finitely generated.

By Nagata's counterexample [118], there exist $n \in \mathbb{Z}^+$ and a connected unipotent algebraic subgroup N of $\mathrm{GL}_n(k)$ such that for the natural action of N on the k-algebra of polynomials $C = k[x_1, \dots, x_n]$ the subalgebra C^N is not finitely generated. Let $\{e\} = N_0 \subset N_1 \subset \cdots \subset N_r = N$ be a chain of normal subgroups for which $\dim N_{i+1}/N_i = 1\ \forall i$. In the chain of algebras $C = C^{N_0} \supset C^{N_1} \supset \cdots \supset C^{N_r}$ the leftmost term is finitely generated whereas the rightmost is not. Therefore, there exists an i such that C^{N_i} is an affine algebra and $C^{N_{i+1}} = (C^{N_i})^{N_{i+1}/N_i}$ is not finitely generated. It is clear now that one can take C^{N_i} for B and define an action of H_u on B by the formula $h(b) = \alpha(h)(b)$, $h \in H_u$, $b \in B$, where $\alpha\colon H_u \to N_{i+1}/N_i$ is an arbitrary surjective rational homomorphism (which is known to exist).

We conclude the proof by constructing an affine k-algebra A and a rational action of H on A by k-automorphisms for which A^H is not finitely generated. The construction is as follows.

Let Z be an affine algebraic variety with $k[Z] = B$. The action of H_u on B gives rise to a regular action of H_u on Z. Consider the variety $H \times Z$ and the regular action of H_u on it defined by the formula $(h, z) \xrightarrow{s} (hs^{-1}, sz)$, $s \in H_u$. It is known (see [138]) that the geometric quotient $H \times_{H_u} Z$ by this action exists; let $\pi\colon H \times Z \to H \times_{H_u} Z$ be the corresponding quotient morphism. The group H acts regularly on $H \times Z$, where the action is defined by $(h, z) \overset{t}{\mapsto} (th, z)$, $t \in H$. This action commutes with the action of H_u mentioned before and, therefore, gives rise to a regular action of H on $H \times_{H_u} Z$. This action induces, in turn, an action of H on $k[H \times_{H_u} Z]$ by k-algebra automorphisms. Now we prove that:

a) $H \times_{H_u} Z$ is an affine variety;

b) k-algebra $k[H \times_{H_u} Z]^H$ is isomorphic to k-algebra $k[Z]^{H_u}$.

It is clear that if assertions a) and b) are proved then one can take algebra $k[H \times_{H_u} Z]$ (with the above action of H) for algebra A.

First we prove b). Let $p_2: H\times Z \to Z$, $p_2(h, z) = z$, be the projection onto the second factor. Using the comorphisms defined, respectively, by π and p_2 we identify the algebras $k[H\times_{H_u} Z]$ and $k[Z]$ with subalgebras of $k[H\times Z]$. Then:

$$k[Z]^{H_u} = \{f\in k[H\times Z] \mid f(h, z) = f(e, sz)\ \forall h\in H,\ s\in H_u,\ z\in Z\}, \quad (1)$$

where e is the identity of H and, by the definition of the quotient variety,

$$k[H\times_{H_u} Z] = \{f\in k[H\times Z] \mid f(h, z) = f(ghs^{-1}, sz) \quad \forall g, h\in H,\ s\in H_u,\ z\in Z\}. \quad (2)$$

It now follows that

$$k[H\times_{H_u} Z]^H = k[Z]^{H_u}. \quad (3)$$

We shall now prove a). Let $H\to H/H_u$ be the canonical morphism. The triple $(H, H/H_u, H_u)$ is a principle bundle with base H/H_u and structure group H_u (see [138]). Because H_u is a unipotent group, this bundle is locally trivial [138] and, because H_u is a normal subgroup of H, the base H/H_u is affine. It follows that this bundle is in fact trivial [138]. Thus H and $H/H_u\times H_u$ are isomorphic as varieties with H_u-action (here H_u acts on H and H_u by right translations and on $H/H_u\times H_u$ via the second factor). It follows that $H\times_{H_u} Z$ is isomorphic to $H/H_u\times(H_u\times_{H_u} Z)$, where $H_u\times_{H_u} Z$ denotes the geometric quotient of $H_u\times Z$ by the action $(h, z)\overset{s}{\mapsto}(hs^{-1}, sz)$ of H_u (see [130]). But varieties $H_u\times_{H_u} Z$ and Z are isomorphic. Indeed, let $p: H_u\times Z\to H_u\times_{H_u} Z$ be the quotient morphism and $q: H_u\times Z\to Z$ and $r: Z\to H_u\times Z$ be the morphisms defined, respectively, by $q(h, z) = hz$ and $r(z) = (e, z)$. Because the fibers of p are contained in the fibers of q, the universal property of the quotient variety implies that there exists a morphism $s: H_u\times_{H_u} Z\to Z$ for which $s\circ p = q$. It is easy to see that $p\circ r\circ s = \mathrm{Id}$ and $s\circ p\circ r = \mathrm{Id}$; hence, s is an isomorphism. Therefore, $H\times_{H_u} Z$ is isomorphic to $H/H_u\times Z$ and hence is affine. □

REMARK. When $\operatorname{char} k = 0$, the proof can be made more direct by using the existence of a reductive Levi subgroup S of H. More precisely, define a regular action of H on $S\times Z$ by

$$s_1 h\cdot(s_2, z) = (s_1 s_2, s_2^{-1} h s_2\cdot z) \qquad \forall s_1, s_2\in S,\ h\in H_u,\ z\in Z.$$

Clearly the restriction of functions from $k[S\times Z]$ to the subvariety $e\times Z$ of $S\times H$ defines an isomorphism between $k[S\times Z]^H$ and $k[Z]^{H_u}$.

§2. Quasihomogeneous varieties of reductive groups and the original Hilbert's 14th problem

1. The theorem just proved does not imply that the algebra A^H cannot be affine for some rational action of a nonreductive group H on an affine algebra A. For example, A^H is affine if H acts on A trivially. A more meaningful example comes from the classical result of Weitzenböck according to which the algebra of invariants of a linear action of a one-dimensional unipotent group is finitely generated (see [48, 134], $\operatorname{char} k = 0$).

It was exactly this special case that was considered in the original Hilbert's 14th problem: A was the algebra of polynomials $k[x_1, \dots, x_n]$, H was a subgroup of $\mathrm{GL}_n(k)$ acting naturally on A, and the problem was to describe those H for which condition 1) is satisfied. At first glance it may seem that the answer depends on some special properties of A and the action of H. However, this is not quite so. The truly important fact here is that the action of H on A can be extended to a rational action of a reductive group G containing H (in this case $G = \mathrm{GL}_n(k)$). It turns out that when such an extension is possible the answer to the question whether A^H is affine depends only on algebro-geometric properties of the homogeneous space G/H (but not on the properties of A).

More precisely, by a quasihomogeneous variety of a linear algebraic group P we will understand any algebraic variety X on which P acts regularly and with a dense, hence open (see [151]), orbit O_X. The subvariety $\Gamma_X = X - O_X$ will be called the boundary of X and O_X will be the principal orbit of X. Now let Y be an arbitrary irreducible affine variety on which a connected reductive algebraic group G acts regularly and let H be an algebraic subgroup of G. We want to know if the algebra $k[Y]^H$ is finitely generated. Let $\widehat{H}$ be the smallest observable algebraic subgroup of G containing H (by definition, S is *observable* in G if G/S is quasiaffine, [58]). Then $k[Y]^{\widehat{H}} = k[Y]^H$ (see [77]), and without loss of generality we may (and will) assume that H is observable. In this case the following conditions are equivalent (see [78]):

a) $k[Y]^H$ is affine;
b) there exists a quasihomogeneous affine variety X of G such that the dense orbit in X is isomorphic to G/H and the boundary is of codimension ≥ 2.

By Nagata's counterexample to the original Hilbert's 14th problem (see [118]), subgroups H (of $G = \mathrm{GL}_n(k)$) not satisfying b) do exist. On the other hand, there are some classes of nonreductive groups H satisfying b) (for example, unipotent radicals of parabolic subgroups of G when $\operatorname{char} k = 0$ (see [84]). According to the aforementioned result, the classification problem for subgroups H for which the original 14th problem is solved positively is directly related to the construction of quasihomogeneous varieties and bounds on the dimensions of their boundaries. Until now no such bounds have been known in general. In what follows we shall obtain one such bound, and using it we draw some consequences for later use in Chapters 6 and 7.

2. Let X be a quasihomogeneous variety of a linear algebraic group G, let O_X be the dense orbit in X, let B be a Borel subgroup of G, and let U be the unipotent radical of B.

THEOREM 2. *Suppose that* $X - O_X \neq \varnothing$ *and* $O_X \cong G/H$. *Then there exists a point* x *in* O_X *such that* $\dim Bx \leq \dim(X - O_X) + 1$. *If* X *is affine or if* H *does not contain a maximal torus of* G *and* $\operatorname{char} k = 0$, *then* $\dim Ux \leq \dim(X - O_X)$.

PROOF. Consider the closure $\overline{By}$ in X of the orbit By of a point $y \in O_X$. Then $By \neq \overline{By}$ otherwise O_X would be closed in X (see [151]) contrary to the assumption that $X \neq O_X$. We claim that By is an affine variety. Indeed, because B is a connected solvable group, the quotient B/B_y is an affine variety (see [138]). On the other hand, we have the bijective canonical morphism $B/B_y \to By$ and therefore (because By is smooth) the variety By is also affine. Therefore $\overline{By} - By$ is an unmixed subvariety of codimension 1 in $\overline{B}y$ (see [30]). Thus if $\dim By > \dim(X - O_X) + 1$, then $\dim(\overline{By} - By) > \dim(X - O_X)$ and therefore $(\overline{By} - By) \cap O_X \neq \varnothing$. Now consider the orbit Bz of a point z from $(\overline{By} - By) \cap O_X$. Because $\overline{By} - By$ is a B-invariant subvariety, we have $\dim Bz \leq \dim(\overline{By} - By) = \dim By - 1$. Replacing y with z we can repeat the procedure several times and in a finite number of steps we find a point $x \in O_X$ satisfying the condition $\dim Bx \leq \dim(X - O_X) + 1$.

Now consider the U-orbit Ux of x. Clearly $Ux \subset Bx$. We shall now prove that, under the assumptions of this theorem, $Ux \neq Bx$, and therefore $\dim Ux \leq \dim Bx - 1 \leq \dim(X - O_X)$. Indeed, if X is affine then Ux is closed in X (see [151]), and because Bx is not closed, we have $Ux \neq Bx$. If H does not contain a maximal torus of G and $\operatorname{char} k = 0$, then Ux is isomorphic to $\mathbb{A}^s$ and Bx is isomorphic to $\mathbb{A}^r \times (\mathbb{A}^1_*)^t$ for some $s, r, t, t > 0$ (see [138]). Those varieties are not isomorphic (one of them has nonconstant invertible regular functions, the other does not), hence $Ux \neq Bx$. □

REMARKS. a) If X is affine and G is reductive, then H contains no maximal tori of G, otherwise O_X would be closed in X (see [110]).

b) Under the assumptions of a), if H is not reductive then G/H is not affine (see [126]), hence $X - O_X \neq \varnothing$.

COROLLARY 1. $\dim(X - O_X) \geq b_G - b_H - 1$ *and if the assumptions of Theorem* 2 *on either* X *or* H *are satisfied then* $\dim(X - O_X) \geq u_G - u_H$, *where* b_p *and* u_p *are, respectively, the dimensions of a Borel subgroup of an algebraic group* P *and the unipotent radical of this subgroup.*

COROLLARY 2. *Let* X *be affine and* $\dim(X - O_X) = 0$. *Then* H *contains a maximal unipotent subgroup of* G, *i.e.,* X *is an* S*-variety of* G *in the sense of* [16] (*see also part* 4 *of* §1, *Chapter* 6). *If* G *is connected then the boundary of* X *is reduced to a point.*

The last assertion of this corollary can be proved as follows. The group $G_u = \operatorname{Rad}_u G$ is contained in H and therefore lies in the noneffectivity kernel of the action of G on X. If the boundary of X contained more than one point then, because those points are fixed under the action of the reductive group G/Gu (G/Gu is connected!), condition 2) from §1 would imply that $k[X]^{G/Gu} \neq k$ contrary to $k[X]^{G/G_u} = k[X]^G = k$.

3. This corollary yields yet another bound on the dimension of the boundary of a quasihomogeneous variety. More precisely, assume that $\operatorname{char} k = 0$, X is affine, and $G = G^0$ is reductive. According to [77], Condition b) for the pair $G \supset H$ is satisfied if and only if it is satisfied for the pair $R \supset H^0$, where R is the smallest reductive subgroup of G containing H^0. In other words, when checking Condition b) one can assume that H is connected and is not contained in any proper reductive subgroup of G. Let us assume this.

Now, X has a unique closed orbit. Its existence follows from the Noetherian property of X and from the fact that any orbit is open in its closure (see [151]); the uniqueness follows from Condition 2) of §1 and from the equality $k[X]^G = k$. Being closed, this orbit is affine and therefore, by Matsushima's criterion (see [126]), the stabilizer of any of its points is reductive. By [108], this point can be chosen so that its stabilizer would contain H. This means that the only closed orbit of G in X is a fixed point o. If H contains a maximal unipotent subgroup of G, then X is an S-variety and the general theory of such varieties (see [16]) provides complete information about the orbital structure of X (in particular, it follows that for such a variety H Condition b) is satisfied and therefore the original 14th problem is solved positively). Therefore, we can assume that H does not contain maximal unipotent subgroups of G. Then, by Corollary 2, there exists a point $x_1 \in X - O_X$, $x_1 \neq o$. If $\overline{Gx_1} - Gx_1 \neq o$ then consider a point $x_2 \in \overline{Gx_1} - Gx_1$, $x_2 \neq o$, etc. After a finite number of steps we will find a two-orbit S-variety of G contained in the boundary of X. The dimension of any such subvariety is equal to $1 + \dim P_u$, where P_u is the unipotent radical of a parabolic subgroup P of G (see [16]). This yields the following bound on the dimension of the boundary of X: $\dim(X - O_X) \geq 1 + \min_P \dim P_u$.

4. A similar argument proves

COROLLARY 3. *Suppose we are given a rational representation of a connected reductive group G in a vector space V. Let B be a Borel subgroup of G and $v \in V$, $v \neq 0$, a nonsemistable point (i.e., $0 \in \overline{Gv}$). Then $\overline{Gv}$ contains a point of the form $v_{\Lambda_1} + \cdots + v_{\Lambda_n}$, where v_{Λ_i} is a B-weighted vector of weight $\Lambda_i \neq 0$. Moreover the weights $\Lambda_1, \ldots, \Lambda_n$ are proportional in* $\operatorname{Hom}(B, k^*) \otimes_{\mathbb{Z}} \mathbb{Q}$ *with positive coefficients of proportionality.*

The proportionality of the weights $\Lambda_1, \ldots, \Lambda_n$ follows from [16] (the number of orbits of the S-variety $\overline{Gu}$, where $u = v_{\Lambda_1} + \cdots + v_{\Lambda_n}$, is equal to the number of faces of the rational polyhedral cone in $\operatorname{Hom}(B, k^*) \otimes_{\mathbb{Z}} \mathbb{Q}$, spanned by $\Lambda_1, \ldots, \Lambda_n$). □

CHAPTER 2

Constructive Invariant Theory

§1. Formulation and reduction of the problem

1. First we are going to fix notation. Let V denote the n-dimensional coordinate linear space (of columns vectors) over k, Mat_n be the space of all $n \times n$ matrices with coefficients in k, and GL_n be the group of all nondegenerate matrices in Mat_n. An element of the space Mat_n will be denoted by (a_{ij}); this is a matrix with element a_{ij} at the intersection of the ith row and the jth column, $1 \leq i, j \leq n$. Let x_i, $1 \leq i \leq n$, and x_{ij}, $1 \leq i, j \leq n$, be the coordinate functions on V and, respectively, on Mat_n relative to the canonical bases, i.e.,

$$x_i \begin{pmatrix} a_1 \\ \vdots \\ a_n \end{pmatrix} = a_i \quad \text{and} \quad x_{ij}((a_{pq})) = a_{ij}.$$

Let G be a reductive algebraic subgroup of GL_n. The group GL_n and therefore G act naturally on V (via multiplication of a matrix by a column) and on $k[V] = k[x_1, \dots, x_n]$ by $(g \cdot f)(v) = f(g^{-1} \cdot v)$, $f \in k[V]$, $g \in \mathrm{GL}_n$, $v \in V$.

The goal of constructive invariant theory is to find explicitly a minimal (i.e., irreducible) system of homogeneous invariants of the finitely-generated (by Hilbert's theorem) algebra $k[V]^G$. From the theoretical point of view this means finding an algorithm that would allow us, in principle, to determine such a system after a finite number of explicitly performable operations. Of course, from the practical point of view we immediately face the question of complexity, but this is a different aspect of the problem, related to the choice of the most efficient algorithm, improving the corresponding bounds, etc., but not to the existence of the algorithm itself. A good illustration is the case of a finite group G, where Noether found an algorithm that later prompted the following remark from Weyl [11, p. 370] "one cannot ask for anything more specific"; in practice, however, this algorithm is too cumbersome to be used in computations with concrete groups. But the main problem of constructive invariant theory is also important from the conceptual point of view; in fact, it emerged simultaneously with invariant theory itself. Hilbert's classical paper [81] deals exactly with this problem (see part 3 of the Introduction).

2. Working on the problem of constructive invariant theory, one can assume that the group G "is known," i.e., certain questions about its structure can be constructively answered. Henceforth we will assume that we know the following:

a) the Lie algebra $\mathfrak{g}$ of G (we view $\mathfrak{g}$ as a linear subspace of Mat_n and for our purposes it would be sufficient to know a set of matrices forming a basis of this subspace);

b) a maximal torus T of G (we view T as the image of the standard torus $(k^*)^r = \underbrace{k^* \times \cdots \times k^*}_{r}$ under the homomorphism

$$\varphi : (k^*)^r \to \mathrm{GL}_n, \qquad \varphi((a_1, \dots, a_r)) = (f_{ij}(a_1, \dots, a_r)), \tag{1}$$

where $f_{ij}(t_1, \dots, t_r)$, $1 \le i, j \le n$ are known rational functions of the coordinate functions $t_1, \dots, t_r$ on $(k^*)^r$, i.e., the functions defined by the equalities $t_s((a_1, \dots, a_r)) = a_s)$.

For example, a typical situation in classical invariant theory is when G is the image of a known standard group $\widetilde{G}$ under a known explicitly-defined homomorphism (say, $\widetilde{G} = \mathrm{SL}_d$ and G is the image of $\widetilde{G}$ under the natural representation in the space of forms of a fixed degree in d variables or, more generally, in the space of tensors of a fixed type on the d-dimensional space). It is clear that if we know the Lie algebra and a maximal torus for $\widetilde{G}$, we also know $\mathfrak{g}$ and T for G.

3. Suppose we know an algorithm that can explicitly determine a minimal system of homogeneous generators of $k[V]^G$. Then, in principle, we can explicitly compute a constant $M \in \mathbb{N}$ such that $k[V]^G$ is generated (as an algebra) by the subspace $\bigoplus_{d=0}^M k[V_d]^G$ (from now on, R_d will denote the subspace of all homogeneous elements of degree d of an arbitrary $\mathbb{N}$-graded k-algebra R). Conversely, if we know that for some $M \in \mathbb{N}$ the subspace $\bigoplus_{d=0}^M k[V]_d^G$ generates the algebra $k[V]^G$ then there is an algorithm that would explicitly determine a minimal system of generators of this algebra. Such a system can be constructed as follows.

First we remark that for each fixed d the subspace $k[V]_d^G$ of $k[V]_d$ can be described explicitly. Indeed, if G is finite then $k[V]_d^G$ is the linear span of the polynomials $(1/|G|)\sum_{g\in G} g \cdot f$, where f runs over a basis of $k[V]_d$ over k (say, all the monomials of degree d in $x_1, \dots, x_n$). If G is connected then $k[V]_d^G = \{f \in k[V]_d \mid X_i f = 0,\ i = 1, \dots, s\}$, where $X_1, \dots, X_s$ is a basis of $\mathfrak{g}$ (here $\mathfrak{g}$ acts naturally on $k[V]$ by differentiations). Hence in this case the problem reduces to solving a finite system of linear equations. The general case obviously reduces to the cases just considered.

Let us now determine the subspaces $k[V]_d^G$ for $d = 1, \dots, M$. Let d_1 be the minimal integer such that $0 < d_1 \le M$ and $k[V]_{d_1}^G \ne 0$ (if no such integer exists then $k[V]^G = k$). Let $f_1, \dots, f_q$ be a basis of $k[V]_{d_1}^G$ and let d_2 be the minimal integer such that $d_1 < d_2 \leq M$ and $k[V]^G_{d_2}$ is different from the linear span of those monomials in $f_1, \ldots, f_q$ that are contained in $k[V]^G_{d_2}$ (if no such integer exists then $f_1, \ldots, f_q$ is a minimal system of generators of $k[V]^G$). Let $g_1, \ldots, g_p$ be a basis of an arbitrary linear subspace of $k[V]^G_{d_2}$ complementary to the aforementioned linear span. Similarly, let d_3 be the minimal integer such that $d_2 < d_3 \leq M$ and $k[V]^G_{d_3}$ is different from the linear span of those monomials in $f_1, \ldots, f_q, g_1, \ldots, g_p$ that are contained in $k[V]^G_{d_3}$ (if no such integer exists then $f_1, \ldots, f_q, g_1, \ldots, g_p$ is a minimal system of generators of $k[V]^G$) and let $h_1, \ldots, h_r$ be a basis of a linear subspace of $k[V]^G_{d_3}$ complementary to this linear span, etc. This process will stop after a finite number of steps, and at the end we will obtain a finite set of polynomials $f_1, \ldots, f_q, g_1, \ldots, g_p, h_1, \ldots, h_r, \ldots$ (it is clear that in principle those polynomials can be found explicitly). It is not difficult to see that this set of polynomials will be a minimal generating set of $k[V]^G$ and that any such set can be obtained by this procedure. Therefore, in some sense a minimal system of homogeneous generators of $k[V]^G$ is determined uniquely; in particular the number of those generators (and their degrees) is determined uniquely and is equal to $\dim I_+/I_+^2$, where I_+ is the ideal of $k[V]^G$ generated by the elements of positive degree (i.e., this number is equal to $\operatorname{ed} k[V]^G$).

Thus the main problem of constructive invariant theory is equivalent to the problem of determining number M. As was mentioned in the Introduction, Noether proved that for a finite group G one can take $M = |G|$, which is a satisfactory solution of the main problem in this particular case. Because of that, from now on we will assume that G is a connected semisimple group. We will reduce the problem of determining M to the problem of finding an upper bound to degrees of some system of homogeneous parameters of $k[V]^G$; the latter problem will be solved in the next section. To accomplish this reduction we are going to establish some remarkable symmetry properties of the Poincaré series of $k[V]^G$. Those properties will also be heavily used in Chapter 3.

4. According to the theorem of Hochster and Roberts (see [85]), $k[V]^G$ is a Cohen-Macaulay algebra. This means that if $\theta_1, \ldots, \theta_m$ is a system of homogeneous parameters of $k[V]^G$, i.e., a system of algebraically independent elements for which $k[V]^G$ is integral over $P = k[\theta_1, \ldots, \theta_m]$, then $k[V]^G$ is a free P-module. Let $\eta_1, \ldots, \eta_l$ be a system of homogeneous generators of this module, so that $k[V]^G = P\eta_1 \oplus \cdots \oplus P\eta_l$. We will assume that

$$\deg \theta_i = d_i, \quad d_1 \leq \cdots \leq d_m; \qquad \deg \eta_j = e_j, \quad 0 = e_1 \leq \cdots \leq e_l. \tag{2}$$

It follows from (2) that the Poincaré series of $k[V]^G$ is

$$F(k[V]^G, x) = \sum_{d=0}^{\infty} \dim k[V]_d^G x^d = \frac{\sum_{i=1}^{l} x^{e_i}}{\prod_{j=1}^{m}(1 - x^{d_j})}. \tag{3}$$

Because G is a connected semisimple group, it follows from [85] that the algebra $k[V]^G$ is Gorenstein. This implies that (and in fact, by a theorem of Stanley, is equivalent to) $F(k[V]^G, x)$ satisfies the following functional equation:

$$F(k[V]^G, x^{-1}) = (-1)^m x^q F(k[V]^G, x), \quad \text{where } q \in \mathbb{Z}. \tag{4}$$

Utilizing (2) and (3) we see that (4) is equivalent to the condition

$$d_1 + \cdots + d_m - e_{l-i+1} = q + e_i, \qquad i = 1, \dots, l. \tag{5}$$

In particular it follows from (5) that

$$e_l = d_1 + \cdots + d_m - q \tag{6}$$

and that $e_i + e_{l-i+1} = e_l$, $i = 1, \dots, l$.

Note that in principle we can calculate the rational function $F(k[V]^G, x)$. Indeed, because $\operatorname{char} k = 0$ we can assume, without loss of generality, that $k = \mathbb{C}$ (say, by the Lefshetz principle). In this case the classical formula of Molien–Weyl along with Weyl's integral formula reduces the calculation of $F(k[V]^G, x)$ to the evaluation of an integral of a certain function over a maximal torus of a maximal compact subgroup in G (see Chapter 3, §1, Equation (6)). This integral, in turn, can be reduced to the evaluation of integrals over the unit circle in $\mathbb{C}$, which can be accomplished using residues.

Because in principle we can explicitly find $F(k[V]^G, x)$, we can also find q. In fact, for our purposes we need q and not $F(k[V]^G, x)$. Several results are now known about q; for example, $q \geq 0$ (see [97]). In Chapter 3 we will obtain much more detailed information about the meaning and properties of this important constant (for example, we will show that if V is an irreducible G-module, then almost always $q = \dim V$).

Obviously $\theta_1, \dots, \theta_l, \eta_1, \dots, \eta_l$ are homogeneous generators of $k[V]^G$. Therefore, by (2) and (6), we can set $M = \max(d_1 + \cdots + d_m - q, d_m)$.

In the general case, however, we do not know the numbers $d_1, \dots, d_m$. Moreover we cannot directly extract upper bounds on those numbers from the function $F(k[V]^G, x)$ (for example, instead of $\theta_1, \dots, \theta_m$ we can take $\theta_1^{s_1}, \dots, \theta_m^{s_m}$ with arbitrary $s_i \in \mathbb{Z}^+$, $i = 1, \dots, m$, and it will also be a system of homogeneous parameters of $k[V]^G$). Thus we have to find an a priori upper bound on the numbers $d_1, \dots, d_m$.

For example, if we know that $d_m \leq D$, then we can set $M = \max(mD - q, D) \leq mD$. The number m is equal to the order of the pole of $F(k[V]^G, x)$ at $x = 1$, and, therefore, in principle it can be explicitly determined. The number m can also be found from the relation $m = n - \dim Gv = n - \dim G + \dim\{X \in \mathfrak{g} \mid Xv = 0\}$, where v is a point in general position in V (see [30]). For connected simple groups G and those connected semisimple G for which the G-module V is irreducible, the number m was found in [26–28, 52, 53] (we note however that for our purposes the inequality $m < n$ will be sufficient). In the next section we will determine the number D.

§2. A bound on the degree of a system of parameters and the main theorem

1. Let $\mathfrak{N}$ denote the null-variety $\pi_{V,G}^{-1}(\pi_{V,G}(0))$. Thus $\mathfrak{N}$ is the zero set of the maximal homogeneous ideal I_+ of $k[V]^G$ and $\mathfrak{N} = \{v \in V \mid 0 \in \overline{Gv}\}$. The points of $V - \mathfrak{N}$ are called semistable [114, 117, 21].

Let us assume for a moment that we found an integer $C > 0$ such that

$$\text{for any semistable point } v \in V \text{ there exists a homogeneous polynomial } f_v \in k[V]^G \text{ of degree } \leq C \text{ such that } f_v(v) \neq 0. \tag{7}$$

Let A be the subalgebra of $k[V]^G$ generated by the subspace $\bigoplus_{d=0}^{C} k[V]_d^G$. Let $h_1, \dots, h_d$ be a minimal system of homogeneous generators of A. In principle we can find such a system explicitly using the algorithm described in §1. Clearly, $\deg h_i \leq C$ for all i. Because of (7) we have $\mathfrak{N} = \{v \in V \mid h_i(v) = 0 \ \forall i\}$. By a theorem of Hilbert (see [81]), this means that $k[V]^G$ is integral over A. Hence any system of homogeneous parameters of A will also be a system of homogeneous parameters of $k[V]^G$. By Noether's normalization lemma [5] we can choose m elements of A, which are linear combinations of the generators $h_1, \dots, h_d$, so that A will be integral over the subalgebra generated by these elements. If the degrees of the h_i were the same then the above elements would be homogeneous and we would get a system of homogeneous parameters of A. However, in general the degrees of h_i need not coincide and we have to pass from A to its "Veronese subalgebra" $B = k[h_1^{\widetilde{C}/\deg h_1}, \dots, h_d^{\widetilde{C}\deg h_d}]$, where

$$\widetilde{C} = \operatorname*{LCM}_{1\leq i\leq d}(\deg h_i) \leq \operatorname*{LCM}_{a\in\mathbb{Z},\ 0<a\leq C} a. \tag{8}$$

Because A is integral over B and $\deg(h_i^{\widetilde{C}/\deg h_i}) = \widetilde{C}$ for any $i = 1, \dots, d$, the normalization lemma applied to B yields a system of homogeneous parameters (of degree $\widetilde{C}$) of B, which is also a system of homogeneous parameters of $k[V]^G$. We now see that for the constant D from §1 one can take $\widetilde{C}$ or, less precisely (but avoiding the explicit determination of the polynomials $h_1, \dots, h_d$), the right-hand side of (8). Thus we need to find a constant C satisfying (7).

2. We shall now solve this problem. Let $\dim G = s$. From now on we will identify k^* with the subgroup of scalar matrices of GL_n. Consider the subgroup $H = k^*G$ of GL_n. This is an $(s+1)$-dimensional connected reductive subgroup with a one-dimensional center. We also consider its closure

$\overline{H}$ in Mat_n. The group H acts on Mat_n by left multiplication and both H and $\overline{H}$ are invariant under this action.

Let $\overline{x}_{ij}$ be the restriction of x_{ij} to $\overline{H}$. The algebra $k[\overline{H}]$ is generated by the elements x_{ij}, $1 \le i,\ j \le n$, $k[\overline{H}] = k[\dots, \overline{x}_{ij}, \dots]$, and is $\mathbb{N}$-graded ($\deg \overline{x}_{ij} = 1$, $1 \le i,\ j \le n$).

Let v be a point of V. Both the orbit Hv and its closure $\overline{Hv}$ in V are cones, i.e., they are invariant under the action of k^*. Let $\overline{x}_i$, $1 \le i \le n$, be the restriction of x_i to $\overline{Hv}$. The algebra $k[\overline{Hv}]$ is generated by the elements $\overline{x}_i$, $1 \le i \le n$, $k[\overline{Hv}] = k[\overline{x}_1, \dots, \overline{x}_n]$ and is $\mathbb{N}$-graded ($\deg \overline{x}_i = 1$, $1 \le i \le n$). We remark that in both cases the structure of the $\mathbb{N}$-graded algebra in question is determined by its weight decomposition with respect to the action of the torus k^*, i.e., a function f is homogeneous of degree d if and only if $f(tx) = t^d f(x)$ for each $t \in k^*$ and x.

PROPOSITION 1. *A point $v \in V$ is semistable if and only if $k[\overline{Hv}]^G \ne k$.*

PROOF. Suppose v is semistable. Then $0 \notin \overline{Gv}$. Hence 0 and $\overline{Gv}$ can be separated by G-invariants (see [114]) and therefore $k[\overline{Hv}]^G \ne k$.

Conversely, suppose that $k[\overline{Hv}]^G \ne k$. It is clear that $k[\overline{Hv}]^G$ is a graded subalgebra of $k[\overline{Hv}]$. Let $f \in k[\overline{Hv}]^G$ be an element of positive degree. Being homogeneous, f satisfies the following conditions: $f(0) = 0$ and $f(tgv) = t^{\deg f} f(gv) = t^{\deg f} f(v)$. Because Hv is dense in $\overline{Hv}$, we have $f(v) \ne 0$, and therefore $\overline{Gv}$ is contained in the closed subset $\{u \in \overline{Hv} \mid f(u) = f(v)\} \not\ni 0$. □

Consider the morphism $\alpha : \overline{H} \to \overline{Hv}$ defined by the formula $\alpha(h) = hv$ (multiplication of a matrix by a column). It commutes with the action of H. It is clear that the corresponding comorphism of the $\mathbb{N}$-graded algebras $\alpha^* : k[\overline{Hv}] \to k[\overline{H}]$ is a degree-preserving monomorphism. If $x_i(v) = a_i$, $1 \le i \le n$, then $\alpha^*(\overline{x}_i) = a_1\overline{x}_{i1} + \dots + a_n\overline{x}_{in}$. Therefore

$\alpha^*(k[\overline{Hv}])$ is an H-invariant graded subalgebra of $k[\overline{H}]$ generated by the n homogeneous elements $a_1\overline{x}_{i1} + \dots + a_n\overline{x}_{in}$ of degree 1. (9)

From Proposition 1 we now have

PROPOSITION 2. *A point $v \in V$ is semistable if and only if $\alpha^*(k[\overline{Hv}]) \cap k[\overline{H}]^G \ne k$.*

An advantage of this reformulation is that much can be said about the structure and properties of the algebra $k[\overline{H}]^G$. More precisely, consider an arbitrary rational character f of H, i.e., a homomorphism $H \to k^*$. We will identify it with the corresponding rational function on $\overline{H}$, which is regular on H. This function is G-invariant because G is semisimple. The set S of all characters of H regular on the entire set $\overline{H}$ is a semigroup (with unit) under multiplication, which is a subsemigroup in the group of all characters of H. The elements of S are homogeneous elements of $k[\overline{H}]^G$.

PROPOSITION 3. 1) *$k[\overline{H}]^G$ is a vector space generated by S over k,*

2) $\dim k[\overline{H}]_d^G \leq 1$ *for any $d \in \mathbb{N}$,*

3) $k[\overline{H}]^G \neq k$,

4) *S is a finitely generated semigroup and the map $f \mapsto \deg f$ is an isomorphism between S and a subsemigroup of $\mathbb{N}$.*

PROOF. 1) Let f be a nonzero homogeneous element of $k[\overline{H}]^G$, $\deg f = d$. Then $f(tg) = t^d f(g) = t^d f(1_n)$ for any $t \in k^*$ and $g \in G$ (here 1_n is the identity of G). Because H is dense in $\overline{H}$ we have $f(1_n) \neq 0$. Multiplying f by a suitable constant we may assume that $f(1_n) = 1$. Therefore, $f(tg) = t^d$, i.e., $f \in S$.

2) Let f and h be nonzero elements of $k[\overline{H}_d]^G$. They are semi-invariants of the torus k^* of the same weight. Hence the rational function f/h is an H-invariant and, therefore, $f/h \in k$.

3) It is clear that the function det: $(a_{ij}) \mapsto \det(a_{ij})$ is a nonconstant homogeneous element of $k[\overline{H}]^G$ of degree n.

4) $k[\overline{H}]^G$ is finitely generated because G is reductive. It follows from 1) and 2) that S is finitely generated. The algebra $k[\overline{H}]$ is $\mathbb{N}$-graded and the elements of S are homogeneous. Hence $\deg f \in \mathbb{N}$ for any $f \in S$, whence 4). □

Note that if for some reason it is known that $\overline{H}$ is a normal algebraic variety, then S can be described more precisely.

PROPOSITION 4. *If $\overline{H}$ is normal then S is a semigroup with one generator. This generator is, in turn, a generator of the group of characters of H.*

PROOF. The group of characters of H is isomorphic to $\mathbb{Z}$ because the center of H is one dimensional. Take any (of the two possible) generator of this group, say, the generator Φ uniquely determined by the condition $\det = \Phi^d$ for some $d > 0$. But $\det \in k[\overline{H}]$. Therefore Φ is integral over $k[\overline{H}]$. Because $\overline{H}$ is normal we have $\Phi \in S$. Thus $\Phi^h \in S$, $h = 0, 1, \ldots$. On the other hand, $\Phi(0) = 0$ and therefore $\Phi^h \notin S$ if $h < 0$. □

REMARK. It is easy to see that $\Phi(tg) = t^{|k^* \cap G|}$, $t \in k^*$, $g \in G$ and therefore $\deg \Phi = |k^* \cap G|$. In particular, if the center of G is trivial then $\Phi(tg) = t$.

In general we do not know if $\overline{H}$ is normal. Therefore, in the explicit constructions we can use only those elements that are definitely contained in S, and those elements are $\det^d$, $d \in \mathbb{N}$. However they are sufficient for our purposes.

The condition from Proposition 2 is equivalent to the condition $\alpha^*(k[\overline{Hv}]) \cap S \neq 1$. When is it satisfied? The answer is given by the following:

PROPOSITION 5. *Let F be a graded H-invariant k-subalgebra of $k[\overline{H}]$, $k \subset F$. Then the following conditions are equivalent:*

1) $F \cap S \neq 1$.
2) *There exists* $f \in S$, $f \neq 1$, *such that* f *is integral over* F.
3) *Any* $h \in S$ *is integral over* F.
4) $k[\overline{H}]^G$ *is integral over* F.

PROOF. 2) $\Leftrightarrow$ 3). It follows from Proposition 3, part 4) that $f^{\deg h} = h^{\deg f}$. Because f is integral over F, $f^{\deg h}$ is also integral over F. Hence $h^{\deg f}$ is integral over F, and the inequality $\deg f > 0$ implies that h is also integral over F.

3) $\Leftrightarrow$ 4). This follows from Proposition 3, part 4).

1) $\Leftrightarrow$ 2). Let $f^d + f^{d-1}b_1 + \cdots + fb_{d-1} + b_d = 0$, $b_i \in F$, $1 \leq i \leq d$ be an integral dependence equation for f over F. Because f is homogeneous, we may assume that all b_i, $1 \leq i \leq n$, are homogeneous. We now average both sides of this equation over G by applying the Reynolds operator $\natural$ (see [9]). Because f is G-invariant we have $f^d + f^{d-1}b_1^\natural + \cdots + fb_{d-1}^\natural + b_d^\natural = 0$, where $b_i^\natural \in k[\overline{H}]^G$, $1 \leq i \leq d$. It is clear that F is invariant under $\natural$, and therefore $b_i^\natural \in F \cap k[\overline{H}]^G$, $1 \leq i \leq d$. Moreover, we have $\deg b_i^\natural = i \deg f \leq d \cdot \deg f$, $1 \leq i \leq d$.

Because $f \neq 1$, we can find i for which $b_i^\natural$ is a nonconstant homogeneous G-invariant. By Proposition 3, statements 1) and 2) we can multiply this invariant by a suitable constant from k and obtain an element of S. Hence $F \cap S \neq 1$. □

COROLLARY 1. $F \cap k[\overline{H}]^G \neq k$ *if and only if* det *is integral over* F.

From the proof of Proposition 5 we obtain

COROLLARY 2. *If* $f \in S$, $f \neq 1$, *is an element integral over* F, *then there exists a nonconstant homogeneous G-invariant element in* F *of degree* $\leq d \cdot \deg f$, *where* d *is the minimal degree of an integral dependence equation of* f *over* F.

Applying this to the case $F = \alpha^*(k[\overline{Hv}])$ and utilizing (9) along with Proposition 2 (and the proof of Proposition 1) we obtain

PROPOSITION 6. *A point*

$$v = \begin{pmatrix} a_1 \\ \vdots \\ a_n \end{pmatrix} \in V$$

is semistable if and only if the function det $\in k[\overline{H}]$ *is integral over the graded subalgebra* A_v *of* $k[\overline{H}]$ *generated by the elements* $a_1\overline{x}_{i1} + \cdots + a_n\overline{x}_{in}$, $1 \leq i \leq n$, *of degree* 1. *If* d_0 *is the minimal degree of an integral dependence equation of* det *over* A_v *then there exists a nonconstant homogeneous polynomial of degree* $\leq d_0 n$ *that is contained in* $k[V]^G$ *and does not vanish at* v.

Thus we see that a constant C satisfying (7) will be determined if we can find a number that is bigger than or equal to the minimal degree of an integral dependence equation of det over A_v for any semistable point v. We shall now show how to find such a number.

3. Let $v \in V$ be a semistable point. Because A_v is generated by homogeneous elements of degree 1, there exists a system of homogeneous parameters $b_1, \dots, b_h$ of A_v with $\deg b_i = 1$, $1 \le i \le h$. Because det is integral over A_v, it is also integral over $B_v = k[b_1, \dots, b_h]$. Therefore, the minimal degree of an integral dependence equation of det over A_v is not bigger than that over B_v. But B_v is integrally closed as $b_1, \dots, b_h$ are algebraically independent. Therefore, the coefficients of a minimal equation for det over $k(b_1, \dots, b_h)$ belong to B_v. We now choose elements $b_{h+1}, \dots, b_{s+1}$ among the elements $\overline{x}_{ij}$, $1 \le i, j \le n$, in such a way that $b_1, \dots, b_{s+1}$ are algebraically independent over k (which can always be done). It is easy to see that a minimal polynomial for det over $k(b_1, \dots, b_h)$ will automatically be a minimal polynomial for det over $k(b_1, \dots, b_{s+1})$. Consequently, for our purposes it suffices to find an upper bound (which would not depend on v) on the degree of a minimal polynomial for det over $k(b_1, \dots, b_{s+1})$. This can be done as follows.

For each $p \in \mathbb{N}$ we consider the set of all monomials of the form $b_1^{i_1} \cdots b_{s+1}^{i_{s+1}} \det^i$, $i_1 + \cdots + i_{s+1} + in = p$. If for some $p = p_0$ these monomials happen to be linearly dependent over k then the degree of a minimal polynomial for det over $k(b_1, \dots, b_{s+1})$ is not bigger than p_0/n. Indeed, if for $p = p_0$ a nontrivial linear combination

$$\sum_{i_1 + \cdots + i_{s+1} + ni = p} \alpha_{i_1 \cdots i_{s+1} i} b_1^{i_1} \cdots b_{s+1}^{i_{s+1}} \det^i, \tag{10}$$

where $a_{i_1 \cdots i_{s+1} i} \in k$, is equal to zero then there exists at least one nonzero coefficient $\alpha_{i_1 \cdots i_{s+1} i}$, with $i \ne 0$ (this follows from the algebraic independence of $b_1, \dots, b_{s+1}$). Because $i \le p_0/n$ for each $\alpha_{i_1 \cdots i_{s+1} i}$, we see that det is a root of an equation of degree $\le p_0/n$ with coefficients in $k(b_1, \dots, b_{s+1})$.

Thus we have to find p_0. From the technical point of view it is more convenient to consider only the monomials for which

$$p = nc, \qquad i_1 = nj_1, \dots, i_{s+1} = nj_{s+1}, \tag{11}$$

where $c, j_1, \dots, j_{s+1}$ are integers. We shall find a constant $c = c_0$ such that the monomials satisfying (11) are linearly dependent over k. In this case the degree of a minimal polynomial for det over $k(b_1, \dots, b_{s+1})$ will not exceed c_0.

4. Let $\mathfrak{n}^+$ and $\mathfrak{n}^-$ be a couple of "opposite" maximal nilpotent subalgebras of $\mathfrak{g}$ corresponding to a certain choice of positive and negative roots with respect to T. Let $Y_1, \dots, Y_d$ and $Z_1, \dots, Z_d$ be bases of the linear spaces $\mathfrak{n}^+$ and $\mathfrak{n}^-$, where $d = (s-r)/2$, where $y_1, \dots, y_d$ and $z_1, \dots, z_d$ are the corresponding coordinate functions on those spaces.

Consider the variety $\mathfrak{n}^+ \times (k^*)^r \times \mathfrak{n}^-$. The functions $y_1, \dots, y_d, z_1, \dots, z_d, t_1, t_1^{-1}, \dots, t_r, t_r^{-1}$ can be naturally identified with regular functions on this variety. It is clear that $k[\mathfrak{n}^+ \times (k^*)^r \times \mathfrak{n}^-]$, its generated by those functions.

For an arbitrary nilpotent matrix $X \in \mathrm{Mat}_n$, its exponential $\exp X$ is the matrix

$$1_n + \frac{1}{1!}X + \frac{1}{2!}X^2 + \dots + \frac{1}{(n-1)!}X^{n-1}, \tag{12}$$

because $X^p = 0$ for $p \geq n$.

It is well known (see [50]) that the morphism $\beta : \mathfrak{n}^+ \times (k^*)^r \times \mathfrak{n}^- \to \overline{H}$ given by the formula (see (1))

$$\beta(X^+, (a_1, \dots, a_r), X^-) = \exp X^+(f_{ij}(a_1, \dots, a_r)) \exp X^-,$$

is an isomorphism between $\mathfrak{n}^+ \times (k^*)^r \times \mathfrak{n}^-$ and an open subset of G (the "big cell").

Now consider the matrices (with coefficients in $k[\mathfrak{n}^+ \times (k^*)^r \times \mathfrak{n}^-]$) : $Y = \sum_{i=1}^d y_i Y_i$, $Z = \sum_{i=1}^d z_i Z_i$. Let $\exp Y$ and $\exp Z$ be the results of substitution of Y and, respectively, Z into (12). Then

$$\begin{array}{l}\beta^*(\overline{x}_{ij}) \text{ is the element of the matrix } \exp Y(f_{pq}(t_1, \dots, t_r)) \exp Z \\ \text{at the intersection of the } i\text{th row and the } j\text{th column.}\end{array} \tag{13}$$

Let t denote the maximum of the numbers $|m_l|$ over all l, $1 \leq l \leq r$ and over all monomials $t_1^{m_1} \cdots t_r^{m_r}$ entering at least one rational function $f_{ij}(t_1, \dots, t_r)$, $1 \leq i, j \leq n$ from (1) with a nonzero coefficient. (For example, if G is the image of SL_d under the natural representation in the space of degree p forms in d variables then one can take $t = p$.)

It follows from (13), (12), and the definition of t, that $\beta^*(\overline{x}_{ij})$ is a k-linear combination of monomials $y_1^{r_1} \cdots y_d^{r_d} z_1^{q_1} \cdots z_d^{q_d} t_1^{m_1} \cdots t_r^{m_r}$, where $r_i, q_i \in \mathbb{N}$, $1 \leq i \leq d$, $m_j \in \mathbb{Z}$, $1 \leq j \leq r$, and $0 \leq r_1 + \dots + r_d \leq n-1$, $0 \leq q_1 + \dots + q_d \leq n-1$, $\max_{1 \leq i \leq r} |m_i| \leq t$.

Notice that function (10) is homogeneous, and therefore it vanishes if and only if its restriction to G vanishes. The latter condition is satisfied if and only if the image of this restriction under the comorphism β^*, i.e., the function

$$\sum_{i_1 + \dots + i_{s+1} + ni = p} \alpha_{i_1 \cdots i_{s+1} i} \beta^*(b_1^{i_1} \cdots b_{s+1}^{i_{s+1}} \det{}^i) \tag{14}$$

vanishes.

The function $b_1^{i_1} \cdots b_{s+1}^{i_{s+1}} \det^i$, $i_1 + \dots + i_{s+1} + ni = p$ is homogeneous of degree p and, therefore, (14) is a linear combination of monomials $y_1^{r_1} \cdots y_d^{r_d} z_1^{q_1} \cdots z_d^{q_d} t_1^{m_1} \cdots t_r^{m_r}$, where $r_i, q_i \in \mathbb{N}$, $1 \leq i \leq d$, $m_j \in \mathbb{Z}$, $1 \leq j \leq r$ and $0 \leq r_1 + \dots + r_d \leq p(n-1)$, $0 \leq q_1 + \dots + q_d \leq p(n-1)$,

$\max_{1\leq i\leq r}|m_i| \leq pt$, and the coefficients of this linear combination are k-linear combinations of the elements $\alpha_{i_1\cdots i_{s+1}i}$, $i_1+\cdots+i_{s+1}+ni=p$.

It is not difficult to see that the total number of such monomials in $y_1, \dots, y_d, z_1, \dots, z_d, t_1, \dots, t_r$ does not exceed $\binom{d+p(n-1)}{d}^2(2pt+1)^r$. On the other hand, assume that conditions (11) are satisfied. Then the total number of the coefficients $\alpha_{i_1\cdots i_{s+1}i}$ in (14) is $\binom{s+c+1}{c}$. Therefore if the inequality

$$\binom{d+cn(n-1)}{d}^2(2cnt+1)^r < \binom{s+c+1}{c} \tag{15}$$

holds for some c, then the monomials $b_1^{i_1}\cdots b_{s+1}^{i_{s+1}}\det^i$, $i_1+\cdots+i_{s+1}+in=p$ (for which (11) also holds) are linearly dependent over k. We shall now show that (15) holds if

$$c=\left[\frac{2^{s+r}n^s(n-1)^{s+r}t^r(s+1)!}{3^s((\frac{s-r}{2})!)^2}\right]. \tag{16}$$

Indeed, because $d=(s-r)/2\leq n(n-1)/2$, we have

$$\begin{aligned}\binom{d+cn(n-1)}{d}^2(2cnt+1)^r &< \frac{(cn(n-1)+d)^{2d}}{(d!)^2}(2cnt+1)^r\\ &= \frac{(c+\frac{d}{n(n-1)})^{s-r}n^{s-r}(n-1)^{s-r}}{((\frac{s-r}{2})!)^2}\left(c+\frac{1}{2nt}\right)^r(2nt)^r\\ &\leq \left(c+\frac{1}{2}\right)^s\frac{n^s(n-1)^{s-r}2^rt^r}{((\frac{s-r}{2})!)^2}.\end{aligned}$$

Moreover $\binom{s+c+1}{c} > (c^{s+1})/(s+1)!$. It now follows easily that (15) is satisfied if

$$\frac{n^s(n-1)^{s-r}2^rt^r(s+1)!}{((\frac{s-r}{2})!)^2}\leq c\left(\frac{2}{1+\frac{1}{2c}}\right)^s.$$

This inequality, in turn, is satisfied for c given by (16), because $1+1/2c\leq 3/2$.

5. Summarizing the results of §1 and parts 1–4 of §2 we obtain the main theorem of this chapter:

THEOREM. *Let $G\subset \mathrm{GL}_n$ be a connected semisimple algebraic group, $s=\dim G$ and $r=\operatorname{rk}G$. Let m be the transcendence degree of $k[V]^G$ over k and let t be the maximum of the numbers $|m_l|$ taken over all $l=1,\dots,r$ and all monomials $t_1^{m_1}\cdots t_r^{m_r}$ entering at least one rational function $f_{ij}(t_1,\dots,t_r)$, $1\leq i, j\leq n$, from (1). Then the degrees of the elements of a minimal system of homogeneous generators of $k[V]^G$ do not exceed the number*

$$m\cdot \mathrm{LCM}\left\{a\in\mathbb{Z},\ 0<a\leq\frac{2^{r+s}n^{s+1}(n-1)^{s-r}t^r(s+1)!}{3^s((\frac{s-r}{2})!)^2}\right\}.$$

§3. The radical of the ideal I and the approach suggested by Dieudonné and Carrell

1. In this section we shall discuss the approach to the main problem of constructive invariant theory suggested by Dieudonné, Carrell, and Mumford in [21] for the case when G is the image of SL_h under a polynomial representation. We shall show that this approach is flawed and does not lead to the goal even theoretically. Its drawback turns out to be related to a rather subtle property of the algebra $k[V]^G$: the ideal I generated by all homogeneous G-invariant polynomials of positive degree is not radical (in other words, the null-variety $\mathfrak{N} = \pi_{V,G}^{-1}(\pi_{V,G}(0))$ is not reduced).

2. The general strategy in [21] is aimed at finding a constant M' such that the ideal J of $k[V]$ generated by the nonconstant homogeneous invariants of degree $\leq M'$ defines in V exactly the null-variety $\mathfrak{N}$. To find M' the authors basically follow Hilbert's ideas from [81]. (We should remark however that the argument [21, p. 67] is false. Namely, the ring B' depends on point x_0, and therefore the minimal equations for $\det(Y_{ij})$ over the fields of quotients of the rings B' constructed for x_0 and $s \cdot x_0$ are a priori different. Therefore, the equality $Q_k((Y_{ij} \cdot (s \cdot x_0)) = Q_k((Y_{ij})) \cdot x_0)$ remains unproved. However the statement itself is correct, and one only has to change the proof using, for example, the Reynolds operator $\natural$ the way it was done in the proof of Proposition 5). Having found M', the authors then remark: "the Nullstellensatz shows that some power of every invariant is in J. As we may (at least theoretically) pick a finite system of generators of J of degree $\leq M'$, we are then reduced to forming a system of generators of the 'root' of J; Hilbert's own proof of his Nullstellensatz shows how this (again theoretically) may be done explicitly." However, this assertion leads to a solution only if $\sqrt{J}$ coincides with I (it is easy to see that the set of homogeneous invariants, which is a system of generators of I, is also a system of generators of $k[V]^G$). In general we only have $\sqrt{J} = \sqrt{I}$. Therefore $\sqrt{J} = I$ if and only if $\sqrt{I} = I$. It seems as if the equality $\sqrt{I} = I$ is considered in [21] as self-evident. However, it is not true in general, as the examples given below in part 3 show. We only remark here that if we know a basis of $\sqrt{I}$, but $\sqrt{I} \neq I$, we can say nothing about a possible basis of I (and therefore the approach suggested in [21] does not lead to a solution).

3. We shall now give some examples. Consider the general case when G is a reductive subgroup of GL_n. First of all, it is clear that if G is finite then $\sqrt{I}$ always coincides with the ideal of $k[x_1, \dots, x_n]$ generated by $x_1, \dots, x_n$. Therefore $\sqrt{I} \neq I$, if G is nontrivial. It is more interesting (and more difficult) to construct an example when G is a connected semisimple group. In [79] one can find such an example with G isomorphic to SO_{2l+1}. But [21] deals with the classical case of polynomial representations of SL_d. Therefore, it would be more instructive to find an example related to this case. We shall now show that if G is the image of SL_2 under the natural representation in the space of degree 6 binary forms, then $\sqrt{I} \neq I$.

Let S_d denote the unique (up to isomorphism) irreducible SL_2-module of dimension $d+1$. It is known that the multiplicity $c(d, e)$ of S_d in $k[V]_e$ is equal to the dimension of the space of covariants of degree e binary form that are homogeneous of degree 6 in the coefficients and are of order d in the variables (i.e., $c(d, e) = m(e, 6, d)$ in the notation of [48]). Computing those numbers one way or another (for example, using the tables of Poincaré series from Sylvester's paper [153]) we obtain Table 1.

TABLE 1

e	1	2	3	4	5	6	8	10
d	2	4	2	4	2	4	4	4
$c(d, e)$	0	1	1	2	2	4	7	11

It is also known (see [48]) that $k[V]^G$ has a minimal system of homogeneous generators $f_2, f_4, f_6, f_{10}, f_{15}$, $\deg f_i = i$, with f_2, f_4, f_6, f_{10} algebraically independent and $f_{15}^2 \in k[f_2, f_4, f_6, f_{10}]$.

Clearly $I_5 = k[V]_3 f_2 \oplus k[V]_1 f_4$. It follows from the table above that the multiplicity of S_2 in I_5 is 1. Because $c(2, 5) = 2$, there exists a submodule S_2 of $k[V]_5$ that is not contained in I_5. Moreover, we have $I_{10} = k[V]_8 f_2 + k[V]_6 f_4 + k[V]_4 f_6 + k[V]_0 f_{10}$. It is also clear that the polynomials f_2, f_4, f_6 are irreducible. Therefore $k[V]_8 f_2 \cap k[V]_6 f_4 \cap k[V]_4 f_6 = 0$. However $k[V]_8 f_2 + k[V]_6 f_4 \cong (k[V]_8 f_2 \oplus k[V]_6 f_4)/(k[V]_8 f_2 \cap k[V]_6 f_4)$ and it is easy to see that $k[V]_8 f_2 \cap k[V]_6 f_4 = k[V]_4 f_2 f_4$. It now follows from the table above that the multiplicity of S_4 in I_{10} is $c(4, 8) + c(4, 6) - c(4, 4) + c(4, 4) = 11 = c(4, 10)$. Therefore any irreducible submodule S_4 of $k[V]_{10}$ is in fact contained in I_{10}. Consider now the submodule S_2 of $k[V]_5$ not contained in I_5, and let f be an element of this submodule, which is a highest vector (relative to a fixed maximal torus and a Borel subgroup B of SL_2). Then f is a semi-invariant of B of weight 2. But $f^2 \in k[V]_{10}$ is obviously a semi-invariant of B of weight 4. Therefore, f^2 is a highest vector of $S_4 \subset k[V]_{10}$. It follows that $f^2 \in I_{10}$. Thus $f \in k[V]$, $f \notin I$, but $f^2 \in I$, i.e, $\sqrt{I} \neq I$.

4. In conclusion we remark that the property of I to be radical could, probably, be considered "good" in the sense discussed in the Introduction.

CHAPTER 3

The Degree of the Poincaré Series of the Algebra of Invariants and a Finiteness Theorem for Representations with Free Algebra of Invariants

In this chapter the characteristic of k is assumed to be zero (in fact, it will be convenient to set $k = \mathbb{C}$, which would not result in loss of generality).

§1. The degree of the Poincaré series and a functional equation

1. Throughout this chapter G will denote a connected semisimple group. Depending on the context, we use either the language of G-modules or the language of representations (which are always assumed to be rational, see §1 of Chapter 1). Our aim is to prove that there exist only finitely many G-modules with free algebra of invariants and without nonzero fixed vectors (see Theorem 6 of §3); this is the case $d = 0$ of the finiteness theorem from part 10 of the Introduction. Our proof will be constructive: we will explicitly describe (by specifying the numerical labels of the highest weights) a finite list of G-modules containing all modules with free algebra of invariants. Notice that the methods developed in Chapter 2 in principle allow us to extract the desired modules from that list.

For our purposes we may (and will) assume that G is simply connected.

The proof relies heavily on several important properties of the Poincaré series of the algebra of invariants, and a significant part of this chapter is devoted to these properties. Besides the aforementioned finiteness theorem, we also use these properties in other applications, like the theory of Poincaré series, as well as invariant theory itself (see §§3 and 1).

2. Let V be an n-dimensional vector space and $\rho\colon G \to \mathrm{GL}(V)$ be a nontrivial representation. Let $k[V]^G$ denote $k[V]^{\rho(G)}$. Then all of the assertions of part 4 of §1, Chapter 2, hold. In particular, we have the functional equation for the Poincaré series $F(k[V]^G, x)$:

$$F(k[V]^G, x^{-1}) = (-1)^m x^q F(k[V]^G, x) \tag{1}$$

as well as formula (3) of Chapter 2, which represents this series as a rational function. Henceforth we will use the notation from part 4 of §1, Chapter 2, without further reference.

Let $m_{\rho(G)} = \max_{v\in V} \dim \rho(G)v$. Then the number m in (1) is equal to $n - m_{\rho(G)}$ (see [30]). Our main effort will be to obtain information about number q.

We first show that q has a simple meaning. Namely, by the degree of a rational function $f \in \mathbb{Q}(x)$ we will understand $\deg f = \deg S - \deg T$, where S and T are polynomials from $\mathbb{Q}[x]$ such that $f = S/T$ (this number does not depend on the choice of S and T).

THEOREM 1. *The number q defined by* (1) *is equal to* $-\deg F(k[V]^G, x)$.

PROOF. Considering the numerator and the denominator in the right-hand side of formula (3), Chapter 2, as S and T for $F(k[V]^G, x)$ we have $\deg T = d_1 + \cdots + d_m$, and, by inequalities (2) of Chapter 2, $\deg S = e_l$. Therefore $\deg F(k[V]^G, x) = e_l - (d_1 + \cdots + d_m)$, and the desired assertion follows from formula (6) of Chapter 2. □

COROLLARY 1. $\deg F(k[v]^{\rho(G)}, x) \leq 0$ *for any representation* $\rho\colon G \to \mathrm{GL}(V)$ *of a connected semisimple group* G.

PROOF. It follows from the fact that $q \geq 0$ (see [97]). □

From this corollary one can deduce (see, e.g., [148]) a formula for the dimension of the space of homogeneous invariants of a fixed degree:

COROLLARY 2. *There exist polynomials* $P_1, \ldots, P_t \in \mathbb{Q}[x]$ *and distinct complex numbers* $\nu_1, \ldots, \nu_t$ *that are roots of unity such that for any integer* $d \geq 0$ $\dim k[V]_d^{\rho(G)} = \sum_{i=1}^t P_i(d)\nu_i^d$.

For example, if $G = \mathrm{SL}_2$ and ρ is the adjoint representation then $t = 2$, $P_1 = P_2 = 1/2$ and $\nu_1 = \nu_2 = -1$.

3. Next we shall relate q to other important invariants of $k[V]^G$ given by its Poincaré series, the coefficients of its Laurent series.

It follows from (3), Chapter 2, that $F(k[V]^G, x)$ has a pole of order m at $x = 1$. Therefore the Laurent expansion near $x = 1$ is as follows:

$$F(k[V]^G, x) = \frac{\gamma}{(1-x)^m} + \frac{\tau}{(1-x)^{m-1}} + \cdots. \tag{2}$$

It now follows from formula (3), Chapter 2, that

$$\gamma = \lim_{x\to 1}(1-x)^m \frac{\sum_{j=1}^l x^{e_j}}{\prod_{i=1}^m (1-x^{d_i})} = \frac{l}{\prod_{i=1}^m d_i},$$

and

$$\begin{aligned}\tau &= -\lim_{x\to 1}\frac{d}{dx}\frac{\sum_{j=1}^{l}x^{e_j}}{\prod_{i=1}^{m}(1-x^{d_i})}(1-x^m)\\ &= -\lim_{x\to 1}\frac{d}{dx}\frac{\sum_{j=1}^{l}x^{e_j}}{\prod_{i=1}^{m}(1+x+\cdots+x^{d_i-1})}\\ &= \frac{-\sum_{j=1}^{l}e_j\prod_{i=1}^{m}d_i+\sum_{i=1}^{m}((\frac{d_i(d_i-1)}{2})\prod_{j\neq i}d_j)}{\prod_{i=1}^{m}d_i^2}\\ &= \frac{l\sum_{i=1}^{m}(d_i-1)-2\sum_{j=1}^{l}e_j}{2\prod_{i=1}^{m}d_i}.\end{aligned}$$

Therefore

$$\frac{2\tau}{\gamma}=\sum_{i=1}^{m}(d_i-1)-\frac{2}{l}\sum_{j=1}^{l}e_j. \tag{3}$$

Summing up equalities (5), Chapter 2, for all $i=1,\ldots,l$, we obtain

$$\sum_{i=1}^{m}d_i-q=\frac{2}{l}\sum_{j=1}^{l}e_j. \tag{4}$$

From (3) and (4) we now deduce

THEOREM 2. *$2\tau/\gamma=q-m$ and $q=2\tau/\gamma+\dim V-m_{\rho(G)}$.*

COROLLARY. 1) *$2\tau/\gamma$ is always an integer;*
2) *$q=\dim V\Leftrightarrow 2\tau/\gamma=(\dim\rho(G)v$, where $v\in V$ is a point in general position).*

To deduce this we only used the facts that $k[V]^G$ is Gorenstein and $m=n-m_{\rho(G)}$. Therefore, if these conditions are satisfied for a reductive (and not necessarily connected semisimple) group, then the same assertion is also true. In particular, if G is finite then the condition $m=n-m_{\rho(G)}$ is always satisfied (see [108]), and $k[V]^G$ is Gorenstein if, for example, $\rho(G)\subset \mathrm{SL}_n$ (see [149, 156]). It is a remarkable fact that for a finite G both γ and τ can be described entirely in terms of internal properties of G itself: $\gamma=1/|\rho(G)|$ and $\tau=$ (the number of pseudoreflections in $G)/2|\rho(G)|$. Hence in this case $2\tau/\gamma=$ (the number of pseudoreflections in $\rho(G)$).

If G is finite and $k[V]^G$ is Gorenstein then it follows from Theorem 2 that $q=\dim V+$ (the number of pseudoreflections in $\rho(G)$). If, for example, $\rho(G)\subset\mathrm{SL}_n$ then $q=\dim V$ (and conversely, if $k[V]^G$ is Gorenstein and $q=\dim V$ then $\rho(G)\subset\mathrm{SL}_n$ (see [149])).

Assume again that G is connected and semisimple. Unlike the case of a finite G, no internal characterization of γ and τ is known in this case. Nevertheless these coefficients still play an important role in invariant theory (see [48]). They were already subjects of interest for Hilbert who found an explicit formula for γ in the case of the (unique) irreducible $(d+1)$-dimensional representation ρ_d of SL_2. $(d \geq 3)$:

$$\gamma = \left\{ \begin{array}{ll} -\frac{1}{4}(d!)^{-1} \sum\limits_{0 \leq e < d/2} (-1)^e \binom{d}{e} \left(\frac{d}{2} - e\right)^{d-3}, & \text{if } d \equiv 1 \mod 2, \\ -\frac{1}{2}(d!)^{-1} \sum\limits_{0 \leq e < d/2} (-1)^e \binom{d}{e} \left(\frac{d}{2} - e\right)^{d-3}, & \text{if } d \equiv 0 \mod 2, \end{array} \right\} \tag{5}$$

as well as the asymptotic behavior of γ as $d \to \infty$ (see [48]). In §3, as an application of the aforementioned results about q, we will also determine τ.

The case of a finite G with $\rho(G) \subset \mathrm{SL}_n$ provides a hint that for a connected semisimple group G, the equality $q = \dim V$ is in a sense "typical" (although it is not always true as is seen from the examples in part 7 of §2 and part 2 of §3). In §2 we will show that this is indeed the case: $q = \dim V$ if ρ is "sufficiently large" (more precisely, if the so-called zonohedron of weights of ρ is sufficiently large, see part 6 of §2). The exact meaning of this assertion will become clear later (see Theorem 4 and the results of §2); here we only remark that it does not mean in general that $q = \dim V$ for all ρ except for a finite number (although this is the case when ρ is irreducible).

4. We shall now prove the above results about q. First we show that if ρ is "sufficiently large" then the functional equation (1) is a consequence of certain other functional equations.

Let K be a maximal connected compact subgroup of G and T be a maximal torus of K. Fixing a basis $\chi_1, \dots, \chi_r$ of the free abelian group $\mathrm{Hom}(T, k^*)$ of finite rank r, we set $\chi^a = \chi_1^{a_1}, \dots, \chi_r^{a_r}$ for any $a = (a_1, \dots, a_r) \in \mathbb{Z}^r$. Then $\chi\colon \mathbb{Z}^r \to \mathrm{Hom}(T, k^*)$, $a \mapsto \chi^a$ is a group isomorphism.

Let $\alpha_i = \chi^{c_i}$, $i = 1, \dots, p$ be the system of roots of K relative to T, and let W be the Weyl group of K. Because $k[V]^{\rho(G)} = k[V]^{\rho(K)}$, the Molien–Weyl formula (see, for example [48]), combined with Weyl's integral formula (see [11]), yields

$$\begin{aligned} F(k[V]^G, x) &= \int_K \frac{d\mu(g)}{\det(1 - x\rho(g))} \\ &= \frac{1}{|W|} \int_T \frac{(1-\chi^{c_1}(t)) \cdots (1-\chi^{c_p}(t))}{\det(1 - x\rho(t))} \, d\mu(t) \\ &= \frac{1}{|W|} \sum_{s=0}^{p} \sum_{1 \leq i_1 < \cdots < i_s \leq p} \int_T \frac{\chi^{c_{i_1} + \cdots + c_{i_s}}(t)}{\det(1 - x\rho(t))} \, d\mu(t), \qquad |x| < 1, \end{aligned} \tag{6}$$

where $d\mu(g)$ and $d\mu(t)$ are the Haar measures on K and T normalized by the conditions $\int_K d\mu(g) = \int_T d\mu(t) = 1$ (we set $c_{i_1} + \cdots + c_{i_s} = 0$ when $s = 0$).

For an arbitrary $a \in \mathbb{Z}^r$ let $k[V]^a$ denote the isotypic component of type χ^α of the T-module $k[V]$ (with respect to the action defined by ρ): $k[V]^a = \{f \in k[V] \| f(\rho(t^{-1})v) = x^a(t)f(v) \text{ for all } t \in T,\ v \in V\}$. The Poincaré series of this component is given by

$$\begin{aligned} F(k[V]^a, x) &= \sum_{d=0}^{\infty} \dim k[V]_d^a x^d \\ &= \int_T \frac{\chi^{-a}(t)\, d\mu(t)}{\det(1 - x\rho(t^{-1}))} = \int_T \frac{\chi^a(t)\, d\mu(t)}{\det(1 - x\rho(t))}, \qquad |x| < 1. \end{aligned} \tag{7}$$

This formula implies that the integrals in the right-hand side of (6) can be interpreted as the Poincaré series of certain isotypic components of the T-module $k[V]$, namely

$$F(k[V]^G, x) = \frac{1}{|W|} \sum_{s=0}^{p} (-1)^s \sum_{1 \le i_1 < \cdots < i_s \le p} F(k[V]^{c_{i_1} + \cdots + c_{i_s}}, x). \tag{8}$$

It turns out that this formula allows us to show that for a "sufficiently large" ρ the functional equation (1) is a consequence of the corresponding functional equations for $F(k[V]^{c_{i_1} + \cdots + c_{i_s}}, x)$.

More precisely, let $x_1, \ldots, x_b$ be a basis consisting of eigenvectors of the torus T in $V^* = k[V]_1$ and $\lambda_i = \chi^{b_i}$, $1, \ldots, n$, be the corresponding weights (i.e., $x_i \in k[V]^{b_i}$, $i = 1, \ldots, n$; notice that it may happen that $b_i = b_j$ for $i \neq j$). For any $t \in T$ the determinant of $\rho(t)$ is equal to $\chi^{b_1 + \cdots + b_n}(t)$. It follows from the triviality of $\operatorname{Hom}(G, k^*)$ that

$$b_1 + \cdots + b_n = 0. \tag{9}$$

It is now clear that for any $a \in \mathbb{Z}^r$ and $d \in \mathbb{N}$ the set of all monomials $x_1^{j_1} \cdots x_n^{j_n}$ with $j_1 + \cdots + j_n = d$, $j_1 \ge 0, \ldots, j_n \ge 0$, and

$$j_1 b_1 + \cdots + j_n b_n = a, \tag{$*$}$$

is a basis of $k[V]_d^a$. Consider now all n-tuples $j_1, \ldots, j_n$ of integers satisfying

$$j_1 b_1 + \cdots + j_n b_n = -a, \tag{$**$}$$

and the formal power series $\Phi_a(t_1, \ldots, t_n) = \sum t_1^{j_1} \cdots t_n^{j_n}$ (summation over all integers $j_1 \ge 0, \ldots, j_n \ge 0$ satisfying $(*)$) and $\overline{\Phi}_a(t_1, \ldots, t_n) = \sum t_1^{j_1} \cdots t_n^{j_n}$ (summation over all integers $j_1 > 0, \ldots, j_n > 0$ satisfying $(**)$). It is easy to see that both series are rational functions of $t_1, \ldots, t_n$ (see [147]). It follows easily from $(*)$, $(**)$, and (9) that

$$\begin{aligned} \overline{\Phi}_a(t_1, \ldots, t_n) &= t_1 \cdots t_n \Phi_{-a-b_1-\cdots-b_n}(t_1, \ldots, t_n) \\ &= t_1 \cdots t_n \Phi_{-a}(t_1, \ldots, t_n). \end{aligned} \tag{10}$$

The set of integral solutions $(j_1, \dots, j_n)$ of the system $j_1 b_1 + \dots + j_n b_n = 0$ is a subgroup of $\mathbb{Z}^n$ of rank $n - \dim \rho(T)$ (this number, in turn, is equal to the transcendence degree of the field of $\rho(T)$-invariant rational functions on V over k (see [131])). According to Stanley's terminology (see [147]), the system $(*)$ of diophantine equations has the R-property if the rational functions $\overline{\Phi}_a$ and Φ_a satisfy the following functional equation $\overline{\Phi}_a(t_1, \dots, t_n) = (-1)^{n-\dim \rho(T)} \Phi_a(t_1^{-1}, \dots, t_n^{-1})$. In view of (10), we can rewrite this equation as $\Phi_a(t_1^{-1}, \dots, t_n^{-1}) = (-1)^{n-\dim \rho(T)} t_1 \cdots t_n \Phi_{-a}(t_1, \dots, t_n)$. Because, obviously, $F(k[V]^a, x) = \Phi_a(x, \dots, x)$, it follows from the last equation that if for a given a the system $(*)$ has the R-property then $F(k[V]^a, x)$ satisfies the following functional equation:

$$F(k[V]^a, x^{-1}) = (-1)^{n-\dim \rho(T)} x^n F(k[V]^{-a}, x). \tag{11}$$

Because $\alpha_i^{-1} = \chi^{-l_i}$ is a root of K relative to T for each i, we deduce from (8) that if the system $(*)$ has the R-property for all $a = c_{i_1} + \dots + c_{i_s}$, $0 \le s \le p$, $1 \le i_1 < \dots < i_s \le p$, then the functional equation (1) is a consequence of the functional equations (11) corresponding to such a, and moreover $q = n$ and $m_{\rho(G)} \equiv \dim \rho(T) \bmod 2$.

5. We shall now utilize the Monster Reciprocity Theorem from [147], which gives a sufficient condition for a system to have the R-property. By this theorem, with each system $(*)$ one can associate (through the method described in [147]) a finite set of equations (in $j_1, \dots, j_n$) of the form

$$j_1 p_1 + \dots + j_n p_n = d, \qquad p_1, \dots, p_n, \ d \in \mathbb{Z}, \tag{12}$$

such that: a) each of the equations of (12) is a consequence of $(*)$, i.e., any (not necessarily integral) solution of $(*)$ is also a solution of each equation in (12), b) if each of the equations (12) has the R-property then the whole system $(*)$ has this property. On the other hand, let p_+ (resp., p_-) be the sum of positive (resp., negative) coefficients p_i of equations (12). Then, according to [147], (12) has the R-property if $-p_+ < d < -p_-$. It is easy to see that the existence of such inequalities is equivalent to the existence of a rational solution $(j_1, \dots, j_n) \in \mathbb{Q}^n$ of (12) with $-1 < j_1 \le 0, \dots, -1 < j_n \le 0$.

Thus, in view of a) and b), if $(*)$ has a rational solution $(j_1, \dots, j_n)$ satisfying $-1 < j_1 \le 0, \dots, -1 < j_n \le 0$, then it has the R-property.

At this point it is convenient to identify $\mathrm{Hom}(T, k^*)$ with $\mathbb{Z}^r$ via χ and use the additive notation for the operations in $\mathrm{Hom}(T, k^*)$.

Let us consider the $\mathbb{Q}$-vector space $X = \mathrm{Hom}(T, k^*) \otimes_{\mathbb{Z}} \mathbb{Q}$ and the set $Z_\rho^0 = \{\sum_{i=1}^n t_i \lambda_i \mid t_i \in \mathbb{Q}, \ 0 \le t_i < 1 \text{ for all } i\} \subset X$. Because the root system $\alpha_1, \dots, \alpha_p$ is invariant under the multiplication by -1, it follows that if each of the vectors $\alpha_{i_1} + \dots + \alpha_{i_s}$, $0 \le s \le p$, $1 \le i_1 < \dots < i_s \le p$ is in

Z_ρ^0 then $q = n$ and $m_{\rho(G)} \equiv \dim \rho(T) \bmod 2$ (compare with the notion of "critical character" in [150]). Now we are going to show that this assertion has an elegant geometric reformulation that is amenable to further investigation.

6. Let us consider the set $Z_\rho = \{\sum_{i=1}^n t_i\lambda_i \mid t_i \in \mathbb{Q},\ 0 \le t_i \le 1 \text{ for all } i\}$. It is a convex polyhedron, or more precisely a zonohedron, in X (we call it the *zonohedron of weights of* ρ). Evidently, it has the property

$$Z_{\gamma\oplus\cdots\oplus\mu} = Z_\gamma + \cdots + Z_\mu \quad \forall\gamma, \dots, \mu. \tag{13}$$

Notice that Z_ρ is the image of the n-cube $Q = \{\sum_{i=1}^n t_i\varepsilon_i | t_i \in \mathbb{Q},\ 0 \le t_i \le 1;\ i = 1, \dots, n\} \subset \mathbb{Q}^n$, where $\varepsilon_i = (0, \dots, 0, 1, 0, \dots, 0)$ (the unit is at the ith place), under the linear map $\varphi\colon \mathbb{Q}^n \to X$ defined by $\varphi(\varepsilon_i) = \lambda_i$, $i = 1, \dots, n$. The vertices of Q are all vectors $\varepsilon_{i_1} + \cdots + \varepsilon_{i_s}$, $0 \le s \le n$, $1 \le i_1 < \cdots < i_s \le n$. Because Q is the convex hull of the set of its vertices, Z_ρ is the convex hull of the vectors $\lambda_{i_1} + \cdots + \lambda_{i_s}$, $0 \le s \le n$, $1 \le i_1 < \cdots < i_s \le n$. Therefore, Z_ρ is contained in a convex set if and only if all these vectors belong to this set.

Notice that the set Z_ρ^0 is in fact the interior of Z_ρ (henceforth the topology is induced by the usual Hausdorff topology of $\mathbb{Q}$). Indeed, let $Y = \varphi(\mathbb{Q}^n)$. The map $\varphi\colon \mathbb{Q}^n \to Y$, being linear, is open and the set $Q^0 = \{\sum_{i=1}^n t_i\varepsilon_i | t_i \in \mathbb{Q},\ 0 < t_1 < 1;\ i = 1, \dots, n\}$ is the interior of the cube Q. From this we easily see that $\varphi(Q^0) = \{\sum_{i=1}^n t_i\lambda_i | t_i \in \mathbb{Q},\ 0 < t_i < 1$ for all $i\}$ is the interior of Z_ρ. Clearly, $\varphi(Q^0) \subset Z_\rho^0$, hence we must show only that each point of Z_ρ^0 is an interior point of Z_ρ.

Later we will show that 0 is an interior point of Z_ρ (see Proposition 3 and formula (13)). This implies that $Q^0 \cap \operatorname{Ker}\varphi \ne \varnothing$. Indeed, assuming the contrary, we get that there exists a support plane L for Q in $\mathbb{Q}^n$ containing $\operatorname{Ker}\varphi$. Let B_ε be the ball in $\mathbb{Q}^n$ with center 0 and radius $\varepsilon > 0$ (relative to the standard metric in $\mathbb{Q}^n$). Because $\dim Z_\rho = \dim Y$ and 0 is an interior point of Z_ρ, the set $\varphi(B_\varepsilon)$ is contained in $\varphi(Q^0)$ for a sufficiently small ε. Let L^+ and L^- be the closed half-spaces of $\mathbb{Q}^n$ associated with L, and assume that $Q \subset L^+$. Obviously, $B_\varepsilon \cap (L^- - L)$ is not empty for any $\varepsilon > 0$ and let β be a vector from this intersection. Then for a sufficiently small $\varepsilon > 0$ we have $\varphi(\beta) \subset \varphi(Q^0)$ and therefore $\varphi^{-1}(\varphi(\beta)) \cap Q \ne \varnothing$. But $\varphi^{-1}(\varphi(\beta)) = \beta + \operatorname{Ker}\varphi$, hence in view of the inclusion $\operatorname{Ker}\varphi \subset L$, we also have $\varphi^{-1}(\varphi(\beta)) \subset L^- - L$ and therefore $\varphi^{-1}(\varphi(\beta)) \cap Q = \varnothing$, a contradiction.

Thus $Q^0 \cap \operatorname{Ker}\varphi \ne \varnothing$. It follows that $\varepsilon Q^0 \cap \operatorname{Ker}\varphi \ne \varnothing$ for any rational $\varepsilon > 0$. Let now α be an element of Z_β^0, i.e., $\alpha = \sum_{i=1}^n t_i\lambda_i$, $t_i \in \mathbb{Q}$, $0 \le t_i < 1$, $i = 1, \dots, n$. Take any rational ε, $0 < \varepsilon < \min_{1 \le i \le n}(1 - t_i)$ and let $\beta \in \varepsilon Q^0 \cap \operatorname{Ker}\varphi$ be an arbitrary vector. Because $\varphi(\beta) = 0$, φ is open, $\dim Z_\rho = \dim Y$, and 0 is an interior point of Z_ρ, we can find an open ball U with center β such that $U \subset \varepsilon Q^0$ and $\varphi(U)$ is an open subset of Z_ρ containing 0. By the linearity of φ we have $\varphi(\alpha + U) = \varphi(\alpha) + \varphi(U)$. Therefore $\varphi(\alpha + U)$ is an open set containing $\varphi(\alpha)$. We want to show that this set is contained in Z_ρ. Let $u \in U$. Then, by construction, $u = \sum_{i=1}^n u_i\varepsilon_i$, $u_i \in \mathbb{Q}$, $0 < u_i < \varepsilon$; $i = 1, \dots, n$. Therefore $\varphi(\alpha + u) = \sum_{i=1}^n (t_i + u_i)\lambda_i$ is contained in Z_ρ because $t_i + u_i \leq 1$ for each i. This proves that Z_ρ^0 is the interior of Z_ρ.

Taking into account that Z_ρ^0 is convex, we deduce from Theorem 1 the following

THEOREM 3. *If the zonohedron Z_{Ad} is contained in the interior of the zonohedron Z_ρ,*

$$Z_{\mathrm{Ad}} \subset Z_\rho^0, \tag{14}$$

then $\deg F(k[V]^{\rho(G)}, x) = -\dim V$ *and* $m_{\rho(G)} \equiv \dim \rho(T) \bmod 2$.

7. EXAMPLE. Let $G = \mathrm{SL}_2$. In this case $\dim X = 1$ and we can identify X with $\mathbb{Q}$, the lattice of weights with $\mathbb{Z}$, and the lattice of roots with $2\mathbb{Z}$. Then $r = 2$, $\alpha_1 = 2$, $\alpha_2 = -2$, $Z_{\mathrm{Ad}} = \{a \in \mathbb{Q} \mid |a| \leq 2\}$. The weights of the irreducible representation ρ_d with highest weight $d > 0$ are d, $d-2, d-4, \dots, -d$, so that $Z_{\rho_d} = \{a \in \mathbb{Q} \mid |a| \leq d(d+2)/4\}$ if d is even and $Z_{\rho_d} = \{a \in \mathbb{Q} \mid |a| \leq (d+1)^2/4\}$ if d is odd. Thus, if $\rho = \rho_d$ condition (14) is not satisfied for an even d when $d(d+2)/4 \leq 2$ (i.e., for $d = 2$ only) and for an odd d if $(d+1)^2/4 \leq 2$ (i.e., for $d = 1$ only). In other words, if $d \geq 3$ then condition (14) for $\rho = \rho_d$ is always satisfied. By the property (13), the only representations ρ of SL_2 without nonzero fixed vectors for which condition (14) is not satisfied, are ρ_1, ρ_2, and $\rho_1 \oplus \rho_2$. Their algebras of invariants and Poincaré series are well known (see [48]), and using Theorems 3 and 1 we now obtain complete information about number q for representations of SL_2 (without nonzero fixed vectors): $q = n - 2$ for ρ_1 and $\rho_1 \oplus \rho_1$; $q = n - 1$ for ρ_2, and $q = n$ for other representations (we should remark that number q for the irreducible representations of SL_2 was calculated by a different method in [48]). Thus, by Theorem 1, we also obtain complete information about the degrees of Poincaré series of algebras of invariants of SL_2. Until now such information has only been known, through the explicit formulas for Poincaré series, for ρ_d with $d \leq 16$ (see [65]), and $\rho_d \oplus \rho_1$ with $d \leq 12$ (see [153]).

This information makes it possible to obtain a formula for the second coefficient τ of the Laurent series (2) in the case $\rho = \rho_d$ (the first coefficient γ is given by Hilbert's formula (5)). More precisely, because $m_{\rho(G)} = 3$ when $d \geq 3$ (see [52]), and because of Theorem 2, $\tau = 3\gamma/2$ when $d \geq 3$. Therefore the asymptotic behavior of τ when $d \to \infty$ can be obtained from the asymptotic behavior of γ, which was described by Hilbert.

In §3 we will generalize these results to an arbitrary simple group.

§2. The zonohedron of weights

1. Now we are going to investigate Condition (14) by geometric methods. In this section we will estimate the "size" of the zonohedron of weights or, more precisely, the radius of the ball with center at the origin inscribed in it.

First we introduce some new notation. Let $\alpha_1, \dots, \alpha_r$ be a fixed system of simple roots in the set $\alpha_1, \dots, \alpha_p$; $\omega_1, \dots, \omega_r$ be the corresponding system of weights; $Q = \{\sum_{i=1}^r n_i\alpha_i | n_i \in \mathbb{Z}\}$ be the lattice of roots and $P = \{\sum_{i=1}^r n_i\omega_i | n_i \in \mathbb{Z}\}$ be the lattice of weights. Let

$$C = \left\{ \sum_{i=1}^r t_i\omega_i \mid t_i \in \mathbb{Q},\, t_i \geq 0 \text{ for all } i \right\}$$

be the Weyl chamber corresponding to the above choice of simple roots. Consider also the cone $A = \{\sum_{i=1}^r t_i\alpha_i | t_i \in \mathbb{Q},\ t_i \leq 0 \text{ for all } i\}$ and the hyperplanes $\Gamma_j = \{\sum_{i=1, i\neq j}^r t_i\alpha_i | t_i \in \mathbb{Q}\}$, $j = 1, \dots, r$. We fix an (invariant with respect to W) scalar product $(\ ,\ )$ in X and the metric $\|\ \|$ defined by it. For any $d \geq 0$ let B_d denote the ball of radius d with center at 0 : $B_d = \{\alpha \in X | \|\alpha\| \leq d\}$. Let (c_{ij}) denote the inverse of the Cartan matrix defined by the root system $\alpha_1, \dots, \alpha_r$; thus

$$\omega_j = c_{j1}\alpha_1 + \cdots + c_{jr}\alpha_r$$

for all $j = 1, \dots, r$. We also set $\delta = \omega_1 + \cdots + \omega_r$.

The group G is the product of its connected (and simply connected) simple normal subgroups G_i, $i = 1, \dots, s$. Let K_i be a maximal connected compact subgroup of G_i and T_i be a maximal torus of K_i chosen so that $K = K_1 \times \cdots \times K_s$ and $T = T_1 \times \cdots \times T_s$. Then $X = X_1 \oplus \cdots \oplus X_s$, where $X_i = \operatorname{Hom}(T_i, k^*)\otimes_{\mathbb{Z}}\mathbb{Q}$. The system of roots of K relative to T is the union over all $i = 1, \dots, s$ of the root systems of K_i relative to T_i; the same is true for the systems of simple roots and fundamental weights. Setting $r_0 = 0$ and $r_s = r$ we will assume that $\{\alpha_j | r_{i-1} + 1 \leq j \leq r_i\}$ is a system of simple roots of K_i relative to T_i. The spaces X_i and X_j are orthogonal when $i \neq j$, and W is the direct product of the Weyl groups of K_i, $i = 1, \dots, s$.

For any subset M of X we denote by $\operatorname{conv} M$ its convex hull in X. Let α be an element of X. By $\operatorname{supp}\alpha$ we denote the sum of those subspaces X for which the orthogonal projection of α on X is different from zero. We assume that $\operatorname{supp} 0 = 0$. Obviously $\operatorname{conv} W(\alpha) \subset \operatorname{supp}\alpha$. The numbers t_j in the decomposition

$$\alpha = \sum_{j=1}^r t_j\omega_j, \qquad t_j \in \mathbb{Q} \quad \forall j, \tag{15}$$

are called the numerical labels of the vector α. Obviously, $X_i \subset \operatorname{supp}\alpha$ if and only if $t_j \neq 0$ for at least one j, $r_{i-1} + 1 \leq j \leq r_i$.

2. PROPOSITION 1. *We have*

$$Z_{\mathrm{Ad}} = \operatorname{conv} W(2\delta). \tag{16}$$

PROOF. It is known that 2δ is the sum of all positive roots. Given any vector $\alpha = \alpha_{i_1} + \cdots + \alpha_{i_s}$, $i \leq i_1 < \cdots < i_s \leq p$, we will show that $\alpha \in \operatorname{conv} W(2\delta)$. Because Z_{Ad} is the convex hull of all such vectors and all elements of the orbit $W(2\delta)$ are also of this form, formula (16) would follow. Any root is an integral linear combination of simple roots with either nonpositive (for negative roots) or nonnegative (for positive roots) coefficients. Therefore, $2\delta = \sum_{i=1}^r a_i\alpha_i$ and $\alpha = \sum_{i=1}^r b_i\alpha_i$, where $a_i, b_i \in \mathbb{Z}$, $a_i > 0$, and $a_i \geq b_i$ for all $i = 1, \dots, r$. Hence, $2\delta - \alpha = \sum_{i=1}^r c_i\alpha_i$, $c_i \in \mathbb{Z}$, $c_i \geq 0$ for all $i = 1, \dots, r$. Because both $\operatorname{conv} W(2\delta)$ and the system of all roots are W-invariant, and the W-orbit of any weight contains a dominant weight (because C is a fundamental domain for the action of W on X), we may assume that $\alpha \in C$. It now follows from the above equality that α is a weight of an irreducible K-module with highest weight 2δ (see [10]). Therefore $\alpha \in \operatorname{conv} W(2\delta)$ ([10]). □

COROLLARY. *Condition* (14) *is equivalent to* $2\delta \in Z_p^0$.

This follows from (16) and the fact that Z_p^0 is convex and W-invariant. □

3. LEMMA 1. $C \cap \operatorname{conv} W(\alpha) = C \cap (\alpha + A)$ *for any* $\alpha \in C$.

PROOF. Let $\beta \in C \cap (\alpha + A)$. Then $\beta = \sum_{i=1}^r t_i\alpha_i$, $t_i \in \mathbb{Q}$, $t_i < 0$ for all i. Because P is a lattice in X of rank $\dim X$, there exists $t \in \mathbb{Z}$, $t > 0$, such that $t\beta \in P$ and $tt_i \in \mathbb{Z}$ for all i. We now have the equality $t\beta = t\alpha + \sum_{i=1}^r tt_i\alpha_i$. Because $P \supset Q$, it follows that $t\alpha \in P$. Because $t > 0$ and $\alpha \in C$, it follows that $t\alpha$ is a dominant weight. Consider the irreducible K-module with highest weight $t\alpha$. Because $\beta \in C$ and $t > 0$, the weight $t\beta$ is also dominant. Next, because $tt_i \in \mathbb{Z}$ and $tt_i < 0$ for all i, it follows from the equality above that $t\beta$ is a weight of the aforementioned K-module, and therefore $t\beta \in \operatorname{conv} W(t\alpha)$ (see [10]). Obviously, $\operatorname{conv} W(t\alpha) = t \operatorname{conv} W(\alpha)$, therefore $\beta \in C \cap \operatorname{conv} W(\alpha)$. Thus

$$C \cap (\alpha + A) \subset C \cap \operatorname{conv} W(\alpha).$$

Conversely, let γ be an element of $\operatorname{conv} W(\alpha)$. Because Q is a lattice in X of rank $\dim X$ there exists $t \in \mathbb{Z}$, $t > 0$ such that both $t\gamma$ and $t\alpha$ are in Q. Because $t > 0$ and $\alpha \in C$, it follows that $t\alpha \in C$ and we can consider the irreducible K-module with highest weight $t\alpha$. We have that $t\gamma \in \operatorname{conv} W(t\alpha)$ and $t\alpha - t\gamma \in Q$. Hence $t\gamma$ is a weight of that K-module, and therefore $t\gamma = t\alpha + \sum_{i=1}^r t_i\alpha_i$, $t_i \in \mathbb{Z}$, $t_i \leq 0$ (see [10]), whence $\gamma \in \alpha + A$. Thus $\operatorname{conv} W(\alpha) \subset \alpha + A$. □

LEMMA 2. *Let* α *and* β *be elements of* C *and* $\beta \neq 0$. *Then the intersection of the ray* $\{t\beta \mid t \in \mathbb{Q},\ t \geq 0\}$ *with the polyhedron* $\operatorname{conv} W(\alpha)$ *is the segment* $\{t\beta \mid 0 \leq t \leq t_0;\ t, t_0 \in \mathbb{Q}\}$ *whose endpoint* $t_0\beta$ *is contained in a hyperplane* $\alpha + \Gamma_i$. *Moreover,* t_0 *is the only point of that ray lying on this hyperplane.*

PROOF. It is known that the numbers c_{ij} (see part 1) are rational and nonnegative for any i, j (see [10]). Because both α and β are in C, it follows that $\alpha = \sum_{i=1}^r a_i\alpha_i$ and $\beta = \sum_{i=1}^r b_i\alpha_i$ where $a_i, b_i \in \mathbb{Q}$, $a_i \geq 0$, $b_i \geq 0$ for all i.

Consider the system of inequalities $a_i - tb_i \geq 0$, $i = 1, \ldots, r$. Because the numbers a_i and b_i are nonnegative for any i, the set of its solutions (relative to t) is $\{t \in \mathbb{Q} | t \leq \min_{1\leq i\leq r,\ b_i\neq 0} a_i/b_i = t_0\}$ (the condition $\beta \neq 0$ implies that there is a b_i different from 0; it is also clear that $t_0 \geq 0$). Because β is in C, $t\beta$ is also in C for $t \geq 0$. Lemma 1 and the equality $\alpha - t\beta = \sum_{i=1}^r (a_i - tb_i)\alpha_i$ imply $t\beta \in C \cap (\alpha + A) = C \cap \operatorname{conv} W(\alpha)$ when $t \leq t_0$ and (because $\alpha_1, \ldots, \alpha_r$ are linearly independent) $t\beta \notin C \cap \operatorname{conv} W(\alpha)$ when $t > t_0$. Finally, if $a_i/b_i = t_0$ then, by the mentioned equality, $t_0\beta \in \alpha + \Gamma_i$. □

PROPOSITION 2. *Let α be an element of C, $\alpha \neq 0$ and*

$$d = \min_{1\leq i\leq r,\,(\alpha+\omega_i)\neq 0} \frac{(\alpha, \omega_i)}{\|\omega_i\|}.$$

Then $d > 0$ and $B_d \cap \operatorname{supp} \alpha$ is contained in $\operatorname{conv} W(\alpha)$ *whereas $B_{d+\varepsilon} \cap \operatorname{supp} \alpha$ is not contained in* $\operatorname{conv} W(\alpha)$ *for any $\varepsilon > 0$.*

PROOF. Let the numerical labels of α be given by (15). Then $t_i \geq 0$ for all i. We have $(\alpha, \omega_i) = \sum_{j=1}^r t_j(\omega_j, \omega_i) = \sum_{j=1}^r t_j(\sum_{k=1}^r c_{jk}\alpha_k, \omega_i) = \frac{(\alpha_i, \alpha_i)}{2}\sum_{j=1}^r t_j c_{ji}$. It is known that $c_{ji} > 0$ if there is a k such that both indices i and j are in the set $\{r_{k-1}+1, \ldots, r_k\}$, and $c_{ji} = 0$ otherwise (see [10]). Therefore $(\alpha, \omega_i) > 0$ if $\omega_i \in \operatorname{supp} \alpha$ and $(\alpha, \omega_i) = 0$ if $\omega_i \notin \operatorname{supp} \alpha$. Because ω_i is orthogonal to Γ_i we have $(\alpha, \omega_i) > 0$ if $\Gamma_i \not\supset \operatorname{supp} \alpha$ and $(\alpha, \omega_i) = 0$ otherwise. The vector $((\alpha, \omega_i)/(\omega_i, \omega_i))\omega_i$ is also orthogonal to Γ_i and is contained in $\alpha + \Gamma_i$. Hence $\|((\alpha, \omega_i)/(\omega_i, \omega_i))\omega_i\| = (\alpha, \omega_i)/\|\omega_i\| = \inf_{\gamma\in\alpha+\Gamma_i} \|\gamma\|$ and therefore

$$d = \min_{\substack{1\leq i\leq r,\\ \Gamma_i\not\supset \operatorname{supp}\alpha}} \inf_{\gamma\in\alpha+\Gamma_i} \|\gamma\|, \tag{17}$$

where d is a positive number.

Since the norm $\|\cdot\|$ and the sets $\operatorname{conv} W(\alpha)$ and $\operatorname{supp} \alpha$ are W-invariant, and W acts transitively on Weyl chambers, the inclusion $B_d \cap \operatorname{supp} \alpha \subset \operatorname{conv} W(\alpha)$ follows from inclusion $C \cap B_d \cap \operatorname{supp} \alpha \subset C \cap \operatorname{conv} W(\alpha)$. Let β be an element of $C \cap B_d \cap \operatorname{supp} \alpha$, we want to show that $\beta \in C \cap \operatorname{conv} W(\alpha)$.

First we remark that $0 \in \operatorname{conv} W(\alpha)$: this follows from Lemma 1 because $0 \in C$ and $0 \in \alpha + A$ (the latter follows from the fact that $\alpha \in C$ and therefore α is a linear combination of simple roots with nonnegative coefficients, see the proof of Lemma 2). Thus we may assume that $\beta \neq 0$. Let t_0 and i be defined as in Lemma 2. The vectors α and β are both in $\operatorname{supp} \alpha$. Therefore if $\Gamma_i \supset \operatorname{supp} \alpha$ then $t\beta \in \alpha + \Gamma_i$ for each $t \in \mathbb{Q}$, which would contradict Lemma 2. Hence $\Gamma_i \not\supset \operatorname{supp} \alpha$ and it follows from $t_0\beta \in \alpha + \Gamma_i$ and

(17) that $\|t_0\beta\| = t_0\|\beta\| \geq d$. Because $\beta \in B_d$ we have $\|\beta\| \leq d$. Because $\beta \neq 0$, it follows that $t_0 \geq 1$. Therefore, by Lemma 2, $\beta \in C \cap \operatorname{conv} W(\alpha)$.

We now consider an arbitrary $\varepsilon > 0$ and a number i such that $\Gamma_i \not\supset \operatorname{supp}\alpha$, and $d = \inf_{\gamma\in\alpha+\Gamma_i}\|\gamma\|$. Then $\omega_i \in \operatorname{supp}\alpha$ and vector $\beta = ((\alpha, \omega_i)/(\omega_i, \omega_i))\omega_i$ is of length d and is contained in $\alpha + \Gamma_i$. In the notation of Lemma 2, this means that $t_0 = 1$ for this vector. Therefore the vector $((d+\varepsilon)/d)\beta \in B_{d+\varepsilon} \cap \operatorname{supp}\alpha$ is not in $\operatorname{conv} W(\alpha)$. □

We now set

$$h = \min_{\substack{1\leq i,j\leq r\\ c_{ji}\neq 0}} c_{ji}\frac{(\alpha_i, \alpha_i)}{2\|\omega_i\|} \quad \text{and} \quad c = \min_{\substack{1\leq i,j\leq r\\ c_{ji}\neq 0}} c_{ji}. \tag{18}$$

From the fact that the numerical labels of α are nonnegative, formula (18) and the formula for (α, ω_i) deduced in the beginning of the proof of Proposition 2, we obtain the following

COROLLARY. *Let α be an element of C, $\alpha \neq 0$. Then the polyhedron* $\operatorname{conv} W(\alpha)$ *contains the ball of positive radius* $B_{th} \cap \operatorname{supp}\alpha$, *where t is the minimum of the numerical labels of α and h is given by* (18). *In particular, if α is a dominant weight, i.e., $\alpha \in P \cap C$ then* $\operatorname{conv} W(\alpha)$ *contains the ball of positive radius* $B_h \cap \operatorname{supp}\alpha$. *In any case, if* $\operatorname{supp}\alpha = X$ *then* $\operatorname{conv} W(\alpha)$ *is a "solid," i.e., has the same dimension as X.*

4. Let now $\alpha \in P \cap C$ be an arbitrary nonzero dominant weight with numerical labels defined by (15). Let $S(\alpha)$ denote the irreducible K-module with highest weight α and $S(\alpha)_\gamma$ denote its weight subspace of weight γ. The set of weights of $S(\alpha)$ is $\operatorname{conv} W(\alpha) \cap (\alpha + Q)$ (see [10]). We shall now give a lower bound on the number of such vectors β in the set $M = C \cap \operatorname{conv} W(\alpha) \cap (\alpha + Q)$ that $\operatorname{supp}\beta = \operatorname{supp}\alpha$.

For any subspace $Y \subset X$ we set $M_Y = \{\alpha \in M \mid \operatorname{supp}\gamma = Y\}$. It is clear that $M_Y = \varnothing$ if Y is not the sum of some of those X_i that are contained in $\operatorname{supp}\alpha$. Let π_Y be the orthogonal projection of X on Y.

LEMMA 3. *For any subspace Y of X we have the inequalities*

$$|M_Y| \leq |M_{\operatorname{supp}\alpha}|, \tag{19}$$

$$\sum_{\gamma\in M_Y} \dim S(\alpha)_\gamma \leq \dim S(\alpha - \pi_Y(\alpha))_0 \sum_{\gamma\in M_{\operatorname{supp}\alpha}} \dim S(\alpha)_\gamma. \tag{20}$$

(*if $M_Y \neq \varnothing$ then the left-hand side of* (20) *is assumed to be zero*).

PROOF. We may assume that $M_Y \neq \varnothing$ (henceforth the sum of zero summands X_i is assumed to be 0). The vectors $\pi_Y(\alpha)$ and $\alpha - \pi_Y(\alpha)$ are also dominant and $S(\alpha) = S(\pi_Y(\alpha)) \otimes S(\alpha - \pi_Y(\alpha))$ (see [10]). Because all weights of $S(\pi_Y(\alpha))$ and $S(\alpha - \pi_Y(\alpha))$ are contained, respectively, in

$Y = \operatorname{supp} \pi_Y(\alpha)$ and $Z = \operatorname{supp}(\alpha - \pi_Y(\alpha)) = \operatorname{Ker} \pi_Y \cap \operatorname{supp} \alpha$, and $\operatorname{supp} \alpha = Y \oplus Z$, it follows that any weight subspace of $S(\alpha)$ is of the form $S(\pi_Y(\alpha))_\lambda \otimes S(\alpha - \pi_Y(\alpha))_\mu$ for some uniquely determined weights λ and μ and, conversely, any such tensor product will be a weight subspace of $S(\alpha)$. In particular, if $\gamma \in M_Y$ then $S(\alpha)_\gamma = S(\pi_Y(\alpha))_\gamma \otimes S(\alpha - \pi_Y(\alpha))_0$. Because $S(\alpha)_\gamma \neq 0$, it follows that $S(\pi_Y(\alpha))_\gamma \neq 0$ for all $\gamma \in M_Y$ and $S(\alpha - \pi_Y(\alpha))_0 \neq 0$. As is known [10], the subspace $S(\alpha - \pi_Y(\alpha))_{\alpha - \pi_Y(\alpha)}$ is one-dimensional, and therefore for any $\gamma \in M_Y$ we have $\dim S(\pi_Y(\alpha))_\gamma \otimes S(\alpha - \pi_Y(\alpha))_{\alpha - \pi_Y(\alpha)} = \dim S(\alpha)_{\gamma + \alpha - \pi_Y(\alpha)} = \dim S(\pi_Y(\alpha))_\gamma$ (in particular, $S(\alpha)_{\gamma + \alpha - \pi_Y(\alpha)} \neq 0$). It is not difficult to see that $\gamma + \alpha - \pi_Y(\alpha) \in M_{\operatorname{supp} \alpha}$. This defines the map $M_Y \to M_{\operatorname{supp} \alpha}$, $\gamma \mapsto \gamma + \alpha - \pi_Y(\alpha)$, which is obviously an injection, whence (19). Finally, $\sum_{\gamma \in M_Y} \dim S(\pi_Y(\alpha)) = (\sum_{\gamma \in M_Y} \dim S(\alpha)_\gamma) / \dim S(\alpha - \pi_Y(\alpha))_0 = \sum_{\gamma \in M_Y} \dim S(\alpha)_{\gamma + \alpha - \pi_Y(\alpha)} \leq \sum_{\gamma \in M_{\operatorname{supp} \alpha}} \dim S(\alpha)_\gamma$ whence (20). □

LEMMA 4. *Let* $\operatorname{supp} \alpha = X_{i_1} \oplus \cdots \oplus X_{i_k}$ *and* $\mathfrak{Z}$ *be a (finite) set of all subspaces of* X *that are sums of some of the subspaces* X_j *for* $j = i_1, \ldots, i_k$. *Then we have the inequalities*

$$|M_{\operatorname{supp} \alpha}| \geq \frac{|\operatorname{conv} W(\alpha) \cap (\alpha + Q)|}{2^k |W|}, \tag{21}$$

$$\sum_{\gamma \in M_{\operatorname{supp} \alpha}} \dim S(\alpha)_\gamma \geq \frac{\dim S(\alpha)}{|W| \sum_{Y \in \mathfrak{Z}} \dim S(\pi_Y(\alpha))_0}. \tag{22}$$

PROOF. Being W-invariant, $\operatorname{conv} W(\alpha) \cap (\alpha + Q)$ is a union of W-orbits. Because each W-orbit intersects C at exactly one point, we can assign this point to that orbit and obtain a bijection between the set of W-orbits in $\operatorname{conv} W(\alpha) \cap (\alpha + Q)$ and M. Because M is a disjoint union of all M_Y, where $M_Y = \varnothing$ if $Y \notin \mathfrak{Z}$, we see that (21) follows from (19).

It follows from (20) and from the remarks made above that

$$\begin{aligned} \dim S(\alpha) &= \sum_{Y \in \mathfrak{Z}} \sum_{\gamma \in M_Y} |W(\gamma)| \dim S(\alpha)_\gamma \leq |W| \sum_{Y \in \mathfrak{Z}} \sum_{\gamma \in M_Y} \dim S(\alpha)_\gamma \\ &\leq |W| \sum_{\gamma \in \mathbb{Z}} (\dim S(\alpha - \pi_Y(\alpha))_0 \sum_{\gamma \in M_{\operatorname{supp} \alpha}} \dim S(\alpha)_\gamma) \\ &= |W| \sum_{\gamma \in M_{\operatorname{supp} \alpha}} \dim S(\alpha)_\gamma \sum_{Y \in \mathfrak{Z}} \dim S(\pi_Y(\alpha))_0; \end{aligned}$$

whence (22). □

LEMMA 5. *For any vector* $\alpha \in P \cap C$ *with numerical labels* $t_1, \ldots, t_r$ *(see (15)) we have* $|\operatorname{conv} W(\alpha) \cap (\alpha + Q)| \geq 2c \sum_{i=1}^{s} \operatorname{rk} G_i \sum_{j=r_{i-1}+1}^{r_i} t_j$.

PROOF. Let $w_0 \in W$ be such that $w_0(C) = -C$. As is known (see [10]), there exists a permutation ν of the set of indices $\{1, \ldots, r\}$ such that

$w_0(\alpha) = -\sum_{j=1}^r t_{\nu(j)}\omega_j$. We now have $\alpha - w_0(\alpha) = \sum_{j=1}^r (t_j + t_{\nu(j)})\omega_j = \sum_{i=1}^r \sum_{j=1}^r (t_j + t_{\nu(j)})c_{ji}\alpha_i$. Because $w_0(\alpha) \in \operatorname{conv} W(\alpha) \cap (\alpha + Q)$ [10], there is a sequence of elements $\beta_1, \dots, \beta_k$ of $\operatorname{conv} W(\alpha) \cap (\alpha + Q)$ such that $\beta_1 = \alpha$, $\beta_k = w_0(\alpha)$, and $\beta_{i-1} - \beta_i$ is a simple root for $i = 2, \dots, k$ [10]. Because simple roots are linearly independent, the elements $\beta_1, \dots, \beta_k$ are all distinct and their total number is equal to the sum of the coefficients in the decomposition of the vector $(\beta_1 - \beta_2) + (\beta_2 - \beta_3) + \dots + (\beta_{k-1} - \beta_k) = \alpha - w_0(\alpha)$ by the simple roots. From our formula we have the bound $|\operatorname{conv} W(\alpha) \cap (\alpha + Q)| \geq \sum_{i=1}^r \sum_{j=1}^r (t_j + t_{\nu(j)})c_{ji}$. The assertion of the lemma now follows because ν preserves the connected components of the Dynkin diagram [10] (i.e., $\{r_{i-1} + 1, \dots, r_i\}$ is ν-invariant for any i). □

5. We can now obtain a lower bound on the radius of the ball inscribed in the zonohedron of weights (which is the goal of this section).

For any $\alpha \in P \cap C$ we set $Z_\alpha = \{\sum_{\beta \in \operatorname{conv} W(\alpha) \cap (\alpha + Q)} t_\beta \beta \mid t_\beta \in \mathbb{Q},\ 0 \leq t_\beta \leq 1 \text{ for all } \beta\}$. This is a zonohedron contained in $\operatorname{supp} \alpha$.

PROPOSITION 3. *Let $\alpha \in P \cap C$ be an arbitrary vector with numerical labels $t_1, \dots, t_r$ (see (15)). Then for any number $d > 0$ not exceeding $(hc/2^{s-1}|W|) \sum_{i=1}^s \operatorname{rk} G_i \sum_{j=r_{i-1}+1}^{r_i} t_j$ (where h and c are given by (18)) the ball $B_d \cap \operatorname{supp} \alpha$ is contained in the zonohedron Z_α.*

PROOF. We have $\operatorname{conv} W(\beta) = \{\sum_{w \in W} t_w w(\beta) \mid t_w \in \mathbb{Q},\ t_w \geq 0 \text{ for all } w,\ \sum_{w \in W} t_w = 1\}$. Because $\operatorname{conv} W(\alpha) \cap (\alpha + Q)$ is W-invariant we see that, first, $\operatorname{conv} W(\beta) \subset Z_\alpha$ for any $\beta \in \operatorname{conv} W(\alpha) \cap (\alpha + Q)$ and, second, that $\operatorname{conv} W(\beta_1) + \dots + \operatorname{conv} W(\beta_k) \subset Z_\alpha$, if $\beta_1, \dots, \beta_k \in \operatorname{conv} W(\alpha) \cap (\alpha + Q)$ and $W(\beta_i) \neq W(\beta_j)$ when $i \neq j$. Our assertion now follows easily from (21), Lemma 5, and the corollary of Proposition 2 if one takes into account that any W-orbit meets C exactly at one point and that $B_{kh} \cap \operatorname{supp} \alpha = (B_h \cap \operatorname{supp} \alpha) + \dots + (B_h \cap \operatorname{supp} \alpha)$ (k summands). □

REMARK. Because $\alpha \in P \cap C$ is a dominant weight we can consider an irreducible representation ρ of G whose restriction to K has α as highest weight. The zonohedron of weights of ρ is $Z_\rho = \{\sum_{\beta \in \operatorname{conv} W(\alpha) \cap (\alpha + Q)} t_\beta \beta \mid t_\beta \in \mathbb{Q},\ 0 \leq t_\beta \leq \dim S(\alpha)_\beta\}$. Using the same argument as in the proof of Proposition 3 we can easily deduce from this and from (22) that for any $d > 0$ not exceeding $(h \dim S(\alpha))/|W| \sum_{Y \in 3} \dim S(\pi_Y(\alpha))_0$, the ball $B_d \cap \operatorname{supp} \alpha$ is contained in Z_ρ.

§3. Finiteness theorems

1. With the aid of the results of §§1 and 2 we shall now deduce two types of finiteness theorems: for the degree of the Poincaré series and for representations with free algebra of invariants. We keep the notation of §2 and part 2 of §1 of this chapter; the weights of representations of K are considered to be relative to T; the highest weights are considered with respect to the order defined by a Weyl chamber C.

Clearly, by passing to $G/(\operatorname{Ker}\rho)^0$ we can always reduce the study of the algebra $k[V]^{\rho(G)}$ (and its Poincaré series) to the case of a finite $\operatorname{Ker}\rho$.

2. Let $\rho = \rho_1 \oplus \cdots \oplus \rho_b$ be an irreducible decomposition. Let (P1), (P2), and (P3) denote the following properties of ρ:

(P1) ρ is irreducible (i.e., $b = 1$);

(P2) $\operatorname{Ker}\rho_l$ is finite for any l;

(P3) $(\operatorname{Ker}\rho_1)^0 = \cdots = (\operatorname{Ker}\rho_b)^0$ for $b \geq 2$.

THEOREM 4. *There exist only a finite number of representations* $\rho\colon G \to \mathrm{GL}(V)$ *of a connected semisimple group* G *having one of the properties* (P1), (P2), *and* (P3), *and such that the degree of the Poincaré series of* $k[V]^{\rho(G)}$ *is different from* $-\dim V$. *This number can be explicitly bounded* (*see proof*); *for example, if* ρ *is irreducible,* $\operatorname{Ker}\rho$ *is finite and*

$$\sum_{i=1}^{s} \operatorname{rk} G_i \sum_{j=r_{i-1}+1}^{r_i} t_j > 2^s |W| \, \|\delta\| / hc, \tag{23}$$

where $t_1, \ldots, t_r$ *are the numerical labels of the highest weight* α *of* $\rho|_K$ (*see* (15)), *then the above degree is* $-\dim V$.

PROOF. Suppose ρ has property (P1). Because G has only finitely many connected normal subgroups of positive dimension, it suffices to prove the assertion for those ρ that have finite $\operatorname{Ker}\rho$.

So let us suppose that $\operatorname{Ker}\rho$ is finite and let α be the highest weight of $\rho|_K$. Then $\operatorname{supp}\alpha = X$. Because the set of weights of $\rho|_K$ is $\operatorname{conv} W(\alpha) \cap (\alpha + Q)$, we have $Z_\rho \supset Z_\alpha$. It now follows from Proposition 3 that Z_ρ contains the ball B_d of radius $d = (hc/2^{s-1}|W|)\sum_{i=1}^{s} \operatorname{rk} G_i \sum_{j=r_{i-1}+1}^{r_i} t_j$, where $t_1, \ldots, t_r$ are the numerical labels of α (see (15)). By Theorem 3 and the corollary of Proposition 1, inequality (23) implies that $\deg F(k[V]^{\rho(G)}, x) = -\dim V$. Because $t_i \in \mathbb{N}$ for all i (see [10]), the assertion of the theorem in this case follows from the fact that there are only finitely many r-tuples $(t_1, \ldots, t_r) \in \mathbb{N}^r$ not satisfying (23).

Suppose now that (P2) holds and let α_l be the highest weight of $\rho_l|_K$ with numerical labels $t_1^{(l)}, \ldots, t_r^{(l)}$. Then $\operatorname{supp}\alpha_l = X$ for all l, and, by (13) and the same argument as above, the ball B_d of radius $d = (hc/2^{s-1}|W|) \times \sum_{i=1}^{s} \operatorname{rk} G_i \sum_{j=r_{i-1}+1}^{r_i} \sum_{l=1}^{b} t_j^{(l)}$ is contained in Z_ρ. At this point we replace (23) with inequality $d > 2\|\delta\|$ and use the previous argument to complete the proof.

The case of (P3) is reduced to (P2) if we replace G with $G/(\operatorname{Ker}\rho_1)^0$. □

REMARKS. 1) Inequality (23) can be improved if one takes into account the multiplicities of the weights. Namely, it follows from the remark after Proposition 3 that in the last statement of Theorem 4 inequality (23) can be replaced with $\dim V > 2h^{-1}|W| \, \|\delta\| \sum_{Y \in 3} \dim S(\pi_Y(\alpha))_0$.

2) The following example shows that in Theorem 4 one cannot drop (without changing the assertion) any of the properties (P1), (P2), or (P3): if

$G = \mathrm{SL}_2 \times \mathrm{SL}_2$, then it is not difficult to see that for any d the algebra of invariants of representation $\rho = (\rho_1 \otimes 1) \oplus (1 \otimes \rho_d)$ is isomorphic (as a graded algebra) to the algebra of invariants of representation ρ_d of SL_2. Now the example of part 7 of §1 shows that $\deg F(k[V]^{\rho(G)}, x) = -(d+1) > -\dim V = -(d+3)$ for $d \geq 3$.

3. From Theorem 4, the corollary of Theorem 2, and the fact that there are only finitely many representations ρ having property (P2) and satisfying inequality $m_{\rho(G)} < \dim G$ (see §4) we now deduce

THEOREM 5. *There are only finitely many representations $\rho\colon G \to \mathrm{GL}(V)$ of a connected semisimple group G having property* (P2) *and satisfying condition $2\tau/\gamma \neq \dim G$, where γ and τ are the first and second coefficients of the Laurent expansion* (2) *of the Poincaré series of $k[V]^{\rho(G)}$ at point* 1.

Because (P2) holds automatically for simple groups, from Theorems 4 and 5 we obtain

COROLLARY. *For all, except finitely many, representations $\rho\colon G \to \mathrm{GL}(V)$ of a connected simple group G we have $\deg F(k[V]^{\rho(G)}, x) = -\dim V$ and $2\tau/\gamma = \dim G$.*

For $G = \mathrm{SL}_2$ this was established (along with sharp bounds) in part 7 of §1.

4. The above results will be used later in the proof of the finiteness theorem for representations with free algebra of invariants. We will also need the following

LEMMA 6. *Let V be a finite-dimensional G-module and U be a submodule of V. If $k[V]^G$ is free, then $k[U]^G$ is also free.*

PROOF. This follows immediately from the second monotonicity theorem for the homological dimension of the algebra of invariants, which will be proved in §1 of Chapter 4. □

5. It is easy to see that $k[V \oplus L]^G = k[V]^G \otimes_k k[L]$ for any G-module V and trivial G-module L. It means that $(V \oplus L)/G = V/G \times L$. It now follows from Corollary 2 of Proposition 1.1, Chapter 4, that $k[V]^G$ is free if and only if $k[V \oplus L]^G$ is free. On one hand, this explains why Theorem 6 below is true only up to addition of a trivial direct summand, and, on the other hand, it shows that in the proof of this theorem one can consider only G-modules without nonzero fixed vectors:

THEOREM 6. *There exist only finitely many, up to isomorphism and addition of a trivial direct summand, finite-dimensional G-modules V such that $k[V]^G$ is free.*

REMARK. The assertion of the theorem is also true for finite groups, see §1, Chapter 4.

6. For the proof we will need the following result (we keep the notation of §1).

PROPOSITION 4. *Suppose that* $k[V]^{\rho(G)}$ *is free. Then*: a) *if* $V^{\rho(G)} = 0$, *then* $n \leq q/2 + m_{\rho(G)}$; b) *if* ρ *is irreducible then* $n \leq (q+1)/3 + m_{\rho(G)}$.

PROOF. Because $k[V]^{\rho(G)}$ is free, we can set $P = k[V]^{\rho(G)}$ in part 4 of §1, Chapter 2. In this case $l = 1$, $e_l = 0$, and, as follows from formula (6) of Chapter 2, $q = d_1 + \cdots + d_m$.

Suppose $V^{\rho(G)} = 0$. If $f \in k[V]_1^{\rho(G)}$, $f \neq 0$, then $L = \{v \in V \mid f(v) = 0\}$ is a $\rho(G)$-invariant hyperplane in V. Because G is semisimple we can find in V a $\rho(G)$-invariant line $L^{\perp}$ complementary to L. Because $\mathrm{Hom}(G, k^*) = \{e\}$, it follows that $\rho(G)$ acts trivially on $L^{\perp}$, contrary to the assumption. Hence $d_i \geq 2$ for all i, and assertion a) follows from the above formula for q (and the relation $m = n - m_{\rho(G)}$).

Suppose now that ρ is irreducible. Because ρ is nontrivial, $d_i \geq 2$, for all i, as was shown above. Because the kernel of any $\rho(G)$-invariant symmetric bilinear form on V is $\rho(G)$-invariant, and since ρ is irreducible, we have $\dim k[V]_2^{\rho(G)} \leq 1$. Therefore $d_i \geq 3$ for $i \geq 2$, and, similarly to the above, assertion b) follows. □

COROLLARY. *Let* $q = n$. *Suppose also that* $n > 2m_{\rho(G)}$ *if* $V^{\rho(G)} = 0$, *or* $n > (3m_{\rho(G)} + 1)/2$ *if* ρ *is irreducible. Then* $k[V]^{\rho(G)}$ *is not free.*

7. EXAMPLE. We return to the example from part 7 of §1. Because $m_{\rho(G)} = 3$ if $\rho \neq \rho_1$ or ρ_2, and $m_{\rho(G)} = 2$ otherwise (see [52, 53]), those ρ for which $V^{\rho(G)} = 0$ and at least one of the conditions of the last corollary is not satisfied are exactly the representations ρ_i, $i = 1, 2, 3, 4$, $\rho_1 \oplus \rho_i$, $i = 1, 2, 3$, $\rho_2 \oplus \rho_2$ and $\rho_1 \oplus \rho_1 \oplus \rho_1$. Therefore all ρ different from these have nonfree algebras of invariants (it is well known that all of the representations listed above, except $\rho_1 \oplus \rho_3$, have free algebras of invariants (see [76, 48])).

8. PROOF OF THEOREM 6. As was explained above, it suffices to consider the representations $\rho: G \to \mathrm{GL}(V)$ with $V^{\rho(G)} = 0$.

By Theorems 4 and 1, there are only finitely many irreducible representations ρ with $q \neq n$. On the other hand, it follows easily from Weyl's formula for the dimension of an irreducible representation (see [10]) that there are only finitely many irreducible representations whose dimensions do not exceed any a priori given constant (it also follows from Lemma 5 since $\dim \rho \geq |\operatorname{conv} W(\alpha) \cap (\alpha + Q)|$; in particular, we may set this constant equal to $(3\dim G + 1)/2$. Because $m_{\rho(G)} \leq \dim G$, it follows from the corollary of Proposition 4 that there are only finitely many irreducible representations with free algebra of invariants. Let $\rho_1, \ldots, \rho_N$ be those representations.

By Theorems 4 and 1, for each $i = 1, \ldots, N$ there exists an integer

$b_i > 0$ such that for any integer $b > b_i$ the numbers q and n for the representation $\rho = b\rho_i$ are equal. Moreover, we may (and will) assume that $\dim b\rho_i > 2\dim G$. By the corollary of Proposition 4, in this case the algebra $k[V]^{b\rho_i(G)}$ is not free.

It follows from Lemma 6 and from the complete reducibility of representations of G that any representation of G with free algebra of invariants is contained in the finite set of the representations $d_1\rho_1 \oplus \cdots \oplus d_N\rho_N$, where $d_i \le b_i$ for all i. □

REMARK. Note that our proof is constructive in the sense of part 1 of §1. Namely, the bounds from Theorem 4 and Lemma 5 (with which, for simplicity, we replace the reference to Weyl's formula) yield, in the obvious way, the following assertions for any irreducible representation ρ with finite $\operatorname{Ker}\rho$ and numerical labels $t_1, \dots, t_r$ of the highest weight α of representation $\rho|_K$ (see (15)):

1) if the algebra of invariants of ρ is free then $\sum_{i=1}^{s} \operatorname{rk} G_i \sum_{j=r_{i-1}+1}^{r_i} t_j \le \max\{2^s h^{-1} c^{-1} |W|\,\|\delta\|,\ c^{-1}\dim G\}$ and

$$\dim\rho \le 2\max\left\{\dim G,\ h^{-1}|W|\,\|\delta\| \sum_{Y\in 3} \dim S(\pi_Y(\alpha))_0\right\};$$

2) if the algebra of invariants of $b\rho$ is free then

$$b \le (\max\{2^s h^{-1} c^{-1}|W|\,\|\delta\|,\ c^{-1}\dim G\}) \Big/ \left(\sum_{i=1}^{s} \operatorname{rk} G_i \sum_{j=r_{i-1}+1}^{r_i} t_j\right).$$

CHAPTER 4

Syzygies in Invariant Theory

The assumptions about the field k in this chapter are the same as in Chapter 3.

§0. A description of the results and additional notation

1. In this chapter, for a connected semisimple group G we will prove the following

MAIN THEOREM. *For any integer $d \geq 0$ there are only finitely many, up to isomorphism and addition of a trivial direct summand, finite-dimensional rational G-modules V such that $k[V]^G$ has homological dimension d.*

Obviously, from this we have

COROLLARY. *If, under the assumptions of the main theorem, G is nontrivial then for any integer $d > 0$ there exists a rational G-module V with $\operatorname{hd} k[V]^G > d$. Moreover the set of nonequivalent G-modules satisfying this condition is infinite, and the set of those not satisfying this condition is finite.*

2. Our method of the proof of the main theorem is constructive: it enables us to describe explicitly a finite list of G-modules containing all V with $\operatorname{hd} k[V]^G \leq d$ (from part 5 of §3, Chapter 3 and Corollary 1 of Proposition 1.1 of this chapter we have that $\operatorname{hd} k[V]^G = \operatorname{hd} k[V \oplus L]^G$ for any V and any trivial L; this explains why in the main theorem trivial summands come into play; it also shows that we may, and without further reference will, assume that $V^G = 0$). This allows us in principle to classify explicitly all V with $\operatorname{hd} k[V]^G = d$. Indeed, because $k[V]^G$ is a Cohen–Macaulay algebra (see [85]) we have (see [139, 149]) that

$$\operatorname{hd} k[V]^G = \operatorname{ed} k[V]^G - \dim k[V]^G. \tag{0.1}$$

((0.1) is true for any reductive group G.) Because $\operatorname{ed} k[V]^G$ and $\dim k[V]^G$ in principle can be found explicitly (see Chapter 2), we can, using (0.1), extract from the above list the V's with $\operatorname{hd} k[V]^G = d$. In §10 we will present examples that show how this program can be carried out not only in principle, but also in practice.

3. One of the typical difficulties in the proof of the main theorem is to find closed orbits for whose stabilizers one can control the algebras of invariants of the corresponding slice representations. Among other things, it follows from the proof that for an overwhelming majority of groups G the following property holds: with the exception of finitely many representations, for each locally-exact irreducible representation $G \to \mathrm{GL}(V)$ there exists a point $v \in V$ such that Gv is closed and G_v is a rank 1 reductive group with diagonalizable centralizer of a maximal torus (exceptions can be explicitly described); moreover, for many G's one can assume that $\dim G_v = 1$ (for example, this is true when $G = G_1 \times \cdots \times G_m$, where G_i is a connected simple group of rank > 1). However, there exist groups with infinitely many locally-exact irreducible representations such that the stabilizer of any nonzero point with closed orbit is finite (it is easy to see that SL_2 is one of them; there are also more complicated examples, see §9); such groups should be investigated by special methods. We remark that the assertion of the main theorem, as is easily seen, it not true for an arbitrary reductive group.

4. We shall now describe briefly the logical structure of the proof as well as the other results in this chapter.

In §1 we prove monotonicity theorems, reduce the proof of the main theorem to the irreducible case, and prove the main theorem for finite groups. As an application, we give necessary conditions for $\mathrm{hd}\, k[V]^G = 1$. Section 2 contains explicit bounds on $\mathrm{ed}\, k[V]^G$ and $\mathrm{hd}\, k[V]^G$ in the case when $\mathrm{rk}\, G = 1$ and the centralizer of a maximal torus of G is diagonalizable. From these bounds we deduce the main theorem for SL_2. In §3 we prove a theorem that, under certain conditions, allows us to find points v such that $\overline{Gv} = Gv$ and G_v is a group of the type considered in §2. Under the same conditions we find explicit bounds on $\mathrm{hd}\, k[V]^G$. We also investigate these conditions and give a criterion for the stability of a reductive linear group G as well as sufficient conditions, which are easy to check. Furthermore, we investigate three-dimensional simple subgroups containing a given one-dimensional torus, find sufficient conditions for the generic stabilizer in a reductive linear group with one-dimensional center to be cyclic, and find necessary conditions for $\dim G_v = 3$. In §4 we prove majorizing theorems for multiplicities, generic stabilizers, and stability. They allow us, using certain information about all fundamental G-modules, to obtain the corresponding information about arbitrary irreducible G-modules, which later can be used to verify the conditions of the theorem from §3. The results proved in §4 are of a general nature and can be useful in other problems. In §§5 and 6 we obtain the above-mentioned information for all connected simply connected simple groups G of rank > 1 (§5 deals with the classical groups and §6 deals with the exceptional ones): we choose a one-dimensional root torus T in G, and for each fundamental G-module V we explicitly (in the form of tables) describe the structure of the $Z_G(T)/T$-module V and C-module

V, where C is a three-dimensional root subgroup containing T. In §§7, 8, and 9 we prove the main theorem. In §7 we establish the finiteness of the set of highest weights Λ of locally-effective irreducible G-modules V with $\operatorname{hd} k[V]^G \leq d$, whose numerical labels corresponding to three-dimensional simple normal subgroups of G do not exceed a given constant c. In §§8 and 9 we prove that if c is sufficiently large then the complement of this set (in the set of all highest weights of locally-effective G-modules V with $\operatorname{hd} k[V]^G \leq d$) is empty. In §§7 and 8 the argument is based on the theorem from §3 (with the aid of theorems from §§4, 5, and 6 we find a point v with $\overline{Gv} = Gv$ and $\operatorname{rk} G_v = 1$; in §9 we consider the case when it is impossible to choose such a point: there we work with finite stabilizers). Finally, §10 contains examples for SL_2 and the exceptional simple groups, described in part 10 of the Introduction.

5. Besides the standard notation we will also use the following ones.

The Lie algebras of linear algebraic groups $G, H, T, \ldots$ will be denoted by $\mathfrak{g}, \mathfrak{h}, \mathfrak{t}, \ldots$, respectively.

Unless stated otherwise, modules (representations) of algebraic groups are always assumed to be finite dimensional and rational. A representation $\varphi\colon H \to GL(V)$, and the corresponding H-module V, will often be denoted by (H, V). The linear group $\varphi(H)$ is called the image of φ. Sometimes, if it does not lead to confusion, we identify a representation with its space. A representation of a Lie algebra $\mathfrak{h}$ in V is denoted by $(\mathfrak{h}, V)$.

Let G be an algebraic group and H be a subgroup of G. If V is a G-module then V^H is invariant under $N_G(H)$ and $Z_G(H)$, and when speaking about representations $(N_G(H), V^H)$ and $(Z_G(H), V^H)$ we always mean the representations induced by the action of G on V. The same is assumed about $(\mathfrak{n}_{\mathfrak{g}}(\mathfrak{h}), V^{\mathfrak{h}})$ and $(\mathfrak{z}_{\mathfrak{g}}(\mathfrak{h}), V^{\mathfrak{h}})$. (H, V) will always denote the restriction of (G, V) to H, and $(G, \mathfrak{g})$ will denote the adjoint representation Ad.

The standard notation from the structure theory of semisimple Lie algebras and the theory of root systems, such as X_α, H_α, (1), as well as $\alpha_1, \ldots, \alpha_l$ (a system of simple roots) and $\omega_1, \ldots, \omega_l$ (a system of fundamental weights) will be borrowed, along with the numbering from [9, 10], without further reference. This is not, however, the case with the elements ε_j (and the expressions of α_i and ω_i in terms of those elements): everywhere, except §6, the elements ε_j and the corresponding expressions are taken from [9], and in §6 we use the notation from [15]. This is explained by the fact that the information we need is contained in [15] but not in [9]. We recall this in the beginning of the corresponding sections.

Let G be a reductive group acting regularly on an affine algebraic variety X. If the orbit Gv of a point $v \in X$ is closed in X, then, by Matsushima's criterion, G_v is a reductive group. Therefore the G_v-module $\Theta_{v,X}$ decomposes into the direct sum of $\Theta_{v,Gv}$ and a complementary G_v-module N_v.

The representation (G_v, N_v) is called the slice representation at v. It is clear that $(Gv, \Theta_{v,Gv})$ is equivalent to the representation of G_v in $\mathfrak{g}/\mathfrak{g}_v$ induced by the adjoint representation of G_v in $\mathfrak{g}$.

If X is a linear space V and the action of G on V is linear, then the slice representation is entirely determined by G_v because $(G_v, N_v) \oplus (G_v, \Theta_{v,Gv}) = (G_v, V)$, or

$$(G_v, N_v) \oplus (G_v, \mathfrak{g}) = (G_v, V) \oplus (G_v, \mathfrak{g}_v). \tag{0.2}$$

It is also clear that the subspace $\mathfrak{g}v$ of V is invariant under G_v and that $(G_v, \mathfrak{g}v) = (G_v, \mathfrak{g}/\mathfrak{g}_v)$. Therefore

$$(G_v, N_v) \oplus (G_v, \mathfrak{g}v) = (G_v, V). \tag{0.3}$$

In this case the point $\pi_{V,G}(0)$ is also denoted 0 (and is called the zero of V/G).

In conclusion, we remark that if X is smooth then the stabilizers of points in general position of X are conjugate in G (see [108]). A representative of this class of conjugate subgroups of G is called the generic stabilizer.

§1. Monotonicity theorems

Let G be a reductive group and V be a G-module.

1. Proposition 1.1 $\operatorname{ed} k[V]^G = \dim \Theta_{0,V/G}$.

Proof. This follows from the fact that the (unique) maximal homogeneous ideal I of $k[V]^G$ is the ideal of point $0 \in V/G$, and therefore $I/I^2 = \Theta^*_{0,V/G}$. □

Corollary 1. $\operatorname{hd} k[V]^G = \dim \Theta_{0,V/G} - \dim V/G$.

This follows from Proposition 1.1 and (0.1). □

Corollary 2. *The following conditions are equivalent:*

a) $\operatorname{hd} k[V]^G = 0$;
b) $V/G \cong \mathbb{A}^s$;
c) V/G *is nonsingular;*
d) 0 *is a nonsingular point of* V/G.

The implications a) $\Rightarrow$ b) $\Rightarrow$ c) $\Rightarrow$ d) are obvious; d) $\Rightarrow$ a) follows from Corollary 1 and the definition of a nonsingular point. □

Proposition 1.2. $\dim \Theta_{0,V/G} \geq \dim \Theta_{x,V/G}$ *for any* $x \in V/G$.

Proof. Let $f_1, \dots, f_s$ be a minimal system of homogeneous generators of $k[V]^G$. Then $\dim \Theta_{0,V/G} = s$ in view of Proposition 1.1 and part 3 of §1, Chapter 2. Consider the morphism $\pi\colon V \to k^s$, $\pi(v) = (f_1(v), \dots, f_s(v))$. Then $\pi(V) = \overline{\pi(V)} \cong V/G$, and π can be identified with $\pi_{V,G}$. Therefore $\Theta_{x,V/G} \subset k^s$. □

PROPOSITION 1.3. *Let* $V = W_1 \oplus \cdots \oplus W_n$ *for some G-submodules* W_j, *and* $f_{j1}, \dots, f_{jm_j}$ *be a minimal system of homogeneous generators of* $k[W_j]^G$. *Let* $\pi_j \colon V \to W_j$ *be the natural projection. Then the system of homogeneous polynomials* $\{\pi_j^*(f_{jl})\}$, $1 \le j \le n$, $1 \le l \le m_j$, *can be extended to a minimal system of homogeneous generators of* $k[V]^G$.

PROOF. By induction, we may assume that $n = 2$.

Let I and I_j be the maximal homogeneous ideals in $k[V]^G$ and $k[W_j]^G$, and $\rho_j \colon k[V]^G \to k[W_j]^G$ be the restriction homomorphism. It is not difficult to see that ρ_j and π_j^* are homogeneous homomorphisms of degree 0 and $\rho_j\pi_j^* = \mathrm{Id}$, $\rho_j(I) = I_j$, $\pi_j^*(I_j) \subset I$, $(\rho_i\pi_j^*)(I_j) = 0$ when $i \neq j$, $\pi_1^*(I_1) \cap \pi_2^*(I_2) = 0$. Let $L_j = \langle f_{j1}, \dots, f_{jm_j}\rangle$. Then L_j is a graded subspace of I_j with $I_j = L_j \oplus I_j^2$ (see part 3 of §1, Chapter 2).

The sum of $\pi_j^*(L_j)$, $j = 1, 2$ and I^2 is direct. Indeed $\pi_1^*(L_1) \cap \pi_2^*(L_2) \subset \pi_1^*(I_1) \cap \pi_2^*(I_2)$, therefore the sum of $\pi_1^*(L_1)$ and $\pi_2^*(L_2)$ is direct. If $f \in (\pi_1^*(L_1) \oplus \pi_2^*(L_2)) \cap I^2$, then $f = \pi_1^*(l_1) + \pi_2^*(l_2)$ where $l_j \in L_j$, so that $\rho_j(f) = l_j \in \rho_j(I^2) = I_j^2$. Hence $l_j = 0$ and consequently $f = 0$.

Because π_j^* is a monomorphism, $\{\pi_j^*(f_{jl})\}$, $j = 1, 2$; $l = 1, \dots, m$, is a basis of $\pi_1^*(L_1) \oplus \pi_2^*(L_2)$. It can be extended by the elements of a basis of any (finite-dimensional) graded subspace L of I satisfying the condition $I = L \oplus \pi_1^*(L_1) \oplus \pi_2^*(L_2) \oplus I^2$. The resulting system of polynomials is a minimal system of homogeneous generators of $k[V]^G$. □

2. THEOREM 1.1 (The first monotonicity theorem). *Let* G *be a reductive group and* V *be a G-module.*

a) *If* $v \in V$ *and* $\overline{Gv} = Gv$ *then* $\operatorname{ed} k[N_v]^{G_v} \le \operatorname{ed} k[V]^G$.

b) *If* V *is the direct sum of submodules* $W_1, \dots, W_n$ *then*

$$\sum_{j=1}^{n} \operatorname{ed} k[W_j]^G \le \operatorname{ed} k[V]^G.$$

c) *If* W *is a submodule of* V *then* $\operatorname{ed} k[W]^G \le \operatorname{ed} k[V]^G$.

PROOF. a) By the slices étales theorem (see [108]), there exists a surjective étale morphism of a neighborhood of zero in N_v/G_v onto a neighborhood of $\pi_{V,G}(v)$ in V/G. It now follows from Propositions 1.1 and 1.2 that $\operatorname{ed} k[N_v]^{G_v} = \dim \Theta_{0, N_v/G_v} = \dim \Theta_{\pi_{V,G}(v), V/G} \le \dim \Theta_{0, V/G} = \operatorname{ed} k[V]^G$.

b) This follows from Proposition 1.3.

c) This follows from b) and the complete reducibility of V. □

THEOREM 1.2 (The second monotonicity theorem). *Let* G *be a reductive group and* V *be a G-module.*

a) *If* $v \in V$ *and* $\overline{Gv} = Gv$ *then* $\operatorname{hd} k[N_v]^{G_v} \le \operatorname{hd} k[V]^G$.

b) *If* V *is the direct sum of submodules* $W_1, \dots, W_n$ *then*

$$\sum_{j=1}^{n} \operatorname{hd} k[W_j]^G \leq \operatorname{hd} k[V]^G.$$

c) *If* W *is a submodule of* V *then* $\operatorname{hd} k[W]^G \leq \operatorname{hd} k[V]^G$.

PROOF. a) By the slices étales theorem [108], $\dim N_v/G_v = \dim V/G$. Hence the assertion follows from (0.1) and Theorem 1.1a).

b) Using the induction, we can assume $n = 2$. By Proposition 1.3, there exists a minimal system of homogeneous generators $f_1, \dots, f_l; h_1, \dots, h_s; g_1, \dots, g_r$ of $k[V]^G$, where $h_i = \pi_1^*(h_i')$, $g_j = \pi_2^*(g_j')$, and $h_1', \dots, h_s'$ (resp., $g_1', \dots, g_r'$) is a minimal system of homogeneous generators of $k[W_1]$ (resp., $k[W_2]$).

Suppose that $h_1, \dots, h_p$ (resp., $g_1, \dots, g_q$) is a maximal system of algebraically independent elements among $h_1, \dots, h_s$ (resp., $g_1, \dots, g_r$). Because $k[V] = k[W_1] \otimes k[W_2]$, the polynomials $h_1, \dots, h_p, g_1, \dots, g_q$ are algebraically independent. We can extend this system to a maximal algebraically independent subsystem of all of the polynomials $f_1, \dots, f_l, h_1, \dots, h_s, g_1, \dots, g_r$. It is clear from the construction that the new elements can only be chosen from the polynomials $f_1, \dots, f_l$. Suppose they are $f_1, \dots, f_t$. Then $p = \dim k[W_1]^G$, $q = \dim k[W_2]^G$, $p+q+t = \dim k[V]^G$, $s = \operatorname{ed} k[W_1]^G$, $r = \operatorname{ed} k[W_2]^G$, $s+r+l = \operatorname{ed} k[V]^G$. It now follows from (0.1) that $\operatorname{hd} k[W_1]^G = s-p$, $\operatorname{hd} k[W_2]^G = r-q$ and $\operatorname{hd} k[V]^G = (s+r+l)-(p+q+t) = (s-p)+(r-q)+(l-t)$. Hence the desired result follows from the inequality $l-t \geq 0$.

c) This follows from b) and the complete reducibility of V. □

COROLLARY. *Let* $V = W_1 \oplus W_2$, $v \in W_1$ *and* $\overline{Gv} = Gv$. *Let* N_{v, W_1} *denote the submodule complementary to* $\Theta_{v, Gv}$ *in* $\Theta_{v, W_1} = W_1$. *Then* $\operatorname{hd} k[N_{v, W_1}]^{G_v} + \operatorname{hd} k[W_2]^{G_v} \leq \operatorname{hd} k[V]^G$.

This follows from Theorem 1.2a) and b) and the obvious equality $(G_v, N_v) = (G_v, N_{v, W_1}) \oplus (G_v, W_2)$. □

As is seen from the proof of Proposition 1.2, we can always assume that $V/G \subset k^s$, $s = \operatorname{ed} k[V]^G$. Therefore, by (0.1), $k[V]^G$ is a hypersurface (i.e., $\operatorname{hd} k[V]^G = 1$) if and only if V/G is a hypersurface in k^s.

3. As an application, we shall now prove

PROPOSITION 1.4. *Let* $k[V]^G$ *be a hypersurface,* $v \in V$ *and* $\overline{Gv} = Gv$.

a) *If* $V = W_1 \oplus \cdots \oplus W_n$, *then no more than one of the algebras* $k[W_j]^G$ *is a hypersurface and the remaining ones are free*;

b) $k[N_v]^{G_v}$ *is either free or is a hypersurface*;

c) *if* G *is connected and semisimple and* G_v *is finite then either* (G_v, N_v) *is trivial or* $k[N_v]^{G_v}$ *is a hypersurface. If* (G_v, N_v) *is nontrivial then its image is generated by such* g*'s that* $\operatorname{rk}(g - \mathrm{Id}) = 2$.

d) *if* $V = W_1 \oplus W_2$ *and* $v \in W_1$ *then* $k[W_2]^{G_v}$ *is either free or is a hypersurface;*

e) *if* $V = W_1 \oplus \cdots \oplus W_n$, $n \geq 2$, $v \in W_1$, G *is connected and semisimple, and* G_v *is finite, then no more than one of the representations* (G_v, N_{v,W_1}), (G_v, W_j), $j \geq 2$, *is nontrivial, and the algebra of invariants of the nontrivial one is a hypersurface; the image of this representation is generated by* g*'s such that* $\operatorname{rk}(g - \operatorname{Id}) = 2$.

PROOF. a) and b) follow from Theorem 1.2a) and b); c) follows from the preceding corollary.

c) By b), the algebra of invariants of the image of (G_v, N_v) is either free or is a hypersurface. Because G is connected and semisimple, the images of $(G, \mathfrak{g})$ and (G, V) are, respectively, in $\mathrm{SL}(\mathfrak{g})$ and $\mathrm{SL}(V)$. Therefore the same is true for the images of $(G_v, \mathfrak{g})$ and (G_v, V). Because G_v is finite, by (0.2) we have $(G_v, N_v) \oplus (G_v, \mathfrak{g}) = (G_v, V)$ and therefore the image H of the representation (G_v, N_v) is in $\mathrm{SL}(N_v)$. In particular, H contains no reflections. By the theorem of Chevalley-Shephard-Todd [66, 142], this means that if $k[N_v]^{G_v} = k[N_v]^H$ is free then H is trivial. If H is nontrivial, then $k[N_v]^{G_v}$ is a hypersurface and therefore a complete intersection. Hence (see [19, 95]) H is generated by such g's that $rk(g - \operatorname{Id}) \leq 2$. Because H contains no reflections, we must have $\operatorname{rk}(g - \operatorname{Id}) = 2$.

e) The proof is similar to those of c) and a) if one takes into account that $(G_v, N_v) = (G_v, N_{v,W_1}) \oplus (G_v, W_2) \oplus \cdots \oplus (G_v, W_n)$. □

4. The next assertion reduces the proof of the main theorem to the case of irreducible representations.

LEMMA 1.1. *Let* G *be a connected semisimple group. Suppose that for any* $h \in \mathbb{N}$ *there exist only finitely many irreducible G-modules* V *with* $\operatorname{hd} k[V]^G = h$. *Then there exist only finitely many G-modules (including reducible ones) whose algebras of invariants have any a priori given homological dimension* d.

PROOF. By the assumption, $\{V \in \mathscr{R}(G) \mid \operatorname{hd} k[V]^G \leq d\}$ contains only finitely many elements say, $V_1, \ldots, V_r$. Let $\operatorname{hd} k[V_i]^G = d_i$. Then, by Theorem 1.2c), any G-module V with $\operatorname{hd} k[V]^G = d$ is of the form $V = n_1 V_1 \oplus \cdots \oplus n_r V_r$, $n_i \in \mathbb{N}$. By Theorem 1.2b), we should have $d \geq n_1 d_1 + \cdots + n_r d_r$. Therefore if $d_i \neq 0$ then $n_i \leq d/d_i$. We now consider those V_i for which $d_i = 0$. Suppose these are $V_1, \ldots, V_l$. According to Theorem 6 of Chapter 3, for any $i = 1, \ldots, l$, there exists $m_i \in \mathbb{N}$ such that $k[m_i V_i]^G$ is nonfree, i.e., $\operatorname{hd} k[m_i V_i]^G \geq 1$. It follows from Theorem 1.2b) that $\operatorname{hd} k[q m_i V_i]^G \geq q$ for all $q \in \mathbb{Z}^+$. Because $n_i V_i$ is a submodule of V, it follows from Theorem 1.2c) that $n_i < (d+1) m_i$, $i = 1, \ldots, l$.

5. Our main effort will now be directed toward the verification of the conditions from Lemma 1.1 for any connected semisimple group G.

Repeating, essentially, the argument from the proof of this lemma we can easily prove an analogue of the main theorem and, automatically, its corollaries for finite groups in the following, more precise, form.

THEOREM 1.3 (The finiteness theorem for finite groups). *Let G be a finite group. Then, up to isomorphism and addition of trivial direct summands, for any integer $d \geq 0$ there exist only a finite number $R_G(d)$ of G-modules V with $\operatorname{hd} k[V]^G = d$. Moreover $\sum_{j=0}^{d} R_G(j) \leq \binom{(2d+3)m-1}{m} - 1$, where $m+1$ is the number of conjugacy classes in G.*

PROOF. It is well known that G has exactly m nonequivalent nontrivial irreducible modules, say, $U_1, \ldots, U_m$. Because for any representation R of G the image of the representation pR, $p \geq 2$, does not contain reflections, it follows from the Chevalley-Shephard-Todd theorem that $\operatorname{hd} k[2U_j]^G \geq 1$ for each j. Let $U = n_1 U_1 \oplus \cdots \oplus n_m U_m$ and $n_1 + \cdots + n_m \geq 2(d+1)m$. Then $n_{i_0} \geq 2(d+1)$ for some i_0 and $\operatorname{hd} k[U]^G \geq \operatorname{hd} k[2(d+1)U_{i_0}]^G \geq (d+1)\operatorname{hd} k[2U_{i_0}]^G \geq d+1$, by Theorem 1.2, and the assertion of the theorem follows. □

REMARK. The estimate from Theorem 1.3 shows that the series $S_G(t) = \sum_{d\geq 0} R_G(d)t^d$ converges when $|t| < 1$.

6. EXAMPLES. 1) Let $|G| = 2$. Any G-module V with $V^G = 0$ is uniquely determined by its dimension. If $x_1, \ldots, x_n$ is a basis in V^*, then $\{x_i x_j \mid 1 \leq i \leq j \leq n\}$ is a minimal system of homogeneous generators of $k[V]^G$, hence, in view of (0.1), $\operatorname{hd} k[V]^G = n(n-1)/2$ and therefore $S_G(t) = \sum_{n=1}^{\infty} t^{n(n-1)/2}$. Since in the sequence of the coefficients of this series there are arbitrarily long segments of zeros, $S_G(t)$ is not a rational function.

2) Suppose that $|G| = 3$, g is a generator of G and $\varepsilon \in k$, $\varepsilon^3 = 1$, $\varepsilon \neq 1$. If V is a G-module with $V^G = 0$ then $V = V_1 \oplus V_2$, where $V_i = \{v \in V \mid gv = \varepsilon^i v\}$. Let $x_1, \ldots, x_p$ be a basis in V_1 and $y_1, \ldots, y_q$ be a basis in V_2. Then $\{x_i x_j x_l \mid 1 \leq i \leq j \leq l \leq p\} \cup \{y_i y_j y_l \mid 1 \leq i \leq j \leq l \leq q\} \cup \{x_i y_j \mid 1 \leq i \leq p;\ 1 \leq j \leq q\}$ is a minimal system of homogeneous generators of $k[V]^G$. It now follows from (0.1) that $S_G(t) = \sum_{n=1}^{\infty}\sum_{j=0}^{n} t^{n(n^2-2nj+3j^2+3n-4)/6}$.

§2. Bounds on $\operatorname{hd} k[V]^G$ for certain types of groups

1. Let G be a reductive group, T be a maximal torus of G, and V be a G-module. Throughout this section we will assume that $\dim T = 1$. Thus G^0 is either T or a three-dimensional simple group.

Let χ be a generator of $\operatorname{Hom}(T, k^*)$. Let $V^+ = (T, V)^+ = \bigoplus_{s>0} V_{\chi^s}$, $V^- = (T, V)^- = \bigoplus_{s<0} V_{\chi^s}$ and $V^0 = (T, V)^0 = V^T$. When $\dim G = 3$, we denote by $n_1, \ldots, n_s$ the dimensions of the irreducible direct summands of the G^0-module V. Then (see [48]) $\dim V^+ = \dim V^- = \sum_{i=1}^s [n_i/2]$. Let $q_{(G,V)}$ denote this number. Evidently, $q_{(G,U\oplus V)} = q_{(G,U)} + q_{(G,V)}$.

In addition we also assume that $Z_G(T)$ is a diagonalizable group.

We are going to investigate separately the cases $\dim G = 1$ and $\dim G = 3$.

If $G^0 = T$ then $[G: Z_G(T)] = 1$ or 2, since $Z_G(T)$ diagonalizes. We now examine each possibility.

2. G is a one-dimensional diagonalizable group.

Let $x_1, \dots, x_p$; $y_1, \dots, y_n$; and $z_1, \dots, z_r$ be G-weighted vectors forming respective bases in $(V^+)^*$, $(V^-)^*$, and $(V^0)^*$. (We view these spaces as subspaces of V^*.) Together they give rise to coordinates on V. It follows easily from part 3 of §1, Chapter 2 that in $k[V]^G$ there exists a uniquely determined minimal system $f_1, \dots, f_q$ of homogeneous generators which are monomials in those coordinates, and that a monomial invariant f coincides with one of the elements $f_1, \dots, f_q$ if and only if f is not divisible by any nonconstant monomial invariant of a smaller degree.

PROPOSITION 2.1. *Under the assumptions and in the notation of this section* $\operatorname{ed} k[V]^G \geq pn + r$, *and if* $pn > 0$ *then* $\operatorname{hd} k[V]^G \geq (p-1)(n-1)$.

PROOF. For any x_i, y_j and z_l there exist $a, b, c \in \mathbb{Z}$ such that $x_i^a y_j^b$ and z_l^c are among $f_l, \dots, f_q$. Indeed, let $x_i \in (V^+)^*_{\chi^m}$, $y_j \in (V^-)^*_{\chi^s}$, $m < 0$, $s > 0$. Then $x_i^s y_j^{-m} \in k[V]^T$. Hence $x_i^s y_j^{-m}$ and z_l are semi-invariants of the finite abelian group G/T, and therefore some of their powers, say, $x_i^a y_j^b$ and z_l^c are in $k[V]^G$, where $a, b, c \in \mathbb{Z}^+$. If we now take $a+b$ and c the smallest possible, then our assertion, along with the desired bound for $\operatorname{ed} k[V]^G$, would follow from the factoriality of $k[V]$ (and the remark preceding this proposition). If $pn \neq 0$, then the action of G on V is stable and nontrivial (see also §3 below). Therefore $\dim V/G = p + n + r - 1$, and the bound for $\operatorname{hd} k[V]^G$ follows from (0.1). □

3. $G^0 - T$ is a one-dimensional torus, $Z_G(T)$ is a diagonalizable group, and $G \neq Z_G(T)$.

Let $\bar{g}$ be the class of $g \in G$ in $G/Z_G(T)$. There is a natural action of $G/Z_G(T)$ on $\operatorname{Hom}(T, k^*)$ and $\operatorname{Hom}(Z_G(T), k^*)$; it is clear that $\bar{g}\chi = \chi^{-1}$ if $\bar{g} \neq e$. If λ is a character of T or $Z_G(T)$ then $g(V^*)_\lambda = (V^*)_{\bar{g}\lambda}$. Therefore $g(V_0^\pm)^* = (V_0^\mp)^*$ for $\bar{g} \neq e$ and $= (V_0^\pm)^*$ for $\bar{g} = e$. In particular, $\dim V^+ = \dim V^-$. Suppose now that $(V^*)_\lambda \neq 0$, where $\lambda \in \operatorname{Hom}(Z_G(T), k^*)$ and $v_1, \dots, v_s$ is a basis of $(V^*)_\lambda$. Then $g(V^+)_\lambda = \oplus_{i=1}^s g\langle v_i \rangle$ where $g\langle v_i \rangle$ depends only on $\bar{g}$.

Suppose that $\mu = \bar{g}\lambda \neq \lambda$ for $\bar{g} \neq e$. Then the sum of $(V^*)_\lambda$ and $(V^*)_\mu$ is direct. Let w_i be a nonzero vector from $g\langle v_i \rangle$, where $\bar{g} \neq e$. Then $v_1, \dots, v_s, w_1, \dots, w_s$ is a $Z_G(T)$-weighted basis of $(V^*)_\lambda \oplus (V^*)_r$, with $g\langle v_i \rangle = \langle w_i \rangle$ and $g\langle w_i \rangle = \langle v_i \rangle$ when $\bar{g} \neq e$. Each $(V^*)_{\chi^s} \neq 0$ is a direct sum of $Z_G(T)$-weighted subspaces and for each such subspace $(V^*)_\lambda \neq 0$ we have $g(V^*)_\lambda \subset (V^*)_{\chi^{-s}}$ when $\overline{g} \neq e$. Thus $\overline{g}\lambda \neq \lambda$ for $s \neq 0$. Therefore there exist $Z_G(T)$-weighted bases $x_1, \dots, x_p$; $y_1, \dots, y_p$; and $z_1, \dots, z_r$, of, respectively, $(V^+)^*$, $(V^-)^*$, and $(V^0)^*$ such that for all i

$$\begin{cases} \text{a)} & \text{The } T\text{-weights } x_i \text{ and } y_i \text{ are reciprocal,} \\ \text{b)} & g\langle x_i\rangle = \langle y_i\rangle \text{ and } g\langle y_i\rangle = \langle x_i\rangle \text{ when } \overline{g} \neq e. \end{cases} \tag{2.1}$$

PROPOSITION 2.2. *Under the assumptions and in the notation of this section* $\operatorname{ed} k[V]^G \geq p(p+1)/2 + r$, *and if* $p > 0$, *then* $\operatorname{hd} k[V]^G \geq (p-1)(p-2)/2$.

PROOF. By (2.1), $\langle x_1^{i_1} \cdots x_p^{i_p} y_1^{j_1} \cdots y_p^{j_p}, x_1^{j_1} \cdots x_p^{j_p} y_1^{i_1} \cdots y_p^{i_p}\rangle$ is G-invariant; it is clear that $k[V^+ \oplus V^-]$ is the direct sum of all such subspaces. It follows that in $k[V^+ \oplus V^-]^G$ there is a minimal system $f_1, \dots, f_q$ of generators of the form $x_1^{i_1} \cdots x_p^{i_p} y_1^{j_1} \cdots y_p^{j_p} + \alpha x_1^{j_1} \cdots x_p^{j_p} y_1^{i_1} \cdots y_p^{i_p}$, $\alpha \in k$, where $(i_1 + \cdots + i_p)(j_1 + \cdots + j_p) \neq 0$ and $\alpha \neq 0 \Leftrightarrow (i_1, \dots, i_p) \neq (j_1, \dots, j_p)$. We shall now show that for any n and m, $1 \leq n \leq m \leq p$, we can find an element of the form

$$x_n^a x_m^b y_n^l y_m^s + \alpha x_n^l x_m^s y_n^a y_m^b, \qquad a + l > 0, \quad b + s > 0, \tag{2.2}$$

among $f_1, \dots, f_q$.

Notice that invariants (2.2) do exist. Indeed, as in Proposition 2.1, $x_n^a y_m^b \in k[V^+ \oplus V^-]^{Z_G(T)}$ for some a and b. By (2.1), there exists $\alpha \in k^*$ such that $g(x_n^a y_m^b) = \alpha x_m^b y_n^a$, $g(x_m^b y_n^a) = \alpha^{-1} x_n^a y_m^b$ for all $g \in G - Z_G(T)$. If $m \neq n$ then $x_n^a y_m^b + \alpha x_m^b y_n^a$ is an invariant of the form (2.2), and if $m = n$ then, by (2.1), $a = b$ and $\alpha = \pm 1$, and therefore $(x_n^a y_n^a)^2$ is an invariant of the form (2.2).

Suppose now that f is an invariant of the form (2.2). Then $f = \sum \alpha_{l_1 \dots l_q} f_1^{l_1} \cdots f_1^{l_q}$, $\alpha_{l_1 \dots l_q} \in k$. Under the substitution $x_i = y_i = 0$ for all $i \neq n$ the left-hand side of this equality does not change, hence it follows from the definition of f_i that the right-hand side can be written as a linear combination of those monomials in f_i that depend only on x_n, x_m, y_n, and y_m. In particular, there are such polynomials among $f_1, \dots, f_q$. If $m = n$ then it follows from (2.1) that there is an element (2.2) among $f_1, \dots, f_q$. Suppose now that $m \neq n$ and there are no elements (2.2) among $f_1, \dots, f_q$. Suppose now that $m \neq n$ and there are no elements (2.2) among $f_1, \dots, f_q$. In this case those of the f_i that depend only on x_n, x_m, y_n, and y_m are either of the form $x_n^l y_n^l$ or of the form $x_m^s y_m^s$. We set $f = x_n^a y_m^b + \alpha x_m^b x_n^a$ (which is possible, as was shown above). Repeating the argument with the substitution, we see that f is a linear combination of the monomials $(x_n^l y_n^l)^c (x_m^s y_m^s)^d$. But this is impossible because $m \neq n$.

Because distinct pairs (n, m) yield distinct elements from $f_1, \dots, f_q$, we have $q = \operatorname{ed} k[V^+ \oplus V^-]^G \geq p(p+1)/2$. From this, combined with the inequality $\operatorname{ed} k[V^0]^G \geq r$ and Theorem 1.1, we obtain the desired bound for $\operatorname{ed} k[V]^G$. If $p \neq 0$ then the action of G on V is stable and nontrivial. Hence $\dim V/G = 2p + r - 1$, and the bound for $\operatorname{hd} k[V]^G$ is a consequence of (0.1). □

4. We now consider the second possibility: $\dim G = 3$.

Because $\operatorname{Aut} G^0 = \operatorname{Int} G^0$, we have $G = Z_G(G^0)G^0$. The group $Z_G(G^0)$ is finite and is contained in the (abelian, see part 1) group $Z_G(T)$. Therefore without loss of generality we may assume that

5. $G = K \times \mathrm{SL}_2$, where K is a finite abelian group.

Let S_n be the SL_2-module of binary forms of degree $n \geq 0$ in variables x and y; up to isomorphism this is the unique irreducible SL_2-module of dimension $n+1$. Let $S_n = 0$ for $n < 0$. The action of SL_2 is given by the formula

$$\sum_{j=0}^{n} \alpha_j x^{n-j} y^j \overset{\left(\begin{smallmatrix} a & b \\ c & d \end{smallmatrix}\right)}{\rightarrow} \sum_{j=0}^{n} \alpha_j (dx - by)^{n-j} (-cx + ay)^j. \tag{2.3}$$

We assume that $T = \{\operatorname{diag}(t, t^{-1}) | t \in k^*\}$, $\chi(\operatorname{diag}(t, t^{-1})) = t$. Then $N_G(T) = \{\left(\begin{smallmatrix} t & 0 \\ 0 & t^{-1} \end{smallmatrix}\right), \left(\begin{smallmatrix} 0 & t \\ -t^{-1} & 0 \end{smallmatrix}\right) | t \in k^*\}$ and $\langle x^{n-j} y^j \rangle = (S_n)_{\chi^{-n+2j}}$. Let $a_n = \operatorname{diag}(e^{\pi i/n}, e^{-\pi i/n})$, $b = \left(\begin{smallmatrix} 0 & i \\ i & 0 \end{smallmatrix}\right)$. We will use the following subgroups of G (see [48]): $T_n = \{a_n^{2j} | 0 \leq j < n\}$ the cyclic group of order n; $D_n = \{a_n^l b^j | 0 \leq l < 2n, j = 0, 1\}$ the binary dihedral group of order $4n$.

The formula $(t, g)v = t(gv)$ defines an action of $k^* \times \mathrm{SL}_2$ on any SL_2-module V. For any $\lambda \in \operatorname{Hom}(K, k^*)$ the homomorphism $\lambda \times \mathrm{Id}: G \to k^* \times \mathrm{SL}_2$ makes V a G-module. In particular, for $V = S_n$ we obtain an irreducible G-module, denoted $S_{n,\lambda}$. It is clear that any irreducible G-module is of the form $S_{n,\lambda}$ for some n and λ.

PROPOSITION 2.3. *Under the assumptions and in the notation of this section, let $p \geq 3$ be a prime number and n be an integer. Suppose also that $n \geq 3p + 5$ if n is even and $n \geq 4p + 5$ if n is odd. Then for any λ*

$$\operatorname{ed} k[S_{n-1,\lambda}]^G \geq (p-1) \binom{[n/p] + p - 2}{p} + (p-1)([n/p] - 1)^2/2 + [n/p] - 1,$$

$$\operatorname{hd} k[S_{n-1,\lambda}]^G \geq (p-1) \binom{[n/p] + p - 2}{p} + (p-1)([n/p] - 1)^2/2 + [n/p] - n + 2.$$

PROOF. Suppose first that n is even, so that $n - 1 = 3p + 2l$; $l \in \mathbb{N}$, $l \geq 2$. Let $\alpha, \beta \in k^*$ be such that all of the roots of the form $u = x^{2p} + \alpha x^p y^p + \beta y^{2p}$ are simple and let $v = x^{l+p} y^l u \in S_{n-1,\lambda}$. Then $\dim Gv = 3$ and $\overline{Gv} = Gv$ (see §3, Chapter 6). For any $g \in \mathrm{SL}_2$ and $f \in S_m - \{0\}$ the roots of gf can be obtained, with the same multiplicities, from the roots of f by the fractional linear transformation with matrix g (see [48]). It now follows from this fact and (2.3), that $(\mathrm{SL}_2)_{\langle v\rangle} \subset T$. Let now $g = \mathrm{diag}(t, t^{-1})$. Then $gv = \mu v$ for some $\mu \in k^*$ if and only if $t^{-3p} = t^{-p} = t^p = \mu$. Hence $(\mathrm{SL}_2)_{\langle v\rangle} = T_{2p}$, $(\mathrm{SL}_2)_v = T_p$.

Let H denote the image of the homomorphism $\lambda \times \mathrm{Id}\colon G \to k^* \times \mathrm{SL}_2$. If $(t, g) \in H$ then $g \in T_{2p}$. The projection $H_v \to T_{2p}$ onto the second summand has a trivial kernel and its image contains T_p. Thus H_v is a cyclic group generated by the element h, where either a) $h = (1, a_p^2)$ or b) $h = (-1, a_p)$.

Let $\varphi \in \mathrm{Hom}(H_v, k^*)$, $\varphi(h) = e^{2\pi i/p}$. It is easy to see that $(H_v, S_{n-1,\lambda})$ equals $\bigoplus_{j=0}^{n-1} \varphi^{-n+2j+1}$ in case a) and equals $\bigoplus_{j=0}^{n-1} \varphi^{(p-n+1)/2+j}$ in case b). Because the order of φ is equal to p, for any integer c the sequence $\varphi^{c+1}, \dots, \varphi^{c+p}$ consists of distinct elements; and because p is a prime greater than or equal to 3, the same is true for $\varphi^{c+1}, \varphi^{c+3}, \dots, \varphi^{c+2p-1}$. Hence $(H_v, S_{n-1,\lambda}) = \bigoplus_{j=1}^{p} m(j)\varphi^j$, $[n/p] \le m(j) \le [n/p]+1$ for all j in both cases. Obviously, $(H_v, \mathfrak{g}) = \bigoplus_{j=-1}^{1} \varphi^{sj}$ where $s = 2$ in case a) and $s = 1$ in case b). Because $p \ge 3$ and $1 \le s \le 2$, it follows that the elements φ^{s_j}, $-1 \le j \le 1$, are distinct. It follows now from (0.2) that $(H_v, N_v) = \bigoplus_{j=1}^{p} l(j)\varphi^j$, $m_j - 1 \le l(j) \le m_j$.

Let $z_{1j}, \dots, z_{l(j),j}$ be coordinates on $(N_v)_{\varphi^j}$ and $f_1, \dots, f_q$ be a minimal system of homogeneous generators of $k[N_v]^{H_v}$ consisting of monomials in z_{ij} (see the argument preceding Proposition 2.1). Because p is prime, for any j any monomial $z_{1j}^{n_1} \cdots z_{l(j),j}^{n_{l(j)}}$, where $(n_1, \dots, n_{l(j)}) \in \mathbb{N}^{l(j)}$, $n_1 + \dots + n_{l(j)} = p$ is contained among $f_1, \dots, f_q$. The same is true for all monomials $z_{ij}z_{k,p-j}$, where $j = 1, \dots, p-1$, and all monomials z_{ip}. The desired bound on $\mathrm{ed}\, k[S_{n-1,\lambda}]^G$ now follows easily from the inequalities for $m(j)$ and $l(j)$. According to [30], the action of G on $S_{n-1,\lambda}$ is stable and $\dim S_{n-1,\lambda}/G = n-3$. Now the desired bound on $\mathrm{hd}\, k[S_{n-1,\lambda}]^G$ follows from (0.1).

Suppose now that n is odd, so that $n-1 = 4p+2l$, $l \in \mathbb{N}$, $l \ge 2$. Let $\alpha, \beta, \gamma \in k^*$ be such that all roots of $u = x^{3p} + \alpha x^{2p}y^p + \beta x^p y^{2p} + \gamma y^{3p}$ are simple and let $v = x^{l+p}y^l u \in S_{n-1,\lambda}$. Then $\dim Gv = 3$ and $\overline{Gv} = Gv$ (see §3, Chapter 6). The same arguments as before show that $(\mathrm{SL}_2)_{\langle v\rangle} = (\mathrm{SL}_2)_v = T_{2p}$ and that (in the same notation) H_v is a cyclic group generated by $h = (1, a_p)$, whence $(H_v, S_{n-1,\lambda}) = \bigoplus_{j=0}^{n-1} \varphi^{(1-n)/2+j}$. The argument identical to the one given before now finishes the proof. □

COROLLARY. $\mathrm{hd}\, k[S_{n,\lambda}]^G$ *grows faster than any power of* n *when* $n \to \infty$.

6. In describing modules V with small $\mathrm{hd}\, k[V]^G$ we will need, in §10, bounds that are good for small values of $\dim V$. Such bounds are derived below; although their order by n is worse than the one in Proposition 2.3, they are applicable for certain n not covered by that proposition. We will also use them in Proposition 2.8 to improve the bound on $\operatorname{hd} k[V]^G$ in the general case (when V may be reducible).

LEMMA 2.1. *Let* $G = \mathrm{SL}_2$ *and* $v = x^n - y^n \in S_n$, $n \geq 3$. *Then*: 1) $\overline{Gv} = Gv$, 2) $G_{\langle v\rangle} = D_n$, 3) $G_v = T_n$ *if n is odd*, $G_v = D_{n/2}$ *if* $n \equiv 2 \bmod 4$, *and* $G_v = c^{-1}D_{n/2}c$, *where* $c = \begin{pmatrix} 0 & -ie^{-\pi i/2n} \\ ie^{\pi i/2n} & 0 \end{pmatrix}$ *if* $n \equiv 0 \bmod 4$.

PROOF. 1) is proved in §3, Chapter 6.

The Hessian $H(v)$ of v is equal to $-n^2(n-1)^2(xy)^{n-2}$. If $g \in G_{\langle v\rangle}$, then $g \in G_{\langle H(v)\rangle}$ (see [48]). Because $k[x, y]$ is factorial, this means that $G_{\langle v\rangle} \subset N_G(T)$, whence 2). Because $G_v \subset G_{\langle v\rangle}$, 3) follows from the formulas

$$\left.\begin{array}{r} v \xrightarrow{a_n^l} e^{\pi il}v \\ v \xrightarrow{a_n^l b} -(-i)^n e^{\pi il}v \end{array}\right\}. \quad \square \tag{2.4}$$

PROPOSITION 2.4. *Let* $G = \mathrm{SL}_2$ *and* $l \in \mathbb{N}$, $l \geq 2$. *Then*: $\operatorname{hd} k[S_{2l}]^G \geq (l-2)^2$ *if l is odd*; $\operatorname{hd} k[S_{2l}]^G \geq (l-2)(l-3)/2$ *if l is even*; $\operatorname{ed} k[S_{2l}]^G \geq l^2 - 2l + 2$ *if l is odd*; $\operatorname{ed} k[S_{2l}]^G \geq (l^2 - 2l + 2)/2$ *if l is even.*

PROOF. Let $v = x^l y^l \in S_{2l}$. Then $\overline{Gv} = Gv$, and $G_v = T$ when l is odd and $G_v = N_G(T)$ when l is even, see §3, Chapter 6. It follows from (0.2) that $(T, N_v) \oplus (T, \mathfrak{g}) = (T, S_{2l}) \oplus (T, \mathfrak{g}_v)$ and because $(T, \mathfrak{g}_v) = \chi^0$, $(T, S_{2l}) = \bigoplus_{j=-l}^{l} \chi^{2j}$, $(T, \mathfrak{g}) = \bigoplus_{j=-1}^{1} \chi^{2j}$ we have $(T, N_v) = \bigoplus_{j=-l, j\neq\pm1}^{l} \chi^{2j}$. Combining this with Propositions 2.1 and 2.2 and Theorem 1.2, we obtain the desired bound for $\operatorname{hd} k[S_{2l}]^G$. Because a generic orbit is closed and of dimension three (see §3, Chapter 6), we have $\dim S_{2l}/G = 2l - 2$. The desired bound for $\operatorname{ed} k[S_{2l}]^G$ now follows from (0.1). $\square$

PROPOSITION 2.5. *Let* $G = SL_2$ *and* $l \in \mathbb{N}, l \geq 3$. *Then* $\operatorname{hd} k[S_{2l-1}]^G \geq l^2 - 2l - 2$ *and* $\operatorname{ed} k[S_{2l-1}]^G \geq l^2 - 5$.

PROOF. Let $v = x^{2l-1} - y^{2l-1} \in S_{2l-1}$. By Lemma 2.1, $\overline{Gv} = Gv$ and $Gv = T_{2l-1}$. Let φ be the restriction of χ to T_{2l-1}. Then $(G_v, S_{2l-1}) = \bigoplus_{j=-l+1}^{l} \varphi^{2j-1}, (G_v, \mathfrak{g}) = \bigoplus_{j=-1}^{1} \varphi^{2j}$. Since $(G_v, N_v) \oplus (G_v, \mathfrak{g}) = (G_v, S_{2l-1})$, (see (0.2)) it now follows that $(G_v, N_v) = \varphi^0 \oplus (\bigoplus_{j=-l+3}^{l-2} \varphi^{2j-1})$. Therefore there exists a G_v-weighted basis $\{x_j\}$ in N_v^*, where j runs through all odd numbers from $-2l+5$ to $2l-5$ and zero, and the weight of x_j is equal to φ^j. Let $f_1, \ldots, f_q$ be a minimal system of homogeneous generators of $k[N_v]^{G_v}$ consisting of monomials in x_j. It is easy to see that the following monomials are contained in this basis: $x_j^{m_j}$ for all j with some

$m_j \in \mathbb{Z}^+$; $x_1^j x_{-j}$ and $x_j x_{-1}^j$ for all $j \geq 1$; $x_j x_i x_{-1}^{m_{ji}}$ and $x_{-j} x_{-i} x_1^{m_{ji}}$ for all $i \geq j \geq 3$ with some $m_{ij} \in \mathbb{Z}^+$. The desired bound on $\operatorname{ed} k[S_{2l-1}]^G$ now follows from Theorem 1.1. Finally the desired bound on $\operatorname{hd} k[S_{2l-1}]^G$ is a consequence of the fact that $\dim S_{2l-1}/G = 2l-3$ (see §3, Chapter 1) and Condition (0.1). □

7. We now obtain a more precise statement of the main theorem for SL_2:

THEOREM 2.1 (The main theorem for SL_2). *Let $G = \mathrm{SL}_2$. Then for any integer $d \geq 0$ there are, up to isomorphism and addition of trivial direct summands, only a finite number $R_G(d)$ of G-modules V with $\operatorname{hd} k[V]^G = d$. Moreover $\sum_{j=0}^d R_G(j) \leq \binom{16d+19}{4}\binom{d+h-4}{d} - 1$, where $h = [\sqrt{d+3}] + [(\sqrt{d}+1)/2] + [(\sqrt{8d+1}+1)/4] + 4$.*

PROOF. From Propositions 2.4 and 2.5 one can easily deduce that there are no more than h irreducible G-modules V with $\operatorname{hd} k[V]^G \leq d$. Let $U_1, \ldots, U_m$ be those modules and let $\operatorname{hd} k[U_j]^G = 0$ for $j \leq l$ and $\operatorname{hd} k[U_j]^G \geq 1$ for $j > l$. By Theorem 6 of Chapter 3, there exists an integer $f \geq 2$ such that $\operatorname{hd} k[fU_j]^G \geq 1$ for all $j \leq l$; we assume that f is the smallest such number.

Let U be a G-module with $\operatorname{hd} k[U]^G \leq d$. Then, by Theorem 1.2, $U = n_1 U_1 \oplus \cdots \oplus n_m U_m$.

We shall show that $\sum_{j=1}^l n_j \leq lf(d+1) - 1$ and $\sum_{j=l+1}^m n_j \leq d$. Indeed, by Theorem 1.2, $d \geq \operatorname{hd} k[U]^G \geq \operatorname{hd} k[\oplus_{j>l} h_j U_j]^G \geq \sum_{j>l} n_j \operatorname{hd} k[U_j]^G \geq \sum_{j>l} n_j$. If the first inequality did not hold we would have $n_{j_0} \geq f(d+1)$ for some $j_0 \leq l$, then $d \geq \operatorname{hd} k[U]^G \geq \operatorname{hd} k[f(d+1)U_{j_0}]^G \geq (d+1) \operatorname{hd} k[fU_{j_0}]^G \geq d+1$, a contradiction.

It now follows from those inequalities that

$$\sum_{j=0}^d R_G(j) \leq \binom{d+h-l}{h-l}\binom{l(fd+f+1)}{l} - 1.$$

As was shown in part 7 of §3, Chapter 3, in this case $l = f = 4$ (we may assume that $U_j = S_j$ for $1 \leq j \leq 4$). □

REMARK. It follows from the obtained bound on $\sum_{j=0}^d R_G(j)$ that the series $S_G(t) = \sum_{d \geq 0} R_G(d) t^d$ converges in a neighborhood of zero.

8. PROPOSITION 2.6. *Let $G = K \times \mathrm{SL}_2$, where K is a finite abelian group, and $l \in \mathbb{Z}^+$, $l \geq 2$. Then $\operatorname{hd} k[S_{2l,\lambda}]^G \geq (l-2)(l-3)/2$ and $\operatorname{ed} k[S_{2l,\lambda}]^G \geq (l^2 - l + 2)/2$ for all λ.*

PROOF. Let $v = x^l y^l \in S_{2l,\lambda}$. Then $(G_v)^0 = ((G^0)_v)^0 = T$ and $\overline{Gv} = Gv$. If $(a, s) \in G_v$, then $s \in (G^0)_{\langle v \rangle} = N_{G^0}(T)$. Hence $G_v \subset K \times N_{G^0}(T)$

so that $T \subset Z_{G_v}(T) \subset Z_{K\times N_{G^0}(T)}(T) = K \times T$. But then $Z_{G_v}(T)$ is a diagonalizable group with $Z_{G_v}(T)^0 = T$. The rest of the argument is identical to the argument in the proof of Proposition 2.4. □

PROPOSITION 2.7. *Let $G = K\times \mathrm{SL}_2$, where K is a finite abelian group, and $l \in \mathbb{Z}^+$. Then $\operatorname{hd} k[S_{2l-1,\lambda}]^G \geq [l/2]-1$ and $\operatorname{ed} k[S_{2l-1,\lambda}]^G \geq 2l+[l/2]-4$.*

PROOF. Let $v = x^n - y^n \in S_{n,\lambda}$, $n = 2l-1 \geq 3$. It follows from Lemma 2.1 that $\overline{Gv} = Gv$ and $(\lambda \times \mathrm{Id})(G_v) \subset (k^* \times D_n)_v$. By (2.4), $(k^* \times D_n)_v = \{((-1)^j, a_n^j), ((-1)^j i^{-n}, a_n^j b) | 0 \leq j < 2n\}$. We also have $x^p y^q \overset{a_n^j}{\to} e^{\pi i j(-p+q)/n} x^p y^q$, $x^p y^q \overset{a_n^j b}{\to} (-i)^n e^{\pi i j(p-q)/n} x^q y^p$ (see (2.3)), which implies that $\langle x^p y^q, x^q y^p\rangle$, $p+q=n$, is a two-dimensional $(k^*\times D_n)_v$-invariant subspace of V, and that $g \in (k^* \times D_n)_v$ acts there as a reflection exactly when $g = ((-1)^j i^{-n}, a_n^j b)$. Let φ_p be a representation of G_v in this subspace; then $(G_v, S_{n,\lambda}) = \bigoplus_{p=0}^{l-1} \varphi_p$. Because $\mathfrak{g}v = \langle x^n + y^n, x^{n-1}y, y^{n-1}x\rangle$ we have $(G_v, \mathfrak{g}v) \oplus 1 = \varphi_0 \oplus \varphi_1$. It now follows from (0.3) that

$$(G_v, N_v) = 1 \oplus \left(\bigoplus_{p=2}^{l-1} \varphi_p\right).$$

Because no element of G_v is a reflection in $\varphi_j \oplus \varphi_{j+1}$ for all j, the theorem of Chevalley-Shephard-Todd implies that $\operatorname{hd} k[\varphi_j \oplus \varphi_{j+1}]^{G_v} \geq 1$.

Now the desired bound for $\operatorname{hd} k[S_{2l-1,\lambda}]^G$ becomes a consequence of Theorem 1.2. Taking into account that $\dim N_v = 2l-3$, we obtain the desired bound for $\operatorname{ed} k[S_{2l-1,\lambda}]^G$ from (0.1). □

9. In conclusion we consider the general case (i.e., when V is not necessarily irreducible).

PROPOSITION 2.8. *Let $G = K \times \mathrm{SL}_2$, where K is a finite abelian group, and V is a G-module. Then $\operatorname{hd} k[V]^G \geq (q_{(G,V)} - 21)/6$ and $\operatorname{ed} k[V]^G \geq (q_{(G,V)} - 39)/6 + \dim V$ (see part 1).*

PROOF. Let $V = \bigoplus_{j=1}^r n_j S_{j,\lambda_j}$, where $n_j \in \mathbb{N}$ for all j. Then

$$q_{(G,V)} = \sum_{j=1}^r [(j+1)/2] n_j. \tag{2.5}$$

We set $h_j = \operatorname{hd} k[S_{j,\lambda_j}]^G$. By Proposition 2.6, $h_{10} \geq 3$, $h_{12} \geq 6$, $h_{14} \geq 10$, and by Proposition 2.7, $h_{11} \geq 2$. Assuming that $p = 3$ in Proposition 2.3, we have $h_j > j+2$ when $j = 13$ and $j \geq 15$. It follows from Proposition 7 of [94] that $\operatorname{hd} k[n_j S_{j,\lambda_j}]^G \geq (n_j - 3)/4$ when $j = 1, 2$; $\geq (n_j - 1)/2$ when

$j = 3, 4, 5, 6$; $\geq n_j$ when $j \geq 7$. It follows now from Theorem 1.2 that

$$\begin{aligned} \operatorname{hd} k[V]^G \geq & -7/2 + n_1/4 + n_2/4 + \sum_{j=3}^{6} n_j/2 \\ & + \sum_{j=7}^{9} n_j + 3n_{10} + 2n_{11} + 6n_{12} + 15n_{13} + 10n_{14} \\ & + \sum_{j=15}^{r} (j+2)n_j. \end{aligned}$$

Let $l = l(n_1, \dots, n_r)$ denote the right-hand side of this inequality. If the n_j are viewed as variables (taking real values) then l becomes a linear function of those variables. We shall now find its minimum value for a fixed $q_{(G,V)}$ under the assumptions that $n_j \geq 0$ for all j and equality (2.5) holds. These restrictions on n_j define an $(r-1)$-dimensional simplex. Because l is linear, the minimum must be attained at a vertex, i.e., at one of the points $(0, \dots, 0, q_{(G,V)}/[(j+1)/2], 0, \dots, 0)$. It follows that this minimum is equal to $(q_{(G,V)} - 21)/6$, whence the desired bound on $\operatorname{hd} k[V]^G$. When $\dim V \geq 4$, the generic stabilizer for SL_2 in V is finite (see [52]). Hence $\dim V/G = \dim V - 3$ (see [30]). Now the desired bound on $\operatorname{ed} k[V]^G$ is a consequence of (0.1). □

§3. Estimating $\operatorname{hd} k[V]^G$ with the aid of one-dimensional tori of G

1. Let G be a reductive group and V be a G-module. If V contains a point v such that $\overline{Gv} = Gv$ and G_v is a group of one of the types considered in §2, then, applying the monotonicity theorems and the bounds from §2, one could obtain bounds for $\operatorname{hd} k[V]^G$ and $\operatorname{ed} k[V]^G$. In this section we will give a method for determining such points v.

2. LEMMA 3.1. *Let G be a connected reductive group, $H \triangleleft G$ and H^0 be a torus. Then H is a diagonalizable group contained in Z_G.*

PROOF. Because $G = G^0$, we have $H^0 \triangleleft G$, and because H^0 is a torus we have $H^0 \subset Z_G^0$. In Z_G^0 we can find a torus T such that $TH^0 = Z_G^0$, $T \cap H^0 = \{e\}$. Then $G = G'TH^0$. We now consider the connected reductive subgroup $G'T$. We have $(G'T \cap H)^0 \subset H^0 \subset Z_G^0$ so that $(G'T \cap H)^0$ is a torus contained in $Z_{G'T}^0 = T$. Because $T \cap H^0 = \{e\}$ the group $G'T \cap H$ is finite. It is also clear that $G'T \cap H \triangleleft G'T$. Because $G'T$ is connected and $G'T \cap H$ is finite, it follows that $G'T \cap H$ is a finite subgroup of $Z_{G'T}$. Because $H^0 \subset Z_G$ and $G = G'TH^0$, it now follows that $H^0(G'T \cap H) \subset Z_G$. But it is easy to see that $H^0(G'T \cap H) = H$. □

3. THEOREM 3.1. *Let G be a connected reductive group, V be a G-module, and T be a one-dimensional torus in G. Suppose that:* a) *the action of*

$Z_G(T)$ on V^T is stable; b) *the generic stabilizer for the action of the image of the representation $(Z_G(T), V^T)$ in V^T is a finite cyclic group*; c) $\operatorname{Ker}(Z_G(T), V^T)^0 = T$.

If $v \in V^T$ is a point in general position then: 1) $\overline{Gv} = Gv$, 2) G_v *is a reductive group with T as a maximal torus*, 3) $Z_{G_v}(T)$ *is a diagonalizable group.*

PROOF. 1) $Z_G(T)v$ is closed (by a)), hence the assertion follows from [110].

2) Let $(Z_G(T), V^T) = \varnothing$. Because $\overline{Gv} = Gv$, it follows that G is reductive. Let S be a maximal torus in G_v containing T. Then $S \subset Z_{G_v}(T)$ and, by b), $\varphi(S) \cong S/(\operatorname{Ker}\varphi \cap S)$ is a finite group. By c), $\dim(\operatorname{Ker}\varphi\cap S) = 1$. Therefore $\dim S = 1$, i.e., $S = T$.

3) Because G is connected, it follows that $Z_G(T)$ is a connected reductive group (see [50]). Because $\operatorname{Ker}\varphi \lhd Z_G(T)$, it follows from c) and Lemma 3.1 that $\operatorname{Ker}\varphi$ is a diagonalizable group contained in the center of $Z_G(T)$ and, therefore, in the center of $Z_{G_v}(T)$. By b), $\varphi(Z_{G_v}(T))$ is a finite cyclic group. Therefore $Z_{G_v}(T)$ is an abelian group (see [24]). According to 2), $Z_{G_v}(T)^0 = T$, whence 3). □

REMARK. Assertions 1) and 2) (and their proofs) remain valid if, keeping a) and c), one drops in b) the requirement that the stabilizer be cyclic.

THEOREM 3.2. 1) *Suppose that the conditions of Theorem 3.1 are satisfied and that T is contained in some three-dimensional simple subgroup C of G. Let $q = q_{(C,V)} - q_{(C,\mathfrak{g})}$ (see part 1 of §2). Then $q \geq -1$ and does not depend on the choice of C. Moreover* $\operatorname{hd} k[V]^G \geq (q-20)/6$, $\operatorname{ed} k[V]^G \geq (q-20)/6 + \dim V - \dim G$. 2) *If, in addition, $G_v^0 = T$ for a generic point $v \in V^T$ then $q \geq 0$, and if $q > 0$ then* $\operatorname{hd} k[V]^G \geq (q-1)(q-2)/2$, $\operatorname{ed} k[V]^G \geq (q-1)(q-2)/2 + \dim V - \dim G$.

PROOF. The fact that q does not depend on C follows from part 1 of §2 (note that the numbers n_i from part 1 of §2 do depend on C).

Let $v \in V$ be a point in general position. By Theorem 3.1, $\overline{Gv} = Gv$ and G_v is a group of one of the types considered in §2. It follows from (0.2) that, in the notation of §2, $\dim N_v^{\pm} = q + \dim \mathfrak{g}_v^{\pm}$.

If $\dim G_v = 1$, i.e., $G_v^0 = T$, then $\mathfrak{g}_v^{\pm} = 0$ and therefore $q = \dim N_v^{\pm} \geq 0$. Suppose that $q > 0$. It follows from Proposition 2.1 and 2.2 that $\operatorname{hd} k[N_v]^{G_v} \geq (q-1)(q-2)/2$ and, because $\dim N_v^0 = \dim V - \dim G + 1 - 2q$ also $\operatorname{ed} k[N_v]^G \geq (q-1)(q-2)/2 + \dim V - \dim G$. Assertion 2) now follows from Theorems 1.1 and 1.2.

If $\dim G_v = 3$ then $\dim \mathfrak{g}_v^{\pm} = 1$ and therefore $\dim N_v^{\pm} = q+1$. From Proposition 2.8 we have $\operatorname{hd} k[N_v]^{G_v} \geq (q-20)/6$ and (because $\dim N_v = \dim V - \dim G + 3$) $\operatorname{ed} k[N_v]^{G_v} \geq (q-20)/6 + \dim V - \dim G$, so that 1) follows from Theorems 1.1 and 1.2. □

REMARK. If we do not know a priori that T is contained in a three-dimensional simple subgroup we set $p = \dim V^+ - \dim \mathfrak{g}^+$, $n = \dim V^- - \dim \mathfrak{g}^-$. Then the same argument shows that $p \geq -1$, $n \geq -1$ and $\operatorname{hd} k[V]^G \geq (\max\{p, n\} - 20)/6$.

4. Later we will apply Theorems 3.1 and 3.2 with an explicitly chosen one-dimensional torus T in each connected semisimple group G. To this end we will have to verify conditions a), b), and c). The group $Z_G(T)/T$ is connected and reductive (see [50]). If it is semisimple then, according to [30], a) follows from b) (and the cyclicity condition in b) can be dropped). In general however $Z_G(T)/T$ has a nonzero-dimensional center. Therefore we need a general criterion for stability (see below). The group G_v in the mentioned theorems can be either three or one dimensional and it becomes important (for example, for finding better bounds) to know what exactly G_v is. To this end we will establish several facts about three-dimensional simple groups containing a given one-dimensional torus, which will be used in §4 for a description of the "typical" G_v's.

5. The stability of reductive linear groups.

THEOREM 3.3. *Let G be a connected reductive group and V be a G-module. Then the action of G on V is stable if and only if* a) *the generic stabilizer for G' on V is reductive,* b) *the action of Z_G^0 on V/G' is stable.*

PROOF. According to [30], a) is equivalent to the stability of the action of G' on V.

Suppose that a) and b) are satisfied. Let $\alpha = \pi_{V, G'}$, $\beta = \pi_{V, G}$, $\gamma = \pi_{V/G', Z_G^0}$. A closed orbit of maximal dimension is always a fiber of the orbit morphism. Hence there is a nonempty open subset Ω of V/G' such that $\alpha^{-1}(w)$ is a G'-orbit for any $w \in \Omega$. The set $t\Omega$ has the same property for any $t \in Z_G^0$, and we may assume that Ω is Z_G^0-invariant. Similarly, there is a nonempty open subset Θ of V/G' such that $\gamma^{-1}(u)$ is a Z_G^0-orbit for any $u \in \Theta$. Obviously, we may assume that $\Theta \subset \gamma(\Omega)$. Now Ω is Z_G^0-invariant and therefore $\gamma^{-1}(\Theta) \subset \Omega$. Because $\beta = \gamma \circ \alpha$, we have $\beta^{-1}(u) = \bigcup_{w \in \gamma^{-1}(u)} \alpha^{-1}(w)$ for all $u \in \Theta$. Hence $\beta^{-1}(u)$ is a G-orbit and therefore the action of G on V is stable.

Conversely, suppose the action of G on V is stable. Then there exists a nonempty open subset Θ of V/G such that $\beta^{-1}(u)$ is a G-orbit for any $u \in \Theta$. From $\gamma^{-1}(u) = \alpha(\beta(u))$ we easily deduce that $\gamma^{-1}(u)$ is a closed Z_g^0-orbit for any $u \in \Theta$. Hence b) is satisfied. Because $\beta^{-1}(u)$ is also a G-orbit, it follows that all G'-orbits in $\gamma^{-1}(u)$ are equidimensional and therefore closed in V. Thus the action of G' on V is stable. □

Now we show how to verify condition b) in certain cases (it will be sufficient for our purposes later).

PROPOSITION 3.1. *In the notation of Theorem* 3.3, *let* $V = \bigoplus_{j=1}^n W_j$ *be a decomposition into irreducible components and let* χ_j *be the character of the scalar (by Schur's lemma) action of* Z_G^0 *on* W_j. *Suppose that the action of* G' *on* W_j *is not locally transitive exactly when* $1 \le j \le s$, *and* $\chi_1, \dots, \chi_s$ *are not contained in any closed half-space of* $\operatorname{Hom}(Z_G^0, k^*) \otimes_{\mathbb{Z}} \mathbb{Q}$. *Then the action of* Z_G^0 *on* V/G' *is stable.*

PROOF. We will use the notation from Proposition 1.3 (replacing G with G') and identify $k[W_j]^{G'}$ with $\pi_j^*(k[W_j])^{G'}$.

It follows from [131] and [30] that $k[W_j]^{G'} \neq k$ for $j \le s$. By Proposition 1.3, there exists a minimal system of homogeneous generators of $k[V]^{G'}$ of the form $f_1, \dots, f_l, f_{11}, \dots, f_{1m_1}, \dots, f_{n1}, \dots, f_{nm_n}$. Let $\pi: V \to k^N$ be the morphism defined by those elements (here $N = l + m_1 + \dots + m_n$). Because f_{jl} is a Z_G^0–semi-invariant for any j and l, and because we may (and will) assume that f_j is also a Z_G^0–semi-invariant for any j, there exists a linear action of Z_G^0 on k^N such that π is Z_G^0-equivariant. Therefore the action of Z_G^0 on V/G' coincides (upon the identification of V/G' with $\pi(V)$) with the restriction to $\pi(V)$ of the action of Z_G^0 on k^N. Because the set of weights of Z_G^0 on k^N contains $\chi^{\deg f_{jl}}$ for all j, l and by the assumptions on $\chi_1, \dots, \chi_s$ we have $\overline{Z_G^0 v} = Z_G^0 v$ for all $v \in \Omega$ where Ω is the complement in k^N of the union of all possible proper Z_G^0-invariant subspaces (see [16]). Finally, $\dim \Theta_{0, V/G'} = N$, whence $\pi(V) \cap \Omega \neq \varnothing$. □

This proposition is especially easy to use when $\dim Z_G = 1$.

PROPOSITION 3.2. *In the notation of Theorem* 3.3, *let* W_1 *and* W_2 *be (not necessarily irreducible) submodules of* V *with* $W_1 \cap W_2 = 0$. *Suppose that* $\dim Z_G = 1$, *the action of* G' *on* W_j *is not locally transitive, and the weights of* Z_G^0 *are positive (with respect to a fixed isomorphism* $\operatorname{Hom}(Z_G^0, k^*) \to \mathbb{Z}$) *on* W_1 *and negative on* W_2. *Then the action of* Z_G^0 *on* V/G' *is stable.*

The proof is similar to that of Proposition 3.1; one has to take into account that $\dim \operatorname{Hom}(Z_G^0, k^*) \otimes_{\mathbb{Z}} \mathbb{Q} = 1$. □

REMARK. Suppose that $\dim Z_G = 1$. If in $k[V]^{G'}$ there are two homogeneous Z_G–semi-invariants with weights of opposite signs then these semi-invariants must be contained in any system of homogeneous generators of $k[V]^{G'}$. But then the argument from the proof of Proposition 3.1 shows that the action of Z_G^0 on V/G' is stable.

6. Three-dimensional connected simple groups containing a given one-dimensional torus. Let C be a three-dimensional connected simple group. Then either $C \cong \mathrm{SL}_2$ or $C \cong \mathrm{PSL}_2$. Any maximal torus T of C

defines, canonically up to sign, an element $H_{C,T}$ of the Lie algebra $\mathfrak{c}$. More precisely, if $\alpha\colon k^* \to T$ is one of the two possible isomorphisms then $H_{C,T} = d\alpha(d/dz)$ (z is a coordinate on k^*).

Let $C \cong \mathrm{SL}_2$. Then $H = H_{C,T}$ can be extended to a basis A, B, H of the Lie algebra $\mathfrak{sl}_2$ in which $[H, A] = 2A$, $[H, B] = -2B$, $[A, B] = H$. The irreducible C-modules are exactly S_n, $n \in \mathbb{Z}$ (see §2). S_n has a T-weighted basis $e_0, e_1, \dots, e_n$, such that $He_j = (n-2j)e_j$ for all j. Elements A and B act on $\langle e_j \rangle$ as indicated on the diagram

$$0 \underset{A}{\leftarrow} \langle e_0 \rangle \underset{A}{\overset{B}{\rightleftarrows}} \langle e_1 \rangle \underset{A}{\overset{B}{\rightleftarrows}} \cdots \underset{A}{\overset{B}{\rightleftarrows}} \langle e_{n-1} \rangle \underset{A}{\overset{B}{\rightleftarrows}} \langle e_n \rangle \overset{B}{\to} 0$$

where each arrow is a surjection.

Let $C \cong \mathrm{PSL}_2$ and $\pi\colon \mathrm{SL}_2 \to \mathrm{PSL}_2$ be the canonical epimorphism. If D is a maximal torus of SL_2 then $T = \pi(D)$ is a maximal torus of C. Obviously $d\pi(H_{\mathrm{SL}_2, D}) = 2H_{C,T}$. Because $d\pi$ is an isomorphism, $H = H_{C,T}$ can be extended to a basis A, B, H of the Lie algebra $\mathfrak{psl}_2$ in which $[H, A] = A$, $[H, B] = -B$, $[A, B] = 2H$. The irreducible C-modules are those S_n for which the action of SL_2 can be factored through π, i.e., exactly S_{2m}, $m \in \mathbb{N}$. S_{2m} has a T-weighted basis $e_0, e_1, \dots, e_n$ such that $He_j = (m-j)e_j$ for all j. The actions of A and B on $\langle e_j \rangle$ are also given by the above diagram with $n = 2m$.

LEMMA 3.2. *Let G be a reductive group, let T be a one-dimensional torus of G, and let C_1 and C_2 be three-dimensional simple connected subgroups of G containing T. Then the following conditions are equivalent*: a) *C_1 and C_2 are conjugate by an element of $Z_G(T)$*; b) *C_1 and C_2 are isomorphic.*

PROOF. Let $\alpha\colon k^* \to T$ be a fixed isomorphism, $H = H_{C_1,T} = H_{C_2,T} = d\alpha(d/dz) \in \mathfrak{t} \subset \mathfrak{g}$ and H, A_j, B_j the basis of $\mathfrak{c}_j$ mentioned before, $j = 1, 2$. We have the $\mathbb{Z}$-grading $\mathfrak{g} = \bigoplus_{l \in \mathbb{Z}} \mathfrak{g}(l)$, where $\mathfrak{g}(l) = \{X \in \mathfrak{g} \mid [H, X] = lX\}$. It can also be obtained first by decomposing $\mathfrak{g}$ into a direct sum of irreducible C_j-modules and then decomposing each one of them into a direct sum of T-weighted subspaces. Combining this with the remarks preceding the lemma we see that 1) $\mathfrak{g}(0) = \mathfrak{z}_{\mathfrak{g}}(\mathfrak{t})$; 2) $A_j \in \mathfrak{g}(l_j)$, $B_j \in \mathfrak{g}(-l_j)$, where $l_j = 1$, if $C_j \cong \mathrm{PSL}_2$, and $l_j = 2$, if $C_j \cong \mathrm{SL}_2$; 3) $\mathfrak{g}(0) \overset{A_j}{\to} \mathfrak{g}(l_j)$ is surjective and $\mathfrak{g}(-l_j) \overset{A_j}{\to} \mathfrak{g}(0)$ is injective (the map is defined by $X \mapsto [A_j, X]$). The $Z_G(T)$-orbit of point A_j is contained in $\mathfrak{g}(l_j)$ and $A_j + [\mathfrak{g}(0), A_j]$ is the tangent space to this orbit at point A_j. In view of 3) and 2), it coincides with $\mathfrak{g}(l_j)$ and therefore the mentioned orbit is open in $\mathfrak{g}(l_j)$.

Let now $C_1 \cong C_2$. Then $l_1 = l_2$. Because $Z_G(T)$ can have only one open orbit in $\mathfrak{g}(l_j)$ we may, conjugating C_1 by an element from $Z_G(T)$, assume (without changing H) that $A_1 = A_2$. But then $[A_1, B_1 - B_2] = 0$, hence, in view of 2) and 3), $B_1 = B_2$ and therefore $C_1 = C_2$. □

7. An examination of the hypothesis and conclusions of Theorem 3.1. We have already indicated how to verify Condition a) of this theorem. We now consider Condition b). Because $\{y \in V \mid \dim G_y \geq n\}$ is closed in V for any $n \in \mathbb{N}$, there exists, under the assumptions of this theorem, a nonempty open subset Ω of V^T such that either $(G_v)^0 = T$ for all $v \in \Omega$ or $(G_v)^0$ is a three-dimensional simple group for any $v \in \Omega$ (see assertion 2) of Theorem 3.1).

PROPOSITION 3.3. *Let H be a connected reductive group and W be an H-module. Suppose that* a) $\dim Z_H = 1$, b) *the generic stabilizer for H' in W is trivial. Then the generic stabilizer for H in W is either finite or is a one-dimensional torus. The latter case can be dropped if, in addition, the following condition holds*: c) *there exists an H-submodule U of W such that the action of H' on U is not locally transitive and the weights of Z_H^0 on U (under a fixed isomorphism* $\operatorname{Hom}(Z_H^0, k^*) \to \mathbb{Z}$) *are all different from zero and have the same sign.*

PROOF. If $\alpha\colon H \to H/H'$ is the canonical epimorphism and $H'_v = \{e\}$ for $v \in V$ then $\alpha\colon H_v \to \alpha(H_v)$ is an isomorphism and (because H/H' is a one-dimensional torus) $\alpha(H_v)$ is either finite or is a one-dimensional torus.

Suppose c) is satisfied and v is a generic point in W. Assume that H_v is a one-dimensional torus. By b), $\dim H' = \dim H'v$ and, by a), $\dim Hv = \dim H'$. Because H and H' are connected, we have $\overline{Hv} = \overline{H'v}$. From b) and Theorem 3.3, we have $\overline{H'v} = H'v \subset Hv \subset \overline{Hv} = \overline{H'v}$, whence $H'v = Hv$.

Let now $f \in k[W]^{H'}$, $t \in Z_H^0$. As was shown before, $t^{-1}v = hv$ for some $h \in H'$. Therefore $(tf)(v) = f(hv) = f(v)$, so that $tf = f$. Hence $k[W]^{H'} = k[W]^H$. Viewing $k[U]$ as a subalgebra of $k[W]$ we deduce from c) that $k[U]^{H'} \neq k$ and $k[U]^{Z_H^0} = k$. Therefore $k \neq k[U]^{H'} = k[U]^H = k[U]^{H'} \cap k[U]^{Z_H^0} = k$, a contradiction. □

REMARK. Replacing "triviality" with "finiteness" in b) and using the same argument as before, one can deduce from a) and b) that the generic stabilizer for H in W is either finite or is an extension of a one-dimensional torus by a finite group, and if c) is satisfied, the second possibility can be excluded.

PROPOSITION 3.4. *Suppose that Conditions* a), b), *and* c) *of Theorem* 3.1 *are satisfied* (*one can even drop the cyclicity requirement in* b)) *and $H = (G_v)^0$ is a three-dimensional simple group. Then N_v contains no irreducible H-submodules of odd dimensions bigger than* 1 *and, if (H, N_v) is nontrivial then $H \cong \mathrm{SL}_2$. Moreover, (H, N_v) is necessarily nontrivial if the dimension of the generic stabilizer for G in V is at most* 2.

PROOF. According to [108], in N_v there exists a π_{N_v, G_v}-saturated affine neighborhood U of zero such that the image of the natural morphism

$\varphi\colon G\times_{G_v} U \to U$ is a $\pi_{V,G}$-saturated neighborhood W of v, and the commutative diagram

$$\begin{array}{ccc} G\times_{G_v} U & \xrightarrow{\quad\varphi\quad} & W \\ {\scriptstyle \pi_{G\times_{G_v}U,G}}\big\downarrow & & \big\downarrow{\scriptstyle \pi_{W,G}} \\ U/G_v & \xrightarrow{\quad\varphi/G_v\quad} & W/G \end{array} \tag{3.1}$$

corresponds to a change of base (with φ and φ/G_v being étale). Suppose that N_v has a nontrivial irreducible H-submodule of odd dimension. Because $T\subset H$ we have $N_v^T\neq 0$ and $(H_w)^0 = T$ for any $w\in N_v^T$, $w\neq 0$ (see above and the proof of Proposition 2.4). Therefore $(G_x)^0 = T$ for some point $x\in G\times_{G_v} U$. By the above property of diagram (3.1), $(G_{\varphi(x)})^0 = T$. In particular, $\varphi(x)\in V^T$. But, by the assumption, $V^T\subset\{y\in V\mid \dim G_y\geq 3\}$, a contradiction.

If (H, N_v) is nontrivial, then N_v contains a nontrivial irreducible H-submodule whose dimension, as we showed, is even. Therefore $H\cong \mathrm{SL}_2$.

It follows from the properties of diagram (3.1) that the generic stabilizer for G in W (and therefore in V) is isomorphic to the generic stabilizer for G_v in U (and thus in N_v). If its dimension is at most 2 then the dimension of the generic stabilizer for H in N_v is at most 2 and therefore (H, N_v) is nontrivial. □

§4. Majorizing theorems for multiplicities, generic stabilizers, and stability

1. Let G be a connected reductive group and V be a G-module. Our plan is to choose a one-dimensional torus T in G and then to find a bound for $\mathrm{hd}\, k[V]^G$ using Theorems 3.1 and 3.2 (a posteriori, we will see that this is indeed possible in the majority of cases, although not in all cases). Because there are infinitely many modules, we have to standardize our choices. To do this, in §§5 and 6 we will explicitly describe T, $Z_G(T)^{V^T}$, $(Z_G(T), V^T)$ (or a subrepresentation of it), a connected three-dimensional simple subgroup $C\supset T$ and (C, V) (or a subrepresentation of it) for each fundamental G-module V of each simple group G; in this section we will prove several statements that would enable us to obtain from this information similar information about the other modules, as well as to verify the conditions of the aforementioned theorems. More precisely, these statements are concerned with the behavior of the properties we are interested in (like multiplicities, generic stabilizers, and stability) under the passage from an irreducible module to a module majorizing it (in the sense of the theory of highest weights).

Clearly, when calculating $\mathrm{hd}\, k[V]^G$, we may assume that G is simply connected, and we will do so throughout this chapter. We will also choose a fixed maximal torus of G and, with the aid of a fixed Borel subgroup B, in the additively-written group of its rational characters (naturally identified with $\mathrm{Hom}(B, k^*)$) we fix the semigroup $\mathfrak{X}$ of dominant weights with fundamental weights $\omega_1, \dots, \omega_l$.

2. Some of the proofs in this section are based on the existence of the universal model of the set of all rational representations of G. By this we mean a finitely-generated integral commutative integrally-closed k-algebra S with identity endowed with a rational action (see part 1 of §1, Chapter 1) of G by k-automorphisms satisfying the following conditions: a) for any $\Lambda \in \mathfrak{X}$ there exists a unique G-invariant subspace $S(\Lambda)$ of S with $\dim S(\Lambda) < \infty$ and $(G, S(\Lambda)) = R(\Lambda)$; b) $\langle S(\Lambda) \cdot S(M) \rangle = S(\Lambda + M)$ for all Λ, M; c) S is factorial; d) all invertible elements of S are in $k = k \cdot 1$.

The existence of S can be proved as follows. Let U be the maximal unipotent subgroup of B. Then S is exactly the left-invariant subalgebra $k[G/U] = \{f \in k[G] \| f(gu) = f(g) \text{ for all } g \in G, u \in U\}$ of $k[G]$: for a) and b) see [16], c) follows, by [33], from the simple-connectedness of G and the triviality of $\mathrm{Hom}(U, k^*)$.

We say that $\Lambda \in \mathfrak{X}$ majorizes $M \in \mathfrak{X}$ (or $R(\Lambda)$ majorizes $R(M)$) if $\Lambda - M \in \mathfrak{X}$.

Henceforth we will assume that G is semisimple.

3. Multiplicities.

THEOREM 4.1. *Let L and H be subgroups of a connected simply connected semisimple group G and assume that $H \subset N_G(L)$ and the group $P = H/(H \cap L)$ is connected and reductive. Let $R(\Lambda_j)$, $1 \leq j \leq s$, be a set of irreducible G-modules. Suppose that the P-module $R(\Lambda_j)^L$ contains an irreducible submodule M_j with multiplicity $n_j \geq 1$. Then $M_1 \cdot \dots \cdot M_s$ enters $R(\Lambda_1 + \dots + \Lambda_s)$ with multiplicity at least $\max_j n_j$.*

PROOF. We may assume that $s \geq 2$. We will consider highest vectors and weights of P with respect to a fixed Borel subgroup $\widetilde{B} \subset P$. By assumption, if V_j is the linear hull of highest vectors of the submodules of type M_j in $S(\Lambda_j)^L$ then $\dim V_j = n_j$. Let, say, $n_1 = \max_j n_j$ and $f_1, \dots, f_{n_1}$ be a basis of V_1. Let also $h_j \in V_j$, for all $j \geq 2$, be nonzero vectors and $h = h_2 \cdots h_s$. Then, by property c) of algebra S, we have $f_j h \in S(\Lambda_1 + \dots + \Lambda_s)^L$ for all j and it is clear that for any j the element $f_j h$ is a $\widetilde{B}$-semi-invariant whose weight is the highest weight of $M_1 \cdot \dots \cdot M_s$. Because S is integral, it follows that $f_1 h, \dots, f_{n_1} h$ are linearly independent. □

COROLLARY. *Suppose $R(\Lambda_j)^L \neq 0$ for all j. Then* $\dim R(\Lambda_1 + \dots + \Lambda_S)^L \geq \max_j \dim R(\Lambda_j)^L$. *In particular, if $R(\omega_j)^L \neq 0$ for all j, then $R(\Lambda)^L \neq 0$ for all $\Lambda \in \mathfrak{X}$.*

This follows from Theorem 4.1 if $H = L$. □

4. Generic stabilizers and stability.

PROPOSITION 4.1. *Let H be a connected reductive subgroup of G and $\Lambda \in \mathfrak{X}$. Consider the following properties of a point in general position $v \in R(\Lambda)$:*

(A) $(G_v)^0 = H$; (B) $(G_{\langle v\rangle})^0 = H$. *Then* 1) *if* $\Lambda_1, \Lambda_2 \in \mathfrak{X}$, $R(\Lambda_1)^H \neq 0$ *and* (B) *holds for* $\Lambda = \Lambda_2$ *then* (A) *holds for* $\Lambda = \Lambda_1 + \Lambda_2$, 2) *if the action of* $Z_G(H)$ *on* $R(\Lambda)^H$ *is stable then* (A) $\Rightarrow$ (B), 3) *if* $Z_G(H)/(Z_G(H) \cap H)$ *is semisimple and* (A) *holds then the action of* $Z_G(H)$ *on* $R(\Lambda)^H$ *is stable.*

PROOF. Because for any G-module V and any n the sets $\{v \in V \mid \dim G_v \geq n\}$ and $\{v \in V \mid \dim G_{\langle v\rangle} \geq n\}$ are closed in V, it is easy to see that if one of the Conditions (A) and (B) holds for some point v then it also holds for a point v in general position. Now we can assume that $H \neq G$.

1) Suppose that the conditions of 1) are satisfied. Because $H \neq G$, we have $R(\Lambda_2)^H \neq 0$. Let now $(G_{\langle v\rangle})^0 = H$ for $v \in S(\Lambda_2)^H$, $v \neq 0$, and $u \in S(\Lambda_1)^H$, $u \neq 0$. Then $vu \neq 0$ because S is integral. It follows from Properties c) and d) of algebra S that $(G_{vu})^0 \subset (G_{\langle v\rangle})^0 = H$, and it follows from Property b) that $vu \in S(\Lambda_1 + \Lambda_2)^H$. In particular, it means, in view of the equality $H = H^0$, that $H \subset (G_{vu})^0$. Therefore $(G_{vu})^0 = H$.

2) Suppose that the action of $Z_G(H)$ on $R(\Lambda)^H$ is stable and (A) holds. Then $Z_G(H)v$ is closed. Therefore Gv is also closed (see [47]). If (B) were not satisfied we would have $(G_{\langle v\rangle})^0 \neq (G_v)^0$, and therefore $G_{\langle v\rangle}v = \langle v\rangle - \{0\}$. Hence $0 \in \overline{Gv} = Gv$, contrary to the fact that $v \neq 0$ (because $H \neq G$).

3) Suppose that $Z_G(H)/(Z_G(H) \cap H)$ is semisimple and that (A) holds. Then $(Z_G(H)_v)^0 = (Z_G(H) \cap G_v)^0 = (Z_G(H) \cap (G_v)^0)^0 = (Z_G(H) \cap H)^0$. Therefore $(Z_G(H)/(Z_G(H) \cap H))_v$ is finite. It now follows from part a) of Theorem 3.3 that the action of $Z_G(H)/(Z_G(H) \cap H)$ (and therefore of $Z_G(H)$) on $R(\Lambda)^H$ is stable. $\square$

The following statement (in conjunction with Proposition 4.1) is directly related to part 2) of Theorem 3.1.

PROPOSITION 4.2. *Let* T *be a one-dimensional torus of* G *and* $\Lambda_1, \Lambda_2, \Lambda_3 \in \mathfrak{X}$ *weights such that* a) $R(\Lambda_j)^T \neq 0$ *for* $j = 1, 2, 3$; b) $R(\Lambda_j)^T \neq R(\Lambda_j)$ *for* $j = 1, 2$; c) *the action of* $Z_G(T)/T$ *on* $R(\Lambda_3)^T$ *is stable and has finite generic stabilizer. Then* $(G_v)^0 = T$ *for a point in general position* $v \in R(\Lambda_1 + \Lambda_2 + \Lambda_3)^T$.

PROOF. First we show that there is no connected three-dimensional simple subgroup C of G such that $C \supset T$ and $S^C \supset S(\Lambda_1 + \Lambda_2)^T$. Suppose this is not the case and $p_j \in S(\Lambda_j)^T$, $p_j \neq 0$, $j = 1, 2$. Then $0 \neq p_1p_2 \in S(\Lambda_1 + \Lambda_2)^T \cap S^C$. It is not difficult to deduce from Properties c) and d) of algebra S and from the connectedness and simplicity of C that $p_j \in S^C$, $j = 1, 2$. Hence $S(\Lambda_j)^T \subset S^C$, $j = 1, 2$.

Condition b) and the inclusion $T \subset C$ imply that $S(\Lambda_j) \not\subset S^C$, $j = 1, 2$. Therefore one can find an irreducible submodule M_j of the C-module

$S(\Lambda_j)$, $j = 1, 2$, of dimension $d_j \geq 2$. The number d_j is necessarily even; otherwise $\dim M_j^T = 1$ (see §3) and $M_j^C = 0$, contrary to the inclusion $S(\Lambda_j)^T \subset S^C$. By Theorem 4.1 (with $L = \{e\}$ and $H = C$), the C-module $S(\Lambda_1 + \Lambda_2)$ contains submodule $M_1 \cdot M_2$. But $\dim M_1 \cdot M_2 = d_1 + d_2 - 1$ is an odd number. Therefore $\dim(M_1 \cdot M_2)^T = 1$ and $(M_1 \cdot M_2)^C = 0$ contrary to the inclusion $S(\Lambda_1 + \Lambda_2)^T \subset S^C$.

We shall now prove the desired assertion. Clearly it suffices to find a point $w \in S(\Lambda_1 + \Lambda_2 + \Lambda_3)^T$, such that $\dim(G_w)^0 = 1$. If u is a point in general position in $S(\Lambda_3)^T$ then, according to the remark after Theorem 3.1, $\overline{Gu} = Gu$ and G_u is a rank 1 reductive group containing T. Let $z \in S(\Lambda_1 + \Lambda_2)^T$, $z \neq 0$. Then $0 \neq w = zu \in S(\Lambda_1 + \Lambda_2 + \Lambda_3)^T$. From Properties c) and d) of algebra S and the connectedness of $(G_w)^0$ we deduce that $(G_w)^0 \subset (G_{\langle z \rangle})^0 \cap (G_{\langle u \rangle})^0$, and from the closedness of Gu we have $(G_{\langle u \rangle})^0 = (G_u)^0$. Therefore $(G_w)^0 \subset (G_z)^0 \cap (G_u)^0$. It follows from the construction that the dimension of the right-hand side of this inclusion is between 1 and 3, and the dimension of the left-hand side is at least 1. If $(G_u)^0 = T$ then w is the desired point. Thus we may assume that $(G_u)^0$ is a three-dimensional simple group. If now $\dim(G_z)^0 \cap (G_u)^0 = 3$, then $(G_z)^0 \supset (G_u)^0$. This inclusion, according to the first part of the proof, cannot be true for all $z \in S(\Lambda_1 + \Lambda_2)^T$, $z \neq 0$, with a fixed u. Therefore we may assume that $\dim(G_z)^0 \cap (G_u)^0 \leq 2$. But $\dim(G_z)^0 \cap (G_u)^0 \neq 2$: otherwise $(G_z)^0 \cap (G_u)^0$ would be a Borel subgroup of the connected three-dimensional simple group $(G_u)^0$ and therefore would coincide with $(G_u)^0 \cap G_z$ (see part 8 of §1, Chapter 6 and [50]). But then $(G_u)^0 z \cong (G_u)^0/(G_u)^0 \cap G_z \cong \mathbb{P}^1$, which is impossible. □

PROPOSITION 4.3. *Let L and H be subgroups of G such that $H \subset N_G(L)$ and $P = H/(H \cap L)$ is nontrivial, connected, and semisimple. Assume that:* a) *$R(\omega_j)^L \neq 0$ for all $j = 1, \dots, l$;* b) *for any j there exist $m_j \in \mathbb{N}$ and an irreducible submodule M_j of the P-module $R(m_j\omega_j)^L$ such that the noneffectivity kernel of the action of P on M_j is finite. Then for any given constant $h \in \mathbb{N}$ there exists a (depending on h) finite subset Φ of $\mathfrak{X}$ such that for any $\Lambda \in \mathfrak{X} - \Phi$ the P-module $R(\Lambda)^L$ contains a nontrivial submodule of dimension $\geq h$ with finite noneffectivity kernel and with generic stabilizer contained in this kernel.*

PROOF. Let Q be an arbitrary connected semisimple group. Then there exists $s_Q \in \mathbb{N}$ such that the generic stabilizer of any $U \in \mathscr{R}(Q)$ with $\dim U > s_Q$ is contained in the noneffectivity kernel (for an a priori proof (see [4]) where it was shown that one can take $s_Q = 2(\dim Q - \operatorname{rk} Q) \cdot \max_j(\operatorname{rk} Q_j + 1)$, with Q_j running through the connected simple normal subgroups of Q; a posteriori it also follows from [26, 28], where a complete classification of those $U \in \mathscr{R}(Q)$ whose generic stabilizers are contained in the noneffectivity kernel is given). On the other hand, it follows from Weyl's dimension formula (see [10]) (or from Lemma 5 of Chapter 3), that there are only finitely many elements of $\mathscr{R}(Q)$ whose dimension does not exceed any given number. It is also clear that any finite subset of $\mathscr{R}(Q)$ is contained in the set of all $U \in \mathscr{R}(Q)$ whose sums of the numerical labels of the highest weights do not exceed a certain constant. This set is also finite and has the property that if $U \in \mathscr{R}(Q)$ does not belong to it then any module from $\mathscr{R}(Q)$ majorizing U also does not belong to it.

We now apply all this to $Q = P$. Then there exists a finite subset $\mathscr{M}$ of $\mathscr{R}(P)$ such that: 1) if $M \in \mathscr{R}(P) - \mathscr{M}$ and $N \in \mathscr{R}(P)$ majorizes M then $N \in \mathscr{R}(P) - \mathscr{M}$; 2) if $M \in \mathscr{R}(P) - \mathscr{M}$ then $\dim M \geq h$ and the generic stabilizer for P in M is contained in the noneffectivity kernel.

Because of b), the highest weight of M_j is zero. Because under the Cartan multiplication the highest weights add, it follows that $M_j^{r_j} \in \mathscr{R}(P) - \mathscr{M}$ for some $r_j \geq 1$. We now choose a nonzero irreducible submodule N_j in each P-module $R(\omega_j)^L$ (it is possible by a)) and consider the finite set $\Phi = \{\sum_{j=1}^{l} n_j\omega_j \in \mathfrak{X} \mid n_j < m_jr_j \text{ for all } j\}$. Let $\Lambda = \sum_{j=1}^{l} n_j\omega_j \in \mathfrak{X} - \Phi$. Then $n_s \geq m_sr_s$ for some s. By Theorem 4.1, the P-module $R(\Lambda)^L$ contains the submodule $(\prod_{j=1, j\neq s}^{l} N_j^{n_j}) \cdot N_s^{n_s - m_sr_s} \cdot M_s^{r_s}$, which majorizes $M_s^{r_s}$ (and M_s) and therefore, by construction, belongs to $\mathscr{R}(P) - \mathscr{M}$. Because the finiteness of the noneffectivity kernel is equivalent to the existence of a vertex with nonzero label in each connected component of the Dynkin diagram of the given module, it follows from b) that the noneffectivity kernel of the aforementioned submodule is finite. Therefore Φ has the desired properties. □

PROPOSITION 4.4. *Let G, L, H, and P be defined as in Proposition 4.3. Suppose also that P is nontrivial, connected, and reductive and* $\dim Z_P = 1$. *Fix an isomorphism* $\operatorname{Hom}(Z_p, k^*) \to \mathbb{Z}$ *and assume that:* a) $R(\omega_j)^L \neq 0$ *for all* $j = 1, \ldots l$; b) *for any j there exist $m_j \in \mathbb{N}$ and irreducible submodules M_j' and M_j'' of the P-module $R(m_j\omega_j)^L$ such that the noneffectivity kernels of the action of P' on M_j' and M_j'' are finite and the weights of Z_p on M_j' and M_j'' are different from zero and have different signs. Then for any given constant $h \in \mathbb{N}$ there exists a (depending on h) finite subset Φ of $\mathfrak{X}$ with the following properties: for any $\Lambda \in \mathfrak{X} - \Phi$ the P-module $R(\Lambda)^L$ contains two submodules such that the noneffectivity kernels of the action of P' on them are finite, generic P'-stabilizers are contained in those kernels, and the weights of Z_p on those modules are different from zero and have different signs. Moreover, if $P' \neq \{e\}$ then the dimensions of those submodules are bigger than or equal to h.*

of the form $\{\varepsilon_i - \varepsilon_j\}$, $i, j \neq 1, 2$. There are no other α_1-series. Therefore $(C, \mathfrak{sl}_{l+1}) = (l-1)^2 S_0 \oplus 2(l-1)S_1 \oplus S_2$.

In conclusion, we shall obtain similar information for the G-modules $R(p\omega_1) = S^p V$ (see [15, 10]). We have the direct sum decomposition of $Z \times Z_G(T)'$-modules $(S^p V)^T = S^p W \oplus (e_1 e_2) S^{p-2} W \oplus (e_1^2 e_2^2) S^{p-4} W \oplus \cdots$; we also have $(Z_G(T)', (e_1^r e_2^r) S^{p-2r} W) = R((p-2r)\omega_1)$ and the weight of Z on this module equals $p - r(l+1)$. Moreover we have the direct sum decomposition of C-modules $S^p V = S^p W \oplus \langle e_1, e_2 \rangle S^{p-1} W \oplus \langle e_1^2, e_1 e_2, e_2^2 \rangle S^{p-2} W \oplus \cdots$ and it is clear that the C-module $\langle e_1^r, e_1^{r-1} e_2, \ldots, e_2^r \rangle S^{p-r} W$ is isomorphic to $\binom{p+l-r-2}{p-r} S_r$. Therefore $(C, S^p V) = \binom{l+p-2}{p} S_0 \oplus \binom{l+p-3}{p-1} S_1 \oplus \cdots \oplus \binom{l-2}{0} S_p$.

Thus we have proved the following

LEMMA 5.1. *Let $G = \mathrm{SL}_{l+1}$, and T and C be the one-dimensional torus and, respectively, the connected three-dimensional simple subgroup of G containing T and corresponding to the root α_1. Then $Z_G(T)/T \cong k^* \times \mathrm{SL}_{l-1}$ and we have Table 1 (in the middle column we use the following notation: χ is a fixed generator of the group $\mathrm{Hom}(Z_{Z_G(T)/T}, k^*)$ and, by definition, $R(\omega_0) = R(\omega_{l-1}) = 1$, $R(\omega_{-1}) = R(\omega_l) = 0$ and $R(n\omega_1)^G = 1$ if $l = 2$ and $n \geq 0$).*

TABLE 1

G-module U	Structure of the $Z_G(T)/T$ module U^T	Structure of the C-module U
$R(\omega_p)$ $1 \leq p \leq l$	$(\chi^{p-l-1} \otimes R(\omega_{p-2})) \oplus (\chi^p \otimes R(\omega_p))$	$\left(\binom{l-1}{p-2} + \binom{l-1}{p}\right) S_0 \oplus \binom{l-1}{p-1} S_1$
$\mathrm{Ad} = R(\omega_1 + \omega_l)$	$(\chi^0 \otimes \mathrm{Ad}) \oplus 2(\chi^0 \otimes 1)$	$(l-1)^2 S_0 \oplus 2(l-1) S_1 \oplus S_2$
$R(p\omega_1)$	$\bigoplus_{t=0}^{[p/2]} (\chi^{p-r(l+1)} \otimes R((p-2r)\omega_1))$	$\bigoplus_{r=0}^{p} \binom{l+p-r-2}{p-r} S_r$

3. Groups of type B_l, $l \geq 3$. Here we consider two types of tori T. In both cases T and C correspond to some root, i.e.,

$$\mathfrak{t} = \langle H_\alpha \rangle, \qquad \mathfrak{c} = \langle X_\alpha, H_\alpha, X_{-\alpha} \rangle. \tag{5.1}$$

First we consider the case $\alpha = \varepsilon_1 + \varepsilon_2 = \omega_2$.

Because $\mathfrak{h} \subset \mathfrak{z}_\mathfrak{g}(\mathfrak{t})$ we see that $\mathfrak{z}_\mathfrak{g}(\mathfrak{t})$ is the linear hull of $\mathfrak{h}$ and those X_β for which $(\beta|\alpha) = 0$. Obviously, $(\beta|\alpha) = 0 \Leftrightarrow \beta$ is a linear combination of $\alpha_1, \alpha_3 \ldots, \alpha_l$. Hence $\mathfrak{z}_\mathfrak{g}(\mathfrak{t}) = \mathfrak{t} \oplus \mathfrak{a} \otimes \mathfrak{b}$ (this is a direct sum of Lie algebras), where $\mathfrak{a} = A_1$ and is generated by X_{α_1} and $X_{-\alpha_1}$, and $\mathfrak{b} = B_{l-2}$ and is generated by X_{α_j} and $X_{-\alpha_j}$, $3 \leq j \leq l$ (we assume that $B_1 = A_1$).

Let $V = R(\omega_1)$ be the simplest $\mathfrak{g}$-module. Then $\Lambda^p V = R(\omega_p)$, $1 \leq p \leq l-1$ (see [10]). In V there is an $\mathfrak{h}$-weighted basis $e_1, \ldots, e_l, e_0, e_{-l}, \ldots, e_{-1}$, where e_i is of weight $(\operatorname{sgn} i)\varepsilon_{|i|}$ (see [10]). Let $W = \langle e_3, e_{-3}, \ldots, e_l, e_{-l}, e_0 \rangle$. Then $W \subset V^\mathfrak{a}$ and W is a $\mathfrak{b}$-module with $(\mathfrak{b}, W) = R(\omega_1)$ if $l \geq 4$ and $= R(2\omega_1)$ if $l = 3$. It now follows from tables in [15] that $(\mathfrak{b}, \Lambda^p W) = R(\omega_p)$ if $1 \le p \le l-3$ and $= R(2\omega_{l-2})$ if $p = l-2, l-1$. Moreover $\langle e_1, e_2, e_{-1}, e_{-2}\rangle \subset V^b$ and $X_{\alpha_1}(e_1) = X_{\alpha_1}(e_{-2}) = X_{-\alpha_1}(e_2) = X_{-\alpha_1}(e_{-1}) = 0$, $X_{\alpha_1}(e_2) = e_1$, $X_{-\alpha_1}(e_{-2}) = e_{-1}$, $X_{\alpha_1}(e_{-1}) = -e_2$, $X_{-\alpha_1}(e_{-2}) = e_{-1}$. Therefore $U_1 = \langle e_1 \wedge e_{-2}, e_{-1} \wedge e_2, e_1 \wedge e_{-1}, e_2 \wedge e_{-2}\rangle$ and $U_2 = \langle e_1 \wedge e_{-1} \wedge e_2 \wedge e_{-2}\rangle$ are $\mathfrak{a}$-modules of the form $R(2\omega_1) \oplus 1$ and 1.

Because $H_\alpha(e_i) = (\operatorname{sgn} i)e_i$ for $i = \pm 1, \pm 2$ and $= 0$ for $i \ne \pm 1, \pm 2$, (5.1) implies the direct sum decomposition of $\mathfrak{a} \oplus \mathfrak{b}$-modules $(\Lambda^p V)^{\mathfrak{l}} = \Lambda^p W \oplus (U_1 \wedge (\Lambda^{p-2} W)) \oplus (U_2 \wedge (\Lambda^{p-4} W))$ where $\Lambda^p W = 1 \otimes R(\omega_p)$ if $1 \le p \le l-3$ and $= 1 \otimes R(2\omega_{l-2})$ if $p = l-2; l-1$; $U_1 \wedge (\Lambda^{p-2} W) = (R(2(\omega_1)) \oplus 1) \otimes R(\omega_{p-2})$ where we assume for B_{l-2}-modules that $R(\omega_i) = 1$ if $i = 0$ and $= 0$ if $i < 0$.

By (5.1), $W \subset V^{\mathfrak{c}}$. Because $X_\alpha(e_i) = X_{-\alpha}(e_{-i}) = 0$ when $i = 1, 2$, $X_\alpha(e_{-1}) = -e_2$, $X_\alpha(e_{-2}) = e_1$, $X_{-\alpha}(e_1) = -e_{-2}$, $X_{-\alpha}(e_2) = e_{-1}$, it is not difficult to see that $\Lambda^j L$, where $j = 1, 2, 3, 4$, and $L = \langle e_1, e_{-1}, e_2, e_{-2}\rangle$, and $\mathfrak{r}$-modules of the form, respectively, $2S_1$, $3S_0 \oplus S_2$, $2S_1$, S_0. Because $\Lambda^p V = \bigoplus_{i=0}^4 ((\Lambda^i L) \wedge (\Lambda^{p-i} W))$ we have $(C, \Lambda^p V) = (\binom{2l-3}{p} + 3\binom{2l-3}{p-2} + \binom{2l-3}{p-4})S_0 \oplus 2(\binom{2l-3}{p-1} + \binom{2l-3}{p-3})S_1 \oplus \binom{2l-3}{p-2} S_2$.

Now we consider the same problems for the spinor $\mathfrak{g}$-module $M = R(\omega_l)$. All of the weights of $\mathfrak{h}$ in M are simple weights of the form $(\pm\varepsilon_1 \pm \cdots \pm \varepsilon_l)/2$ (with an arbitrary choice of signs) (see [15]). If λ is a weight, then $\lambda(H_\alpha) = 1$ for $\lambda = (\varepsilon_1 + \varepsilon_2 \pm \cdots \pm \varepsilon_l)/2$, $\lambda(H_\alpha) = -1$ for $\lambda = (-\varepsilon_1 - \varepsilon_2 \pm \cdots \pm \varepsilon_l)/2$ and $\lambda(H_\alpha) = 0$ in all other cases. It now follows from (5.1) that $M^{\mathfrak{t}}$ is the linear hull of all $M_{(\varepsilon_1 - \varepsilon_2 \pm \cdots \pm \varepsilon_l)/2}$ and $M_{(-\varepsilon_1 + \varepsilon_2 \pm \cdots \pm \varepsilon_l)/2}$. Because $\alpha_i + \mu$, where $\mu = (\varepsilon_1 - \varepsilon_2 + \varepsilon_3 + \cdots + \varepsilon_l)/2$ is not a weight for $i \ne 2$, we see that M_μ is the line of highest vectors of some irreducible $\mathfrak{a} \oplus \mathfrak{b}$-submodule of $M^{\mathfrak{t}}$ (with respect to the Borel subalgebra generated by $X\alpha_j$ and $H\alpha_j$, $j = 1, 2$). Because $\mu(H\alpha_j) = 1$ for $j = 1, l$, and $= 0$ otherwise, this submodule is of the form $R(\omega_1) \otimes R(\omega_{l-2})$. By the dimension count, it coincides with $M^{\mathfrak{t}}$. Furthermore, if λ is a weight of M, then $\alpha + \lambda$ is also a weight only for $\lambda = (-\varepsilon_1 - \varepsilon_2 \pm \cdots \pm \varepsilon_l)/2$, and $-\alpha + \lambda$ is a weight only for $\lambda = (\varepsilon_1 + \varepsilon_2 \pm \cdots \pm \varepsilon_l)/2$. Therefore $\langle M_\lambda, M_{\lambda - \alpha}\rangle$, where $\lambda = (\varepsilon_1 + \varepsilon_2 \pm \cdots \pm \varepsilon_l)/2$ is a $\mathfrak{c}$-module of type S_1, and $M_{(\varepsilon_1 - \varepsilon_2 \pm \cdots \pm \varepsilon_l)/2}$ and $M_{(-\varepsilon_1 + \varepsilon_2 \pm \cdots \pm \varepsilon_l)/2}$ and $\mathfrak{c}$-modules of type S_0. Therefore $(C, M) = 2^{l-1} S_0 \oplus 2^{l-2} S_1$.

Now consider the case when $\mathfrak{t}$ and $\mathfrak{c}$ are given by (5.1) but $\alpha = \varepsilon_1 = \omega_1$.

By the argument similar to the one given above, we see that $\mathfrak{z}_{\mathfrak{g}}(\mathfrak{t}) = \mathfrak{t} \oplus \mathfrak{a}$, where the subalgebra $\mathfrak{a} = B_{l-1}$ is generated by X_{α_j} and $X_{-\alpha_j}$, $2 \le j \le l$. Let V be the same as above and $W = \langle e_2, e_{-2}, \ldots, e_l, e_{-l}, e_0\rangle$. Then W is the $\mathfrak{a}$-module $R(\omega_1)$. It now follows from the tables in [15] that $(\mathfrak{a}, \Lambda^p W) = R(\omega_p)$ if $1 \le p \le l-2$ and $= R(2\omega_{l-1})$ if $p = l-1$. Because $\pm\varepsilon_1 \pm \alpha_j$ is not an $\mathfrak{h}$-weight for any $j \ge 2$ and any choice of signs, both e_1 and e_{-1} belong to $V^{\mathfrak{a}}$. Because $H_\alpha(e_i) = 2(\operatorname{sgn} i)e_i$ for $i = \pm 1$ and $= 0$ for $i \neq \pm 1$, we have the direct sum of $\mathfrak{z}_{\mathfrak{g}}(\mathfrak{t})/\mathfrak{t}$-modules $(\Lambda^p V)^{\mathfrak{t}} = \Lambda^p W \oplus (e_1 \wedge e_{-1} \wedge (\Lambda^{p-2} W))$, $1 \leq p \leq l-1$. The tables in [15] show $(\mathfrak{z}_{\mathfrak{g}}(\mathfrak{t})/\mathfrak{t} = \mathfrak{a})$ that $\Lambda^p W = R(\omega_p)$ if $1 \leq p \leq l-2$ and $= R(2\omega_{l-1})$ if $p = l-1$; $e_1 \wedge e_{-1} \wedge (\Lambda^{p-2} W) = R(\omega_{p-2})$ where we assume for B_{l-1}-modules that $R(\omega_i) = 1$ for $i = 0$ and $= 0$ for $i < 0$.

Because $\pm\alpha \pm \varepsilon_i$ is not a weight for any $i \geq 2$ and any choice of signs, we have $u = \langle e_2, e_{-2}, \dots, e_l, e_{-l} \rangle \subset V^{\mathfrak{c}}$. From the formulas $X_\alpha(e_1) = X_{-\alpha}(e_{-1}) = 0$, $X_\alpha(e_0) = 2e_1$, $X_\alpha(e_{-1}) = e_0$, $X_{-\alpha}(e_1) = -e_0$, $X_{-\alpha}(e_0) = -2e_{-1}$ we easily deduce that $\Lambda^j L$, $j = 1, 2, 3$, where $L = \langle e_1, e_0, e_{-1} \rangle$, are $\mathfrak{c}$-modules of types S_2, S_2, S_0 respectively. Together with $\Lambda^p V = \bigoplus_{i=0}^3 (\Lambda^p L) \wedge (\Lambda^{p-i} U)$, this yields $(\mathfrak{c}, \Lambda^p V) = (\binom{2l-2}{p} + \binom{2l-2}{p-3}) S_0 \oplus (\binom{2l-2}{p-1} + \binom{2l-2}{p-2}) S_2$.

Finally we consider the spinor $\mathfrak{g}$-module $M = R(\omega_l)$. If λ is an $\mathfrak{h}$-weight of M, then $\lambda(H_\alpha) = 1$ for $\lambda = (\varepsilon_1 \pm \cdots \pm \varepsilon_l)/2$ and $= -1$ for $\lambda = (-\varepsilon_1 \pm \cdots \pm \varepsilon_l)/2$. Therefore $M^{\mathfrak{t}} = 0$. As before, we see that for $\lambda = (\varepsilon_1 \pm \cdots \pm \varepsilon_l)/2$, $\langle M_\lambda, M_{\lambda-\alpha} \rangle$ is a $\mathfrak{c}$-module of type S_1, whence $(\mathfrak{c}, M) = 2^{l-1} S_1$.

Thus we have proved

LEMMA 5.2. *Let G be a group of type B_l, $l \geq 3$, and T and C be the one-dimensional torus and, respectively, the connected three-dimensional subgroup of G defined by some root α (their Lie algebras are given by* (5.1)*). Then $Z_G(T)/T$ is a connected semisimple group with Lie algebra of type B_{l-1} when $\alpha = \omega_1$ and of type $A_1 \oplus B_{l-2}$ when $\alpha = \omega_2$ ($B_{l-2} = A_1$ if $l = 3$). Moreover we have the following table (in the column describing the structure of $Z_G(T)/T$-module U^T we set $R(\omega_i) = 1$ for $i = 0$, and $R(\omega_i) = 0$ for $i < 0$):*

TABLE 2

α	G-module U	Structure of $Z_G(T)/T$-module U^T	Structure of C-module U
ω_1	$R(\omega_p)$ $1 \leq p \leq l-1$	$R(\Lambda_p) \oplus R(\Lambda_{p-2})$, where $\Lambda_p = \begin{cases} \omega_p & \text{for } p \neq l-1, \\ 2\omega_{l-1} & \text{for } p = l-1 \end{cases}$	$(\binom{2l-2}{p} + \binom{2l-2}{p-3}) S_0 \oplus (\binom{2l-2}{p-1} + \binom{2l-2}{p-2}) S_2$
ω_2	$\mathrm{Ad} = R(\omega_2)$	$(1 \otimes R(\Lambda_p)) \oplus (1 \otimes R(\omega_{p-2})) \oplus (1 \otimes R(\omega_{p-4})) \oplus (R(2\omega_1) \otimes R((\omega_{p-2}))$, where $\Lambda_p = \begin{cases} \omega_p & \text{for } p \neq l-2, l-1, \\ 2\omega_{l-2} & \text{for } p = l-2, l-1 \end{cases}$	$(\binom{2l-3}{p} + 3\binom{2l-3}{p-2} + \binom{2l-3}{p-4}) S_0 \oplus 2(\binom{2l-3}{p-1} + \binom{2l-3}{p-3}) S_1 \oplus \binom{2l-3}{p-2} S_2$
ω_1	$R(\omega_l)$	0	$2^{l-1} S_1$
ω_2		$R(\omega_1) \otimes R(\omega_{l-2})$	$2^{l-1} S_0 \oplus 2^{l-2} S_1$

4. Groups of type C_l, $l \geq 2$. We define T and C by (5.1), where $\alpha = 2\varepsilon_1 = \omega_1$. An argument similar to the one given in part 3 shows that $\mathfrak{z}_{\mathfrak{g}}(\mathfrak{t}) = \mathfrak{t} \oplus \mathfrak{a}$, where $\mathfrak{a} = C_{l-1}$ and is generated by X_{α_j} and $X_{-\alpha_j}$, $2 \leq j \leq l$ (we assume that $C_1 = A_1$).

Let $e_1, \dots, e_l, e_{-l}, \dots, e_{-1}$ be an $\mathfrak{h}$-weighted basis of the simplest $\mathfrak{g}$-module $V = R(\omega_1)$ where weight e_i is equal to $(\operatorname{sgn} i)\varepsilon_{|i|}$ (see [10]). Then $W = \langle e_2, e_{-2}, \dots, e_l, e_{-l}\rangle$ is an $\mathfrak{a}$-module of type $R(\omega_1)$. Because $\pm\varepsilon_1 \pm \alpha_j$ is not a weight of V for any $j \geq 2$ and any choice of signs, both e_1 and e_{-1} belong to $V^{\mathfrak{a}}$. From this and from the fact that $H_\alpha(e_i) = (\operatorname{sgn} i)e_i$ for $i = \pm 1$ and $= 0$ for $i \neq \pm 1$ we obtain the direct sum decomposition of $\mathfrak{a}$-modules $(\Lambda^p V)^{\mathfrak{t}} = \Lambda^p W \oplus (e_1 \wedge e_2 \wedge (\Lambda^{p-2} W))$ notice that $e_1 \wedge e_2 \wedge (\Lambda^{p-2} W)$ is isomorphic to $\Lambda^{p-2} W$.

As is known $\Lambda^p V = R(\omega_p) \oplus \Lambda^{p-2} V$ for $1 \leq p \leq l$ as $\mathfrak{g}$-modules (see [10, 15]). Therefore $(\Lambda^p V)^{\mathfrak{t}} = R(\omega_p)^{\mathfrak{t}} \oplus (\Lambda^{p-2} V)^{\mathfrak{t}}$. Because $\mathfrak{a} = C_{l-1}$ we have a similar decomposition for $\mathfrak{a}$-modules $\Lambda^p W = R(\omega_p) \oplus \Lambda^{p-2} W = \cdots = \bigoplus_{i=0}^{[p/2]} R(\omega_{p-2i})$ for $1 \leq p \leq l-1$ (we assume that $R(\omega_0) = 1$). Moreover, because any $\mathfrak{a}$-module is self-contragradient, we have the $\mathfrak{a}$-module isomorphism $\Lambda^l W \cong (\Lambda^{l-2} W)^* \cong \Lambda^{l-2} W$. Therefore $(\mathfrak{a}, (\Lambda^p V)^{\mathfrak{t}}) = R(\omega_p) \oplus (\bigoplus_{i=1}^{[p/2]} 2R(\omega_{p-2i}))$ for $1 \leq p \leq l-1$ and $(\mathfrak{a}, (\Lambda^l V)^{\mathfrak{t}}) = \bigoplus_{i=1}^{[l/2]} 2R(\omega_{l-2i})$. It now follows that the subspace $R(\omega_p)^{\mathfrak{t}}$ of the $\mathfrak{g}$-module $R(\omega_p)$, viewed as an $\mathfrak{a}$-module, is isomorphic to $R(\omega_{p-2}) \oplus R(\omega_p)$ for $1 \leq p \leq l$ (we assume that $R(\omega_{-1}) = R(\omega_l) = 0$).

Because $X_\alpha(e_1) = X_{-\alpha}(e_{-1}) = 0$, $X_\alpha(e_{-1}) = e_1$, $X_{-\alpha}(e_1) = -e_1$, and $\pm\alpha \pm \varepsilon_j$ is not an $\mathfrak{h}$-weight of V for any $j \geq 2$ and any choice of signs, we see that, first, $W \subset V^{\mathfrak{c}}$, and, second, $L = \langle e_1, e_{-1}\rangle$ and $\Lambda^2 L$ are $\mathfrak{c}$-modules of types S_1 and S_0, respectively. Because $\Lambda^p V = \bigoplus_{i=0}^{2}((\Lambda^i L) \wedge (\Lambda^{p-i} W))$ we have $(\mathfrak{c}, \Lambda^p V) = (\binom{2l-2}{p} + \binom{2l-2}{p-2})S_0 \oplus \binom{2l-2}{p-1}S_1$. It now follows easily that the $\mathfrak{g}$-module $R(\omega_p)$, $1 \leq p \leq l$, viewed as a C-module is isomorphic to $(\binom{2l-2}{p} - \binom{2l-2}{p-4})S_0 \oplus (\binom{2l-2}{p-1} - \binom{2l-2}{p-3})S_1$.

Now we shall obtain similar information about the adjoint $\mathfrak{g}$-module $R(2\omega_1)$. Notice that $Z_G(T)$-modules $\mathfrak{g}^{\mathfrak{t}}$ and $\mathfrak{z}_{\mathfrak{g}}(\mathfrak{t})$ coincide.

Therefore $(\mathfrak{a}, \mathfrak{g}^{\mathfrak{t}}) = R(2\omega_1) \oplus 1$. It is easy to see that all α-series of length 2 are of the form $\langle \varepsilon_1 \pm \varepsilon_j, -\varepsilon_1 \pm \varepsilon_j\}$, $2 \leq j \leq l$, all α-series of length 1 are of the form $\{\pm 2\varepsilon_j\}$, $2 \leq j \leq l$, and there are no other α-series. Hence (see part 2),

$$(C, \mathfrak{g}) = (l-1)(2l-1)S_0 \oplus 2(l-1)S_1 \oplus S_2.$$

Thus we have proved

LEMMA 5.3. *Let G be a group of type C_l, $l \geq 2$, and T and C be the one-dimensional torus and, respectively, a connected three-dimensional simple* *subgroup of* G *corresponding to the root* $\alpha = \omega_1$ (*their Lie algebras are given by* (5.1)). *Then* $Z_G(T)/T$ *is a connected semisimple group with Lie algebra of type* $C_{l-1} (= A_1$ *for* $l = 2)$. *We also have the table* (*in the middle column we set* $R(\omega_i) = 1$ *for* $i = 0$ *and* $R(\omega_i) = 0$ *for* $i = -1, l$)

TABLE 3

G-module U	Structure of $Z_G(T)/T$-module U^T	Structure of C-module U
$R(\omega_p)$ $1 \leq p \leq l$	$R(\omega_{p-2}) \oplus R(\omega_p)$	$\left(\binom{2l-2}{p} - \binom{2l-2}{p-4}\right) S_0 \oplus \left(\binom{2l-2}{p-1} - \binom{2l-2}{p-3}\right) S_1$
$\mathrm{Ad} = R(2\omega_1)$	$R(2\omega_1) \oplus 1$	$(l-1)(2l-1)S_0 \oplus 2(l-1)S_1 \oplus S_2$

5. Groups of type D_l, $l \geq 4$. We define T and C by (5.1), where $\alpha = \varepsilon_1 + \varepsilon_2 = \omega_2$. This case is very much like the case B_l, $\alpha = \omega_2$. The assertions stated below can be proved the same way (and in the same notation) as the ones in part 3.

We have $\mathfrak{z}_{\mathfrak{g}}(\mathfrak{t}) = \mathfrak{t} \oplus \mathfrak{a} \oplus \mathfrak{b}$ where $\mathfrak{a} = A$, and is generated by X_{α_1} and $X_{-\alpha_1}$, and $\mathfrak{b} = D_{l-2}$ and is generated by X_{α_j} and $X_{-\alpha_j}$, $3 \leq j \leq l$; we also assume that D_l is $A_1 \oplus A_1$ and D_3 is A_3 with the following nonstandard ordering of simple roots (and fundamental weights):

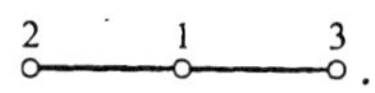

Let $V = R(\omega_1)$ be the simplest $\mathfrak{g}$-module. Then $\Lambda^p V = R(\omega_p)$ for $p = 1, \ldots, l-2$ (see [10]). There is an $\mathfrak{h}$-weighted basis $e_1, \ldots, e_l e_{-l}, \ldots, e_{-1}$ of V, where weight e_i is equal to $(\operatorname{sgn} i)\varepsilon_{|i|}$ [10]. Let $W = \{e_3, e_{-3}, \ldots, e_l, e_{-l}\}$. Then $W \subset V^{\mathfrak{a}}$ and W is a $\mathfrak{b}$-module with $(\mathfrak{b}, W) = R(\omega_1)$ for $l \geq 5$ and $= R(\omega_1 + \omega_2)$ for $l = 4$. It is not difficult to deduce from here and from the tables in [15] that $(\mathfrak{b}, \Lambda^p W) = R(\omega_p)$ for $1 \leq p \leq l-4$; $= R(\omega_{l-3} + \omega_{l-2})$ for $p = l-1$ and $= R(2\omega_{l-3}) \oplus R(2\omega_{l-2})$ for $p = l-2$. For $p = 1, \ldots, l$ we have the direct sum of $\mathfrak{a} \oplus \mathfrak{b}$-modules: $(\Lambda^p V)^{\mathfrak{t}} = \Lambda^p W \oplus (U_1 \wedge (\Lambda^{p-2} W)) \oplus (U_2 \wedge (\Lambda^{p-4} W))$ (see part 3). Their structure is as follows: $\Lambda^p W = 1 \otimes R(\omega_p)$ for $1 \leq p \leq l-4$, $= 1 \otimes R(\omega_{l-3} + \omega_{l-2})$ for $p = l-3$, and $= (1 \otimes R(2\omega_{l-3})) \oplus (1 + R(2\omega_{l-2}))$ for $p = l-2$; $U_1 \wedge (\Lambda^{p-2} W) = (R(2\omega_1) \oplus 1) \otimes R(\omega_{p-2})$ and $U_2 \wedge (\Lambda^{p-4} W) = 1 \otimes R(\omega_{p-4})$ where, for D_{l-2}-modules we set $R(\omega_i) = 1$ for $i = 0$ and $= 0$ for $i < 0$.

The $\mathfrak{c}$-module structure on $\Lambda^p V$, $1 \leq p \leq l-2$, can be described exactly as in the case B_l, $\alpha = \omega_2$; the only difference is that $\dim W = 2l - 4$ (and not $2l - 3$, as it was in part 3).

We shall now obtain the corresponding information for semispinor $\mathfrak{g}$-modules. Let $M = R(\omega_l)$. All of the weights of $\mathfrak{h}$ in M are simple and of the form $(\pm\varepsilon_1 \pm \cdots \pm \varepsilon_l)/2$ with an even number of minuses (see [10, 15]). Applying the arguments from part 3 to $\mu = (\varepsilon_1 - \varepsilon_2 + \varepsilon_3 + \cdots + \varepsilon_{l-1} - \varepsilon_l)/2$ we have $(\mathfrak{a} \oplus \mathfrak{b}, M^{\mathfrak{t}}) = R(\omega_1) \otimes R(\omega_{l-3})$. Proceeding along these lines we see that $(C, M) = 2^{l-2}S_0 \oplus 2^{l-3}S_1$. The module $N = R(\omega_{l-1})$ can be investigated in much the same way; we only consider an odd, rather than even, number of minuses. We then have

$$(\mathfrak{a} \oplus \mathfrak{b}, N^{\mathfrak{t}}) = R(\omega_1) \otimes R(\omega_{l-2}), \qquad (C, N) \cong (C, M).$$

Thus we have proved

LEMMA 5.4. *Let G be a group of type D_l, $l \geq 4$, and T and C be the one-dimensional torus and, respectively, a connected three-dimensional simple subgroup of G corresponding to the root $\alpha = \omega_2$ (their Lie algebras are given by (5.1)). Then $Z_G(T)/T$ is a connected semisimple group with Lie algebra of type $A_1 \oplus D_{l-2}$ (where $D_2 = A_1 \oplus A_1$ and $D_3 = A_3$). We also have the table (in the middle column we set $R(\omega_i) = 1$ if $i = 0$ and $R(\omega_i) = 0$ if $i < 0$, moreover, $D_3 = A_3$ has the following nonstandard ordering of the simple roots and fundamental weights* $\overset{2}{\circ}$——$\overset{1}{\circ}$——$\overset{3}{\circ}$:

TABLE 4

G-module U	Structure of $Z_G(T)/T$-module U^T	Structure of C-module U
$R(\omega_p)$, $1 \leq p \leq l-2$, $\mathrm{Ad} = R(\omega_2)$	$(1 \otimes R_p) \oplus (1 \otimes R(\omega_{p-2})) \oplus (1 \otimes R(\omega_{p-4}))$ $\oplus (R(2\omega_1) \otimes R(\omega_{p-2}))$, where $R_p = \begin{cases} R(\omega_p) & \text{for } 1 \leq p \leq l-4, \\ R(\omega_{l-3} + \omega_{l-2}) & \text{for } p = l-3, \\ R(2\omega_{l-3}) \oplus R(2\omega_{l-2}) & \text{for } p = l-2. \end{cases}$	$\left(\binom{2l-4}{p} + 3\binom{2l-4}{p-2} + \binom{2l-4}{p-4}\right)S_0$ $\oplus 2\left(\binom{2l-4}{p-1} + \binom{2l-4}{p-3}\right)S_1$ $\oplus \binom{2l-4}{p-2}S_2$
$R(\omega_{l-1})$	$R(\omega_1) \otimes R(\omega_{l-2})$	$2^{l-2}S_0 \oplus 2^{l-1}S_1$
$R(\omega_l)$	$R(\omega_1) \otimes R(\omega_{l-3})$	

§6. Torus T for the exceptional simple groups

1. Whenever G is an exceptional simple group we define T and C by (5.1), where α is the highest root. Because $\alpha = \omega_{i_0}$ for some i_0 (see [9]), we have $\mathfrak{z}_{\mathfrak{g}}(\mathfrak{t}) = \mathfrak{t} \oplus \mathfrak{a}$ (a direct sum of Lie algebras), where $\mathfrak{a}$ is the subalgebra generated by all X_{α_i} and $X_{-\alpha_i}$, $i \neq i_0$. For $\mathfrak{g} = G_2, F_4, E_6, E_7, E_8$ we have, respectively, $i_0 = 2, 1, 2, 1, 8$ and therefore $\mathfrak{g} = A_1, C_3, A_5, D_6, E_7$. Because α is the highest root, it is not difficult to see that $\mathfrak{a} = \mathfrak{z}_{\mathfrak{g}}(\mathfrak{c})$ and $\mu_{\mathfrak{g}}(\mathfrak{c}) = \mathfrak{a} \oplus \mathfrak{c}$.

We now remark that in this section for α_i and ω_i we use the construction from [15] and not from [9]; at the same time we keep the old ordering of α_i and ω_i from [9] (in [15] the ordering and the orientation of the Dynkin diagrams are different from those in [9]). More precisely, for $\mathfrak{g} = E_7, E_8$ and G_2 the weights of irreducible $\mathfrak{g}$-modules are expressed in terms of $\varepsilon_1, \dots, \varepsilon_{l+1} \in \mathfrak{h}^*$, where $l = \operatorname{rk}\mathfrak{g}$, $(\varepsilon_i|\varepsilon_j) = -1/(l+1)$ for $i \neq j$ and $= l(l+1)$ for $i = j$. For E_6 they are expressed in terms of $\varepsilon_1, \dots, \varepsilon_6, \eta \in \mathfrak{h}^*$, where $(\varepsilon_i|\varepsilon_j) = -1/6$ for $i \neq j$ and $= 5/6$ for $i = j$, $(\varepsilon_i|\eta) = 0$ for all i, $(\eta|\eta) = 1/2$. For F_4 the elements ε_j are the same as the ones in [9]. Below we list α_i, ω_i, and the roots.

If V is a $\mathfrak{g}$-module, λ is a weight of it, and β is a root, then the set of all weights of V of the form $\lambda + j\beta$, $j \in \mathbb{Z}$ is called the β-series containing λ. If V is the adjoint module and $\lambda \neq \pm\beta$, then the β-series containing λ is also the β-series of roots in the sense of [9], whereas if $\lambda = \pm\beta$ it is not.

2. The group G_2. We have $\alpha_1 = -\varepsilon_2$, $\alpha_2 = \varepsilon_2 - \varepsilon_3$, $\omega_1 = \varepsilon_1$, $\omega_2 = \varepsilon_1 - \varepsilon_3$. The roots are $\varepsilon_i - \varepsilon_j$, $\pm\varepsilon_i$.

a) For the simplest $\mathfrak{g}$-module $V = R(\omega_1)$ all weights are simple and have the form $\pm\varepsilon_i$, 0 (see [15]). Because the α-series of weights are exactly $\{\varepsilon_1, \varepsilon_3\}$, $\{-\varepsilon_1, -\varepsilon_3\}$, $\{\varepsilon_2\}$, $\{-\varepsilon_2\}$, $\{0\}$, we have $(C, V) = 3S_0 \oplus 2S_1$. Because $(\lambda/\alpha) = 0$ only for the weights $\lambda = 0, \pm\varepsilon_2$ and these weights form an α_1-series, we have $(\mathfrak{a}, V^{\mathfrak{t}}) = R(2\omega_1)$.

b) For the adjoint $\mathfrak{g}$-module $V = R(\omega_2)$ we have $V^{\mathfrak{t}}\mathfrak{z}_{\mathfrak{g}}(\mathfrak{t})$, and therefore $(\mathfrak{a}, V^{\mathfrak{t}}) = R(2\omega_1) \oplus 1$. Because the α-series of roots are exactly $\{\varepsilon_1, \varepsilon_3\}$, $\{-\varepsilon_1, -\varepsilon_3\}$, $\{\varepsilon_2\}$, $\{-\varepsilon_2\}$, $\{\varepsilon_1 - \varepsilon_2, \varepsilon_3 - \varepsilon_2\}$, $\{-\varepsilon_1 + \varepsilon_2, -\varepsilon_3 + \varepsilon_2\}$, we have $(C, V) = 3S_0 \oplus 4S_1 \oplus S_2$ (compare with part 2 of §5).

3. The group F_4.

a) If $V = R(\omega_1)$ is the adjoint $\mathfrak{g}$-module then, obviously, $(\mathfrak{a}, V^{\mathfrak{t}}) = R(2\omega_1) \oplus 1$. The α-series of roots are exactly $\langle\varepsilon_1 + \lambda, -\varepsilon_2 + \lambda\}$ and $\langle -\varepsilon_1 + \lambda, \varepsilon_2 + \lambda\}$, where $\lambda = 0, \pm\varepsilon_3, \pm\varepsilon_4$; $\{(\varepsilon_1 + \varepsilon_2 + \mu)/2, (-\varepsilon_1 - \varepsilon_2 + \mu)/2\}$; $\{(\varepsilon_1 - \varepsilon_2 + \mu)/2, (-\varepsilon_1 + \varepsilon_2 + \mu)/2\}$; $\{\mu\}$, and $\mu = \pm\varepsilon_3 \pm \varepsilon_4$; $\{\varepsilon_1 - \varepsilon_2\}$; $\{-\varepsilon_1 + \varepsilon_2\}$, as well as $\{\pm\varepsilon_i\}$, $i = 3, 4$. Therefore

$$(C, V) = 21S_0 \oplus 14S_1 \oplus S_2.$$

b) Let $V = R(\omega_2)$. Then the $\mathfrak{g}$-modules $\Lambda^2 R(\omega_1)$ and $V \oplus R(\omega_1)$ are isomorphic (see [15]). The arguments we are about to go through will also be used several times later. Because of that we will give details only once. Later we will simply skip them and only indicate the key points and the answers. We consider the above relation as an isomorphism of $\mathfrak{c}$-modules; we then deduce from a) that $V \oplus 21S_0 \oplus 14S_1 \oplus S_2 = \Lambda^2(21S_0 \oplus 14S_1 \oplus S_2) = \Lambda^2 S_2 \oplus 14(S_1 \otimes S_2) \oplus 21(S_0 \otimes S_2) \oplus \binom{14}{2}(S_1 \otimes S_1) \oplus 14\Lambda^2 S_1 \oplus 294(S_0 \otimes S_1) \oplus \binom{21}{2}(S_0 \otimes S_0) \oplus 21\Lambda^2 S_0$. From the equality $\Lambda^2 S_p = \bigoplus_{j=0}^{+\infty} S_{2p-2-4j}$ (see [15]), and from the Clebsch–Gordan formula we have $(C, V) = 294S_0 \oplus 294S_1 \oplus 112S_2 \oplus 14S_3$. Similarly we obtain the following isomorphisms of $\mathfrak{a}$-modules $(\Lambda^2 R(\omega_1))^{\mathfrak{t}} = V^{\mathfrak{t}} \oplus R(\omega_1)^{\mathfrak{t}} = V^{\mathfrak{t}} \oplus R(2\omega_1) \oplus 1$. Because the left-hand side contains $\Lambda^2(R(\omega_1)^{\mathfrak{t}}) = \Lambda^2(R(2\omega_1) \oplus 1) = \Lambda^2 R(2\omega_1) \oplus R(2\omega_1)$ and $(\mathfrak{a}, \Lambda^2 R(2\omega_1)) = R(2\omega_1 + \omega_2) \oplus R(2\omega_1)$ (see [15]), we have $(\mathfrak{a}, V^{\mathfrak{t}}) \supset R(2\omega_1) \oplus R(2\omega_1 + \omega_2)$.

c) The weights of the simplest $\mathfrak{g}$-module $V = R(\omega_4)$ are exactly $\pm\varepsilon_i$, $(\pm\varepsilon_1 \pm \varepsilon_2 \pm \varepsilon_3 \pm \varepsilon_4)/2$ (simple) and 0 (double) (see [15]). Therefore the α-series of weights are as follows: $\{(\varepsilon_1 + \varepsilon_2 + \mu)/2, (-\varepsilon_1 - \varepsilon_2 + \mu)/2\}$, $\{(\varepsilon_1 - \varepsilon_2 + \mu)/2, (-\varepsilon_1 + \varepsilon_2 + \mu)/2\}$, where $\mu = \pm\varepsilon_3 \pm \varepsilon_4$; $\{\varepsilon_1, -\varepsilon_2\}$; $\{-\varepsilon_1, \varepsilon_2\}$; $\{\pm\varepsilon_i\}$ for $i = 3, 4$, and $\{0\}$. Hence $(C, V) = 14S_0 \oplus 6S_1$. Because $(\lambda|\alpha) = 0$ only for the weights $\lambda = 0, \pm\varepsilon_3, \pm\varepsilon_4, \pm(\varepsilon_1 - \varepsilon_2 \pm \varepsilon_3 \pm \varepsilon_4)/2$ we see that $\dim V^{\mathfrak{t}} = 14$. Let $\lambda_0 = (\varepsilon_1 - \varepsilon_2 + \varepsilon_3 + \varepsilon_4)/2$ and $v \in V_{\lambda_0}$, $v \neq 0$. Because $\alpha_i + \lambda_0$ is not a weight for $i \neq 1$, and $\lambda_0(H_{\alpha_i}) = 0, 1, 0$ for $i = 2, 3, 4$, respectively, we have $(\mathfrak{g}, V^{\mathfrak{t}}) \supset R(\omega_1)$. But then from the dimension count we see that $(\mathfrak{a}, V^{\mathfrak{t}}) = R(\omega_2)$.

d) Let $V = R(\omega_3)$. Then the $\mathfrak{g}$-modules $\Lambda^2 R(\omega_4)$ and $V \oplus R(\omega_1)$ are isomorphic (see [15]). From this and from a) and c) we deduce, by the same argument as above, that $(C, V) = 91S_0 \oplus 70S_1 \oplus 14S_2$. Now because $(C_3, \Lambda^2 R(\omega_2)) = R(2\omega_1) \oplus R(\omega_1 + \omega_3)$ (see [15]), we have

$$(\mathfrak{a}, V^{\mathfrak{t}}) \supset R(\omega_1 + \omega_3).$$

4. The group E_6. We have $\alpha_1 = \varepsilon_1 - \varepsilon_2$; $d_{i+1} = \varepsilon_i - \varepsilon_{i+1}$ for $i = 2, 3, 4, 5$; $\alpha_2 = \varepsilon_4 + \varepsilon_5 + \varepsilon_6 + \eta$; $\omega_1 = \varepsilon_1 + \eta$; $\omega_{i+1} = \varepsilon_1 + \cdots + \varepsilon_i + \min\{i, 6 - i\}\eta$ for $i = 2, 3, 4, 5$; $\omega_2 = 2\eta$. The roots are $\varepsilon_i - \varepsilon_j$; $\varepsilon_i + \varepsilon_j + \varepsilon_k \pm \eta$; $\pm 2\eta$ (i, j, and k are all different).

a) For the adjoint $\mathfrak{g}$-module $V = R(\omega_2)$ we obviously have $(\mathfrak{a}, V^{\mathfrak{t}}) = R(\omega_1 + \omega_5) \oplus 1$. Because $\{\varepsilon_i + \varepsilon_j + \varepsilon_k + \eta, \varepsilon_i + \varepsilon_j + \varepsilon_k - \eta\}$ and $\{\varepsilon_i - \varepsilon_j\}$ exhaust all α-series of roots, we have $(C, V) = 35S_0 \oplus 20S_1 \oplus S_2$.

b) All weights of the simplest $\mathfrak{g}$-module $V = R(\omega_1)$ are simple and of the form $\varepsilon_i \pm \eta$ and $-\varepsilon_i - \varepsilon_j$, $i \neq j$ (see [15]). Therefore the α-series of weights are exactly $\{\varepsilon_i + \eta, \varepsilon_i - \eta\}$, $\{-\varepsilon_i - \varepsilon_j\}$. Therefore $(C, V) = 15S_0 \oplus 6S_1$. Because $(\lambda|\alpha) = 0$ only for the weights $\lambda = -\varepsilon_i - \varepsilon_j$, $1 \le i < j \le 6$, we see that $\dim V^{\mathfrak{t}} = 15$. Because $\lambda_0 + \alpha_i$ is not a weight when $\lambda_0 = -\varepsilon_5 - \varepsilon_6$ and $i \neq 2$, and $\lambda_0(H_{\alpha_i}) = 0, 0, 0, 1, 0$ when, respectively, $i = 1, 3, 4, 5, 6$, we have $(\mathfrak{a}, V^{\mathfrak{t}}) \supset R(\omega_4)$ and from the dimension count, $(\mathfrak{a}, V^{\mathfrak{t}}) = R(\omega_4)$.

c) Let $V = R(\omega_3)$. Then the $\mathfrak{g}$-modules $\Lambda^2 R(\omega_1)$ and $R(\omega_3)$ are isomorphic (see [15]). This observation together with b) implies that $(C, V) = 126S_0 \oplus 90S_1 \oplus 15S_2$. Because $(A_5, \Lambda^2 R(\omega_4)) = R(\omega_3 + \omega_5)$ (see [15]), we have $(\mathfrak{a}, V^{\mathfrak{t}}) \supset R(\omega_3 + \omega_5)$.

d) Let $V = R(\omega_4)$. Then the $\mathfrak{g}$-modules $\Lambda^2 R(\omega_2)$ and $V \oplus R(\omega_2)$ are isomorphic (see [15]). Hence, as above, $(C, V) = 770S_0 \oplus 700S_1 \oplus 225S_2 \oplus 20S_3$. Because $(A_5, \Lambda^5 R(\omega_1+\omega_5)) = R(2\omega_1+\omega_4)\oplus R(\omega_1+2\omega_4)\oplus R(\omega_1+\omega_5)$ (see [15]), we have $(\mathfrak{a}, V^{\mathfrak{t}}) \supset R(2\omega_1+\omega_4) \oplus R(\omega_2+2\omega_5) \oplus R(\omega_1+\omega_5)$.

e) Because $R(\omega_5) = R(\omega_3)^*$ and $R(\omega_6) = R(\omega_1)^*$ we have isomorphisms of $\mathfrak{c}$-modules $R(\omega_5)$ and $R(\omega_3)$, as well as $R(\omega_6)$ and $R(\omega_1)$. In view of b) and c) $(\mathfrak{a}, R(\omega_5)^{\mathfrak{t}}) \supset R(\omega_1+\omega_3)$, $(\mathfrak{a}, R(\omega_6)^{\mathfrak{t}}) = R(\omega_4)$.

5. The group E_7. We have $\alpha_1 = \varepsilon_6 - \varepsilon_7$; $\alpha_{8-i} = \varepsilon_i - \varepsilon_{i+1}$ for $i = 1, 2, 3, 4, 5$; $\alpha_2 = \varepsilon_5+\varepsilon_6+\varepsilon_7+\varepsilon_8$; $\omega_1 = -\varepsilon_7+\varepsilon_8$; $\omega_{8-i} = \varepsilon_1+\cdots+\varepsilon_i+\min\{i, 8-i\}\varepsilon_8$ for $i = 1, 2, 3, 4, 5$; $\omega_2 = \varepsilon_5+\varepsilon_6+\varepsilon_7+\varepsilon_8$. The roots are $\varepsilon_i - \varepsilon_j$; $\varepsilon_i+\varepsilon_j+\varepsilon_k+\varepsilon_l$ (i, j, k, and l are all distinct).

a) For the adjoint $\mathfrak{g}$-module $V = R(\omega_1)$ we obviously have $(\mathfrak{a}, V^{\mathfrak{t}}) = R(\omega_2)\oplus 1$. Because the α-series of roots are exactly $\{\varepsilon_7+\lambda, \varepsilon_8+\lambda\}$, where $\lambda = -\varepsilon_i, \varepsilon_i+\varepsilon_j+\varepsilon_k$; $\{-\varepsilon_7+\varepsilon_i, -\varepsilon_8+\varepsilon_i\}$; $\{\varepsilon_i-\varepsilon_j\}$; $\{\varepsilon_i+\varepsilon_j+\varepsilon_7+\varepsilon_8\}$ and $\{\varepsilon_i+\varepsilon_j+\varepsilon_k+\varepsilon_l\}$, and $i, j, k, l \neq 7, 8$ we have $(C, V) = 66S_0\oplus 32S_1\oplus S_2$.

b) All of the weights of the simplest $\mathfrak{g}$-module $V = R(\omega_7)$ are simple and of the form $\pm(\varepsilon_i+\varepsilon_j)$, $i \neq j$ (see [15]). Hence the α-series of weights are exactly $\{\varepsilon_7+\varepsilon_i, \varepsilon_8+\varepsilon_i\}$; $\{-\varepsilon_7-\varepsilon_i, -\varepsilon_8-\varepsilon_i\}$; $\{\pm(\varepsilon_i+\varepsilon_j)\}$, where $i, j \neq 7, 8$, and $\{\pm(\varepsilon_7+\varepsilon_8)\}$. Therefore $(C, V) = 32S_0 \oplus 12S_1$. Because $(\lambda|\alpha) = 0$ only for the weights $\lambda = \pm(\varepsilon_7+\varepsilon_8)$, $\pm(\varepsilon_i+\varepsilon_j)$, $i, j \neq 7, 8$, we see that $\dim V^{\mathfrak{t}} = 32$. Because $\lambda_0+\alpha_i$, where $\lambda = \varepsilon_7+\varepsilon_8$ and $i \neq 1$, is not a weight, and $\lambda_0(H_{\alpha_i}) = 1, 0, 0, 0, 0, 0$ when, respectively, $i = 2, 3, 4, 5, 6, 7$, we have $(\mathfrak{a}, V^{\mathfrak{t}}) \supset R(\omega_6)$ (see part 3c)) and, from the dimension count, $(\mathfrak{a}, V^{\mathfrak{t}}) = R(\omega_6)$.

c) Let $V = R(\omega_6)$. The $\mathfrak{g}$-modules $\Lambda^2 R(\omega_7)$ and $V\oplus 1$ are isomorphic (see [15]), hence, similar to part 3b), we have $(C, V) = 573S_0\oplus 384S_1\oplus 66S_2$ and (because $(D_6, \Lambda^2 R(\omega_6)) = R(\omega_4)\oplus 1$ see [15]), $(\mathfrak{a}, V^{\mathfrak{t}}) \supset R(\omega_4)$.

d) Let $V = R(\omega_3)$. The $\mathfrak{g}$-modules $\Lambda^2 R(\omega_1)$ and $V \oplus R(\omega_1)$ are isomorphic (see [15]), hence $(C, V) = 2607S_0 \oplus 2112S_1 \oplus 562S_2 \oplus 32S_3$ and (in view of $(D_6, \Lambda^2 R(\omega_2)) = R(\omega_1+\omega_3) \oplus R(\omega_2)$ see [15]) $(\mathfrak{a}, V^{\mathfrak{t}}) \supset R(\omega_1+\omega_3)\oplus R(\omega_2)$.

e) Let $V = R(\omega_5)$. The $\mathfrak{g}$-modules $\Lambda^3 R(\omega_7)$ and $V\oplus R(\omega_7)$ are isomorphic (see [135]), hence $(C, V) = 7424S_0\oplus 6512S_1\oplus 2112S_2\oplus 220S_3$. Because $\lambda_0 = (3\varepsilon_1+3\varepsilon_2+3\varepsilon_3+\varepsilon_4+\varepsilon_5-\varepsilon_6)/2$ is a weight of $(D_6, \Lambda^3 R(\omega_6))$ and $\lambda_0+\alpha_i$ is not a weight of algebra D_6 for any simple root α_i, $i = 1, 2, 3, 4, 5, 6$, and because $\lambda_0(H_{\alpha_i}) = 0, 0, 1, 0, 1, 0$ for the same range of indices, we deduce from the aforementioned isomorphism of the $\mathfrak{g}$-modules that $(\mathfrak{a}, V^{\mathfrak{t}}) \supset R(\omega_3+\omega_5)$.

f) Let $V = R(\omega_4)$. Arguing as in part e) (with $\lambda_0 = 2(\varepsilon_1+\varepsilon_2+\varepsilon_3)$) and using the isomorphism of the $\mathfrak{g}$-modules $\Lambda^4 R(\omega_7)$ and $V \oplus R(\omega_6)\oplus 1$ (see

[135]) we have $(C, V) = 75790S_0 \oplus 77440S_1 \oplus 34815S_2 \oplus 7040S_3 \oplus 495S_4$ and $(\mathfrak{a}, V^t) \supset R(2\omega_3)$.

g) Let $V = R(\omega_2)$. Then $(\mathfrak{a}\oplus\mathfrak{c}, V) = (R(\omega_1+\omega_5)\otimes 1)\oplus(R(\omega_1)\otimes R(\omega_1)) \oplus (R(\omega_3)\otimes R(\omega_1))\oplus(R(\omega_6)\otimes R(2\omega_1))$ (see [135]), whence

$$(\mathfrak{a}, V^t) = R(\omega_1+\omega_5)\oplus R(\omega_6)$$

and (see [15]) $(C, V) = 352S_0 \oplus 232S_1 \oplus 32S_2$.

6. The group E_8. We have $\alpha_{9-i} = \varepsilon_i - \varepsilon_{i+1}$, $1 \le i \le 6$; $\alpha_1 = \varepsilon_7 - \varepsilon_8$; $\alpha_2 = \varepsilon_6 + \varepsilon_7 + \varepsilon_8$; $\omega_{9-i} = \varepsilon_1 + \cdots + \varepsilon_i - \min\{i, 15-2i\}\varepsilon_9$, $1 \le i \le 6$; $\omega_1 = -\varepsilon_8 - 2\varepsilon_9$; $\omega_2 = -3\varepsilon_9$. The roots are $\varepsilon_i - \varepsilon_j$, $\pm(\varepsilon_i+\varepsilon_j+\varepsilon_k)$ (i, j, and k are all distinct).

a) For the adjoint (= simplest) $\mathfrak{g}$-module $V = R(\omega_8)$ we obviously have $(\mathfrak{a}, V^t) = R(\omega_1)\oplus 1$. The α-series of roots are exactly $\{\varepsilon_1+\lambda, \varepsilon_9+\lambda\}$ and $\{-\varepsilon_1-\lambda, -\varepsilon_9-\lambda\}$ for $\lambda = -\varepsilon_i, \varepsilon_i+\varepsilon_j$, $\{\varepsilon_i-\varepsilon_j\}$, $\{\pm(\varepsilon_i+\varepsilon_j+\varepsilon_k)\}$, $\{\pm(\varepsilon_1+\varepsilon_9+\varepsilon_i)\}$, where $i, j, k \ne 1, 9$. Therefore $(C, V) = 133S_0\oplus 56S_1 \oplus S_2$.

b) Let $V = R(\omega_1)$. Then $(\mathfrak{a}\oplus\mathfrak{c}, V) = (R(\omega_6)\otimes 1)\oplus 1 \oplus (R(\omega_7)\otimes R(\omega_1))\oplus(R(\omega_2)\otimes R(\omega_1))\oplus(R(\omega_1)\otimes R(2\omega_1))$ (see [135]), whence $(\mathfrak{a}, V^t) = 1\oplus R(\omega_1)\oplus R(\omega_6)$ and $(C, V) = 1540S_0\oplus 968S_1\oplus 133S_2$.

c) Let $V = R(\omega_7)$. The $\mathfrak{g}$-modules $\Lambda^2 R(\omega_8)$ and $V\oplus R(\omega_8)$ are isomorphic (see [15]). Hence $(C, V) = 10374S_0\oplus 7504S_1\oplus 1674S_2\oplus 56S_3$. Because $(E_7, \Lambda^2 R(\omega_1)) = R(\omega_3)\oplus R(\omega_1)$ (see [15]), we have $(\mathfrak{a}, V^t) \supset R(\omega_1)\oplus R(\omega_3)$. On the other hand $(\mathfrak{a}, V^t)\supset R(\omega_7)$ (see [135]).

d) Let $V = R(\omega_6)$. Because $(\mathfrak{a}\oplus\mathfrak{c}, V)\supset(R(\omega_6)\otimes 1)\oplus(R(\omega_5)\otimes R(3\omega_1))$, (see [135]), we have $(\mathfrak{a}, V^t)\supset R(\omega_6)$ and $(C, V)\supset 1539S_0\oplus 27664S_3$.

e) Let $V = R(\omega_5)$. Because $(\mathfrak{a}\oplus\mathfrak{c}, V)\supset(R(\omega_5)\otimes 1)\oplus(R(\omega_4)\otimes R(4\omega_1))$ (see [135]), we have $(\mathfrak{a}, V^t)\supset R(\omega_5)\oplus R(\omega_4)$ and $(C, V)\supset 27664S_0\oplus 365750S_4$.

f) Let $V = R(\omega_4)$. Because $(\mathfrak{a}\oplus\mathfrak{c}, V)\supset(R(\omega_4)\otimes 1)\oplus(R(\omega_2+\omega_3)\otimes R(5\omega_1))$ (see [135]), we have $(\mathfrak{a}, V^t)\supset R(\omega_4)$ and $(C, V)\supset 365750S_0\oplus 23760S_5$ (one should remember that $23760 = \dim(E_7, R(\omega_2+\omega_3)))$.

g) Let $V = R(\omega_3)$. Then $(\mathfrak{a}\oplus\mathfrak{c}, V)\supset(R(\omega_3)\otimes 1)\oplus(R(\omega_1+\omega_2)\otimes R(3\omega_1))$, (see [135]). Hence $(\mathfrak{a}, V^t)\supset R(\omega_3)$ and $(C, V)\supset 8645S_0\oplus 86184S_3$ (notice that $86184 = \dim(E_7, R(\omega_1+\omega_2)))$.

h) Let $V = R(\omega_2)$. Then $(\mathfrak{a}\oplus\mathfrak{c}, V)\supset(R(\omega_2)\otimes 1)\oplus(R(\omega_3)\otimes R(2\omega_1))$ (see [135]). Hence $(\mathfrak{a}, V^t)\supset R(\omega_2)\oplus R(\omega_3)$ and $(C, V)\supset 912S_0\oplus 8645S_2$.

We now put the obtained results together into Lemma 6.1.

7. LEMMA 6.1. *Let G be an exceptional simple group and T and C be the one-dimensional torus and, respectively, the connected simple three-dimensional subgroup of G containing it, corresponding to the highest root α (their Lie algebras are given by* (5.1)). *Then $Z_G(T)/T$ is a connected simple group and we have Table 5.*

TABLE 5

G	$\mathfrak{z}_{\mathfrak{g}}(\mathfrak{t})/\mathfrak{t}$	G-module U	Structure of $Z_G(T)/T$-module U^T	Structure of C-module U
G_2	A_1	$R(\omega_1)$	$R(2\omega_1)$	$3S_0 \oplus 2S_1$
		$\mathrm{Ad} = R(\omega_2)$	$R(2\omega_1) \oplus 1$	$3S_0 \oplus 4S_1 \oplus S_2$
F_4	C_3	$\mathrm{Ad} = R(\omega_1)$	$R(2\omega_1) \oplus 1$	$21S_0 \oplus 14S_1 \oplus S_2$
		$R(\omega_2)$	contains $R(2\omega_1) \oplus R(2\omega_1 + \omega_2)$	$294S_0 \oplus 294S_1 \oplus 112S_2 \oplus 14S_3$
F_4	C_3	$R(\omega_3)$	contains $R(\omega_1 + \omega_3)$	$91S_0 \oplus 70S_1 \oplus 14S_2$
		$R(\omega_4)$	$R(\omega_2)$	$14S_0 \oplus 6S_1$
E_6	A_5	$R(\omega_1)$	$R(\omega_4)$	$15S_0 \oplus 6S_1$
		$\mathrm{Ad} = R(\omega_2)$	$R(\omega_1 + \omega_5) \oplus 1$	$35S_0 \oplus 20S_1 \oplus S_2$
		$R(\omega_3)$	contains $R(\omega_3 + \omega_5)$	$126S_0 \oplus 90S_1 \oplus 15S_2$
		$R(\omega_4)$	contains $R(2\omega_1 + \omega_4) \oplus R(\omega_2 + 2\omega_5) \oplus R(\omega_1 + \omega_5)$	$770S_0 \oplus 700S_1 \oplus 225S_2 \oplus 20S_3$
		$R(\omega_5)$	contains $R(\omega_1 + \omega_3)$	$126S_0 \oplus 90S_1 \oplus 15S_2$
		$R(\omega_6)$	$R(\omega_2)$	$15S_0 \oplus 6S_1$
E_7	D_6	$\mathrm{Ad} = R(\omega_1)$	$R(\omega_2) \oplus 1$	$66S_0 \oplus 32S_1 \oplus S_2$
		$R(\omega_2)$	$R(\omega_1 + \omega_5) \oplus R(\omega_6)$	$352S_0 \oplus 232S_1 \oplus 32S_2$
		$R(\omega_3)$	contains $R(\omega_1 + \omega_3) \oplus R(\omega_2)$	$2607S_0 \oplus 2112S_1 \oplus 562S_2 \oplus 32S_3$
		$R(\omega_4)$	contains $R(2\omega_3)$	$75790S_0 \oplus 77440S_1 \oplus 34815S_2 \oplus 7040S_3 \oplus 495S_4$
		$R(\omega_5)$	contains $R(\omega_3 + \omega_5)$	$7424S_0 \oplus 6512S_1 \oplus 2112S_2 \oplus 220S_3$
		$R(\omega_6)$	contains $R(\omega_4)$	$573S_0 \oplus 384S_1 \oplus 66S_2$
		$R(\omega_7)$	$R(\omega_6)$	$32S_0 \oplus 12S_1$
E_8	E_7	$R(\omega_1)$	$R(\omega_1) \oplus R(\omega_6) \oplus 1$	$1540S_0 \oplus 968S_1 \oplus 133S_2$
		$R(\omega_2)$	contains $R(\omega_2) \oplus R(\omega_3)$	$912S_0 \oplus 8645S_2$
		$R(\omega_3)$	contains $R(\omega_3)$	contains $8645S_0 \oplus 86184S_3$
		$R(\omega_4)$	contains $R(\omega_4)$	contains $365750S_0 \oplus 23760S_5$
		$R(\omega_5)$	contains $R(\omega_4) \oplus R(\omega_5)$	contains $27664S_0 \oplus 365750S_4$
		$R(\omega_6)$	contains $R(\omega_6)$	contains $1539S_0 \oplus 27664S_3$
		$R(\omega_7)$	contains $R(\omega_1) \oplus R(\omega_3) \oplus R(\omega_7)$	$10374S_0 \oplus 7504S_1 \oplus 1674S_2 \oplus 56S_3$
		$R(\omega_8)$	$R(\omega_1) \oplus 1$	$133S_0 \oplus 56S_1 \oplus S_2$

§7. Proof of the main theorem: the first case

1. Henceforth we will use the notation of §4 (however $\operatorname{rk} G$ will be denoted by r and not by l). We may (and will) assume that $G \neq \{e\}$ and

$$G = H_1 \times \cdots \times H_s \times G_1 \times \cdots \times G_p, \tag{7.1}$$

where $H_1 = \cdots = H_s = \mathrm{SL}_2$ and, for any i, G_i is a connected, simply connected simple group of rank ≥ 2. We do not exclude the cases $s = 0$ and $p = 0$. We only exclude the case $s = 1, p = 0$ considered in Theorem 2.1. The groups H_j and G_i will be considered as subgroups of G, a fixed maximal torus in G as the direct product of fixed maximal tori in H_j and G_i, $\mathfrak{X}$ as the direct sum of fixed subsemigroups of dominant weights of those maximal tori, and $\{\omega_1, \dots, \omega_r\}$ as the disjoint union of the systems of fundamental weights of those semigroups. Let $\mathfrak{X}$ and $\{\omega_{i1}, \dots, \omega_{ir_i}\}$ be the aforementioned semigroup and, respectively, system of fundamental weights of G; the (only) fundamental weight for H_j will be denoted η_j.

Let $\Lambda \in \mathfrak{X}$; then

$$\Lambda = \sum_{j=1}^{s} M_j + \sum_{i=1}^{p} \Lambda_i, \quad \text{where } M_j = n_j \eta_j, \ \Lambda_i = \sum_{t=1}^{r_i} n_{it}\omega_{it} \in \mathfrak{X}_i. \tag{7.2}$$

Let $\mathfrak{X}^0 = \{\Lambda \in \mathfrak{X} \mid M_j \neq 0, \Lambda_i \neq 0 \text{ for all } i, j\}$. Then $\Lambda \in \mathfrak{X}^0 \Leftrightarrow R(\Lambda)$ the noneffectivity kernel of $R(\Lambda)$ is finite.

We now fix $d \in \mathbb{N}$. By Lemma 1.1, we must prove that the set $\{V \in \mathscr{R}(G) \mid \operatorname{hd} k[V]^G \leq d\}$ is finite. It is clear that using the induction on $s + p$, we must only check this for $V = R(\Lambda)$ with $\Lambda \in \mathfrak{X}^0$.

Let us fix another constant $c \in \mathbb{N}$. The choice of c will be specified in §8 in the course of the proof; it will depend on the group G only. Let (see (7.2)) $\mathfrak{X}_c = \{\Lambda \in \mathfrak{X} \mid \prod_{j=1}^{s}(n_j + 1) < c\}$; if $s = 0$ we set $\mathfrak{X}_c = \mathfrak{X}$. In general both $\mathfrak{X}_c^0 = \mathfrak{X}^0 \cap \mathfrak{X}_c$ and $\mathfrak{X}^0 - \mathfrak{X}_c^0$ are infinite. In this section we will show that the set $\{\Lambda \in \mathfrak{X}_c^0 \mid \operatorname{hd} k[R(\Lambda)]^G \leq d\}$ is finite for any choice of c, and in §§8 and 9 that $\{\Lambda \in \mathfrak{X}^0 - \mathfrak{X}_c^0 \mid \operatorname{hd} k[R(\Lambda)]^G \leq d\}$ is empty for a sufficiently large c. This, along with Theorem 2.1, will give us a proof of the main theorem.

2. THEOREM 7.1. *Let P be a connected group and W_i, $1 \leq i \leq t$ be a set of irreducible P-modules. Suppose that for the action of P on W_1 the generic stabilizer is contained in the noneffectivity kernel K and that $|K| < \infty$. Then the same is true for the action of P on $W = \bigoplus_{i=1}^{t} W_i$.*

PROOF. Because $|K| < \infty$, we have $K \subset Z_P$ and therefore K acts on W_i by scalar multiplication via a character χ_i. Let $w = \sum_{i=1}^{t} w_i$, $w_i \in W_i$ be a generic point in general position in W. Then w_1 is a point in general position in W_1 and therefore $P_w = \bigcap_i P_{\omega_i} \subset P_{\omega_1} \subset K$. Hence $\chi_i(h) = 1$ for all $h \in P_\omega$, for all i so that P_w is contained in the noneffectivity kernel of P on W, which, in turn, is contained in K. □

THEOREM 7.1. *Let G be a connected simply connected simple group of* rank ≥ 2. *Then there is a one-dimensional torus T in G such that:* 1) *$T \subset C \subset G$, where C is a connected three-dimensional simple group;* 2) *$V^T \neq 0$ for any fundamental $V \in \mathscr{R}(G)$;* 3) *$Z_G(T)/T$ is a nontrivial connected reductive group with at most one-dimensional center;* 4) *if $Z_G(T)/T$ is semisimple (resp., not semisimple) then Proposition 4.3 (resp., Proposition 4.4) holds for $L = T$ and $H = Z_G(T)$;* 5) *if $Z_G(T)/T$ is a one-dimensional torus then for any $h \in \mathbb{N}$ there exists a finite subset Φ of $\mathfrak{X}$ (depending on h) such that (in the notation of §2) $\dim(Z_G(T)/T, R(\Lambda)^T)^{\pm} \geq h$ for all $\Lambda \in \mathfrak{X} - \Phi$. Moreover, we can choose T to be a root torus.*

PROOF. We shall explicitly describe T in each of the cases: we choose T and C as was indicated in Lemmas 5.1–5.4 and 6.1; in the case B_l we set $\alpha = \omega_2$. Then assertions 1), 2), and 3) will follow immediately from those lemmas. From the same lemmas we deduce that $Z_G(T)/T$ is a one-dimensional torus only when $G = \mathrm{SL}_3$, and therefore assertion 5) now follows easily from Lemma 5.1 and Theorem 4.1. It remains to prove assertion 4). To do this we will vertify that Conditions a) and b) of Propositions 4.3 and 4.4 are satisfied for the semisimple and, respectively, nonsemisimple group $Z_G(T)/T$. Because 2) has already been proved we will only verify Condition b).

If G is exceptional, then, by Lemma 6.1, $Z_G(T)/T$ is simple and b) is already satisfied for $m_j = 1$ for all j.

Let G be of type C_l. Then, by Lemma 5.3, $Z_G(T)/T$ is simple of type C_{l-1} and b) is satisfied for $m_j = 1$ for all j if $l \geq 3$. If $l = 2$ then we can set $m_1 = 1$. However we cannot set $m_2 = 1$ because $(Z_G(T)/T, R(\omega)^T) = 1$. But it is easy to see that $(Z_G(T)/T, R(2\omega_2)^T) \supset (A_1, R(2\omega_1))$ and therefore we can set $m_2 = 2$.

Suppose G is of type B_l. Then, by Lemma 5.2, $Z_G(T)/T$ is semisimple of type $A_1 \oplus B_{l-2}$ and in Condition b) of Proposition 4.3 we can set $m_j = 1$ for $j \geq 3$. By Lemma 5.2, $(Z_G(T)/T, R(\omega_2)^T) \supset (1 \otimes R(\Lambda_2)) \oplus (R(2\omega_1) \otimes 1)$. Hence, by Theorem 4.1, $(Z_G(T)/T, R(2\omega_2)^T) \supset R(2\omega_1) \otimes R(\Lambda_2)$. Because $\Lambda_2 \neq 0$, this means that we can set $m_2 = 2$. We shall now show that we can set $m_1 = 3$. In the notation of part 3 of §5, the G-modules $S^3 V$ and $R(3\omega_1) \oplus V$ are isomorphic (see [15]) and therefore $(S^3 V)^T = R(3\omega_1)^T \oplus V^T$ as $Z_G(T)/T$-modules. It is easy to see that $(S^3 V)^T = S^3 W \oplus \langle e_{-1}e_1, e_{-1}e_2, e_{-2}e_1, e_{-2}e_2 \rangle W$. Hence

$$(Z_G(T)/T, R(3\omega_1)^T) = (1 \otimes R(3\omega_1)) \oplus (1 \otimes R(\Lambda_1)) \oplus (R(2\omega_1) \otimes R(\Lambda_1))$$

and our assertion follows from the fact that $\Lambda_1 \neq 0$.

Let G be of type D_l. By Lemma 5.4, $Z_G(T)/T$ is semisimple of type $A_1 \oplus D_{l-2}$ and in Condition b) of Proposition 4.3 we can set $m_j = 1$ for

$3 \le j \le l-2$, and $m_{l-1} = m_l = 1$ for $l \ge 5$. By Lemma 5.4, $(Z_G(T)/T, R(\omega_2)^T) \supset (R(2\omega_1) \otimes 1) \oplus (1 \otimes R_2)$. According to Theorem 4.1, $(Z_G(T)/T, R(2\omega_2)^T) \supset R(2\omega_1) \otimes R_2$. Because R_2 is nontrivial and irreducible for $l \ge 5$, we can now set $m_2 = 2$ for $l \ge 5$. Similarly, for $l = 4$ we have $(Z_G(T)/T, R(3\omega_2)^T) \supset R(2\omega_1) \otimes R(2\omega_1 + 2\omega_2)$ so that we can set $m_2 = 3$ for $l = 4$. Finally, we can show, exactly as we did for the case B_l, that $(Z_G(T)/T, R(3\omega_1)^T) = (1 \otimes R(3\omega_1)) \oplus (1 \otimes R_1) \oplus (R(2\omega_1) \otimes R_1)$. Because R_1 is locally faithful we can set $m_1 = 3$. Therefore we can set $m_{l-1} = m_l = 3$ for $l = 4$.

Finally, let $G = \mathrm{SL}_{l+1}$, $l \ge 2$. Then $Z_G(T)/T = k^* \times \mathrm{SL}_{l-1}$. Because the case $l = 2$ has already been considered, we may assume that $l \ge 3$. By Lemma 5.1, in Condition b) of Proposition 4.4 we can set $m_j = 1$ for $3 \le j \le l-2$. By Theorem 4.1, we have $(Z_G(T)/T, R(3\omega_2)^T) \supset (\chi^6 \otimes R(3\omega_2)) \oplus (\chi^{-2l+4} \otimes R(\omega_2))$ and therefore we can set $m_2 = m_{l-1} = 3$. Similarly $(Z_G(T)/T, R(3\omega_1)^T) \supset (\chi^3 \otimes R(3\omega_1)) \oplus (\chi^{-l+2} \otimes R(\omega_1))$ and we can set $m_1 = m_l = 3$. □

3. Proof of the main theorem: the first case. We shall now prove the finiteness of $\{\Lambda \in \mathfrak{X}_c^0 \mid \operatorname{hd} k[R(\Lambda)]^G \le d\}$ for all c. Obviously, we may assume that $p > 0$.

Let T_i be the one-dimensional torus and $C_i \supset T_i$ be the connected simple three-dimensional subgroup of G_i whose existence is guaranteed by Theorem 7.1 (applied to $G = G_i$). It is clear that

$$R = H_1 \times \cdots \times H_s \times G_1 \times \cdots \times G_{i-1} \times Q \times G_{i+1} \times \cdots \times G_p, \qquad (7.3)$$

where $R = Z_G(T_i), Z_G(T_i)/T_i$ and $(Z_G(T_i)/T_i)'$ for, respectively, $Q = Z_{G_i}(T_i), Z_{G_i}(T_i)/T_i$ and $(Z_{G_i}(T_i)/T_i)'$. By [4], for any connected semisimple group P there exists $b \in \mathbb{N}$ (it is explicitly given in [4]) such that for any set of irreducible P-modules $W_1, \dots, W_t$ with finite noneffectivity kernels, the generic stabilizer for P in $\bigoplus_{i=1}^t W_i$ is contained in the (finite) noneffectivity kernel whenever $\dim \bigoplus_{i=1}^t W_i \ge b$. Let b_i be such b for $P = (Z_G(T_i)/T_i)'$.

Let Φ_i be a finite subset of $\mathfrak{X}_i$ whose existence in the case $h = b_i$ is guaranteed by Proposition 4.3 and 4.4 (applied to $G = G_i$, $L = T_i$, $H = Z_{G_i}(T_i)$) when $Z_{G_i}(T_i)/T_i$ is not a one-dimensional torus, and by part 4) of Theorem 4.1 (applied to $G = G_i$, and $T = T_i$) otherwise. Increasing, as necessary, Φ_i we may (and will) assume that $\Phi_i = \{\sum_{j=1}^{r_i} n_{ij}\omega_{ij} \mid 0 \le n_{ij} \le m_{ij}$ for all $j\}$ where $m_{ij} \in \mathbb{N}$ are some fixed constants. Let now $\Phi = \{\Lambda \in \mathfrak{X}_c^0 \mid \Lambda_i \in \Phi_i$ for all $i = 1, \dots, p\}$ (see (7.2)). It is clear that Φ is finite. We shall assume that b_i, Φ_i, and m_{ij} are smallest possible. Then h_i, Φ_i, m_{ij} and Φ depend only on G and c. Clearly, for our purposes it suffices to show the finiteness of the set $\{\Lambda \in \mathfrak{X}_c^0 - \Phi \mid \operatorname{hd} k[R(\Lambda)]^G \le d\}$, which we shall now do.

Let $\Lambda \in \mathfrak{X}_c^0 - \Phi$. Then $\Lambda_i \notin \Phi_i$ for some i. Let $V = (G, R(\Lambda))$, $V_j = (G, R(M_j))$, $U_t = (G, R(\Lambda_t))$. Then (see [10])

$$X = V_1 \otimes \cdots \otimes V_s \otimes U_1 \otimes \cdots \otimes U_{i-1} \otimes Y \otimes U_{i+1} \otimes \cdots \otimes U_p, \tag{7.4}$$

where $X = V$ and V^{T_i} when $Y = U_i$ and, respectively, $U_i^{T_i}$. We shall now show that for G, V, and $T = T_i$ all of the conditions of the first part of Theorem 3.2 are satisfied. We may have the following three cases:

a) $Z_{G_i}(T_i)/T_i$ is semisimple.

From the definition of Φ_i it follows that the $Z_{G_i}(T_i)/T_i$-module $U_i^{T_i}$ contains an irreducible submodule W of dimension $\geq b_i$ with finite noneffectivity kernel. Then, in view of (7.4), the $Z_G(T_i)/T_i$-module V^{T_i} contains the irreducible submodule $(\bigotimes_{j=1}^s V_j) \otimes (\bigotimes_{t=1}^{i-1} U_t) \otimes W \otimes (\bigotimes_{r=i+1}^p U_r)$ of dimension $\geq b_i$ with finite noneffectivity kernel. It follows from the definition of b_i that its generic stabilizer is contained in the noneffectivity kernel. Lemma 7.1 implies now that the noneffectivity kernel of $(Z_{G_i}(T_i)/T_i, V^{T_i})$ is finite and that its generic stabilizer is contained in that kernel. Because $Z_{G_i}(T_i)/T_i$ is semisimple, Theorem 3.3 implies that all of the conditions of the first part of Theorem 3.2 are satisfied.

b) $Z_{G_i}(T_i)/T_i$ is not semisimple and is not a one-dimensional torus.

In view of Theorem 7.1 and the choice of T_i, the centers of $Z_{G_i}(T_i)/T_i$ and $Z_G(T_i)/T_i$ are both one-dimensional. From the definition of Φ_i it follows that the $(Z_{G_i}(T_i)/T_i)'$-module $U_i^{T_i}$ contains two irreducible submodules W_1 and W_2 with finite noneffectivity kernels and of dimensions $\geq b_i$, such that the weights of the center of $Z_{G_i}(T_i)/T_i$ on them are different from 0 and have different signs. By (7.4), the $(Z_G(T_i)/T_i)'$-module V^{T_i} contains two irreducible submodules $(\bigotimes_{j=1}^s V_j) \otimes (\bigotimes_{t=1}^{i-1} U_t) \otimes W_m \otimes (\bigotimes_{r=i+1}^p U_r)$, $m = 1, 2$, with finite noneffectivity kernels and of dimensions $\geq b_i$ on which the weights of the center of $Z_G(T_i)/T_i$ are different from 0 and have different signs. It follows from the definition of b_i that their generic stabilizers are contained in their noneffectivity kernels. In view of Lemma 7.1, the noneffectivity kernel of $((Z_G(T_i)/T_i)', V^{T_i})$ is finite and the generic stabilizer is contained in this kernel. Because the generic stabilizer of a locally transitive representation of a connected semisimple group is at least one-dimensional (see [154]), Propositions 3.3, 3.2, and Theorem 3.3 imply that all of the conditions of the first part of Theorem 3.2 are satisfied.

c) $Z_{G_i}(T_i)/T_i$ is a one-dimensional torus.

Let $W^{\pm} = (Z_{G_i}(T_i)/T_i, U_i^{T_i})^{\pm}$. Then $\dim W^{\pm} \geq b_i$. By (7.4), the $(Z_G(T_i)/T_i)'$-module V^{T_i} contains submodules $(\bigotimes_{j=1}^s V_j) \otimes (\bigotimes_{t=1}^{i-1} U_t) \otimes W^{\pm} \otimes (\bigotimes_{r=i+1}^p U_r)$. They are direct sums of respectively $\dim W^{\pm}$ copies of the irreducible $(Z_G(T_i)/T_i)'$-module $(\bigotimes_{j=1}^s V_j) \otimes (\bigotimes_{t=1}^{i-1} U_t) \otimes 1 \otimes (\bigotimes_{r=i+1}^p U_r)$ whose noneffectivity kernel is finite. Furthermore all of the weights of the center on them are, respectively, bigger than 0 and less than 0. By the definition of b_i, their generic stabilizers are contained in their noneffectivity kernels. By Lemma 7.1, the noneffectivity kernel of $((Z_G(T_i)/T_i)', V^{T_i})$ is finite and the generic stabilizer is contained in this kernel. As in part b), this shows that all of the conditions of the first part of Theorem 3.2 are satisfied.

Now, according to the first part of Theorem 3.2, we have $\operatorname{hd} k[V]^G \geq (q-20)/6$, where $q = q_{(C_i, V)} - q_{(C_i, \mathfrak{g})}$.

Because $C_i \subset G_i$ and G_i is simple, for any t the C_i-module $R(\omega_{it})$ contains at least one nontrivial irreducible submodule, say $S_{a_{it}}$, $a_{it} \geq_1$. It is clear that $R(M_j)$ (resp., $R(\Lambda_j)$, $j \neq i$) is, as a C_i-module, of the form $(\dim R(M_j))S_0$ (resp., $(\dim R(\Lambda_j))S_0$). Hence it follows from Theorem 4.1 (with $L = \{e\}$ and $H = C_i$) that the C_i-module V contains submodule S_m, $m = \sum_{j=1}^{r_i} a_{ij}n_{ij}$. Therefore $q_{(C_i, V)} \geq [(1 + \sum_{j=1}^{r_i} n_{ij})/2]$ and hence, if $\operatorname{hd} k[V]^G \leq d$, then $\sum_{j=1}^{r_i} n_{ij} \leq 12d + 40 + 2q_{(C_i, \mathfrak{g})}$. Because $q_{(C_i, \mathfrak{g})}$ depends only on G and T_i (but not on Λ) the set $\{\Lambda \in \mathfrak{X}_c^0 - \Phi \mid \operatorname{hd} k[R(\Lambda)]^G \leq d\}$ is obviously finite. $\square$

REMARK. Because $\mathfrak{X}_c^0 = \mathfrak{X}^0$ when $s = 0$, we also obtained a complete proof of the main theorem in the case when G does not contain three-dimensional simple normal subgroups.

§8. Proof of the main theorem: the second case

1. We keep the notation of §7. We may now assume that $s \geq 1$ (and $s \geq 2$ when $p = 0$). In this section we shall specify the choice of the constant c, and in all cases, except a rather special one, we show that $\{\Lambda \in \mathfrak{X}^0 - \mathfrak{X}_c^0 \mid \operatorname{hd} k[R(\Lambda)]^G \leq d\} = \varnothing$ for the chosen value of c. The special case just mentioned is as follows: $s = 1$, $p \geq 1$, r_i is even, and $G_i = \mathrm{SL}_{r_i+1}$ for all i. In this case we will only show that $\{\Lambda \in \mathfrak{X}^0 - \mathfrak{X}_i^0 \mid k[R(\Lambda)]^G \leq d$, and either n_1 is even or $\Lambda = \omega_{i1}, \omega_{ir_i}$ for at least one $i\} = \varnothing$. In §9 we will show that the complement of this set in $\{\Lambda \in \mathfrak{X}^0 - \mathfrak{X}_c^0 \mid \operatorname{hd} k[R(\Lambda)]^G \leq d\}$ is also empty (for a sufficiently large c). The reason for investigating these cases by two different methods is that in this section, as it also was in the previous section, the proof is based on a suitable choice of a one-dimensional torus, whereas in §9 no such choice is possible.

At this point we do not explicitly specify the constant c. The first reason is that we would have to introduce lots of new notation. Second, it would be difficult to explain the meaning of our choice. Instead, we will be imposing new restrictions on c (like $c \geq$ some constant) as we go and then, at the end of §9, we will choose a constant that satisfies all of the restrictions.

2. THEOREM 8.1. *Let G be a connected simply connected simple group of rank $l \geq 2$. Then there exists a finite set $\mathfrak{T}$ of tori of G such that:* 1) $Z_G(T)/T$

for all $T \in \mathfrak{T}$ *is a connected reductive group whose center is finite if* $G \neq \mathrm{SL}_{l+1}$, *and at most one-dimensional if* $G = \mathrm{SL}_{l+1}$; 2) *for any* $\Lambda \in \mathfrak{X}$ *different from* ω_1 *and* ω_l *when* $G = \mathrm{SL}_{l+1}$, *there exists a torus* $T \in \mathfrak{T}$ *(depending on* Λ*) with the following properties:* a) $R(\Lambda)^T \neq 0$ b) *if* $Z_G(T)/T$ *is semisimple then* $(Z_G(T)/T, R(\Lambda)^T)$ *contains a nonzero irreducible submodule* W *with finite noneffectivity kernel,* c) *if the center of* $Z_G(T)/T$ *is one-dimensional then* $(Z_G(T)/T, R(\Lambda)^T)$ *contains two nonzero irreducible submodules* W_1 *and* W_2 *such that the noneffectivity kernel of the action of* $(Z_G(T)/T)'$ *on* W_1 *is finite and the weights of the center of* $Z_G(T)/T$ *on* W_1 *and* W_2 *are different from* 0 *and have different signs.*

PROOF. In each of the cases we will describe $\mathfrak{T}$ explicitly. Let $\Lambda = \sum_{i=1}^{l} p_i\omega_i$, $p_i \in \mathbb{N}$ for $\Lambda \in \mathfrak{X}$.

If G is an exceptional group, let $\mathfrak{T}$ consist of only one element, the torus T corresponding to the highest root. This choice is justified by Lemma 6.1 and Theorem 4.1 (applied to $L = T$ and $H = Z_G(T)$).

Let G be of type B_l. Then, by Lemma 5.2 and Theorem 4.1, we easily see that when $p_l = 0$ we may choose T to be the torus corresponding to the root $\alpha = \omega_1$, and, when $p_l \neq 0$ by the root $\alpha = \omega_2$. These two tori form the set $\mathfrak{T}$.

Let G be of type C_l. If either $l \geq 3$ and Λ is arbitrary, or $l = 2$ and $p_1 \neq 0$, then it follows from Lemma 5.3 and Theorem 4.1 that the torus T corresponding to the root $\alpha = \omega_1$ has Properties 1), 2a), and 2b). Suppose now that $l = 2$ and $p_1 = 0$. We are going to keep the notation of part 4 of §5. Let $X \in \mathfrak{t} = \operatorname{Ker}\alpha_1$ and $X \neq 0$. Then $\varepsilon_1(X) = \varepsilon_2(X) \neq 0$ so that $(\Lambda^2 V)^{\mathfrak{t}} = \langle e_1 \wedge e_{-1}, e_1 \wedge e_{-2}, e_2 \wedge e_{-1}, e_2 \wedge e_{-2}\rangle$. We have $\mathfrak{z}_{\mathfrak{g}}(\mathfrak{t}) = \mathfrak{t} \oplus \mathfrak{a}$ where $\mathfrak{a} = A_1$ and is generated by X_{α_1} and $X_{-\alpha_1}$; moreover $X_{\alpha_1}(e_1) = X_{\alpha_1}(e_{-2}) = X_{-\alpha_1}(e_2) = X_{-\alpha_1}(e_{-1}) = 0$, $X_{\alpha_1}(e_2) = e_1$, $X_{\alpha_1}(e_{-1}) = -e_{-2}$, $X_{-\alpha_1}(e_1) = -e_2$, $X_{-\alpha_1}(e_{-2}) = e_{-1}$. Hence $(\mathfrak{a}, R(\omega_2)^{\mathfrak{t}}) = R(2\omega_1)$. It now follows easily from Theorem 4.1 that T can be chosen to be the one with Lie algebra $\mathfrak{t}$. Thus when G is of type C_l the set $\mathfrak{T}$ consists of one torus when $l \geq 3$ and of two tori when $l = 2$.

Let G be of type D_l. If either $l \geq s$ and $p_i \neq 0$ for at least one $i \geq 3$ or $l \geq 4$ and at least two of the numbers p_1, p_3, and p_4 are different from 0 then, in view of Lemma 5.4 and Theorem 4.1, the torus T corresponding to the root $\alpha = \omega_2$ has Properties 1), 2a), and 2b). Suppose now that $l \geq 4$ and $p_i = 0$ for all $i \geq 3$. We shall show that, in the notation of part 5 of §5, T can be chosen to be the one with Lie algebra $\mathfrak{t} = \bigcap_{i\geq 2} \operatorname{Ker}\alpha_i$. We have $\mathfrak{z}_{\mathfrak{g}}(\mathfrak{t}) = \mathfrak{t} \oplus \mathfrak{a}$ where $\mathfrak{a} = D_{l-1}$ and is generated by X_{α_i} and $X_{-\alpha_i}$, $i \geq 2$. If $X \in \mathfrak{t}, X \neq 0$, then $\varepsilon_i(X) = 0$ for each $i \geq 2$, so that $V^{\mathfrak{t}} = \langle e_2, \ldots, e_l, e_{-l}, \ldots, e_{-2}\rangle$ and therefore $(\mathfrak{a}, V^{\mathfrak{t}}) = R(\omega_1)$. Because $\Lambda^2 V = (\mathfrak{g}, R(\omega_2))$ and $(\Lambda^2 V)^{\mathfrak{t}} = \Lambda^2\langle e_2, \ldots, e_l, e_{-l}, \ldots, e_{-2}\rangle$, we have (see [15]) $(\mathfrak{a}, R(\omega_2)^{\mathfrak{t}}) = R(\omega_2)$ for $l \geq 5$, and $= R(\omega_2 + \omega_3)$ for $l = 4$. Our assertion now follows from Theorem 4.1. The remaining case, when $l = 4$, and either $p_i = p_3 = 0$, or $p_1 = p_4 = 0$ can be reduced to the one just considered by applying an outer automorphism.

Let $G = \mathrm{SL}_{l+1}$, $l \geq 2$. First we shall specify the set of all Λ for which the torus T corresponding to the root α_1 has Properties 1), 2a), and 2c). By Lemma 5.1 and the corollary to Theorem 4.1, $Z_G(T)/T = k^* \times \mathrm{SL}_{l-1}$ and Condition 2a) is satisfied for any Λ. By Lemma 5.1 and Theorem 4.1, the $Z_G(T)/T$-module $R(\Lambda)^T$ contains submodules $\chi^{a_i} \otimes R(M_i)$, $i = 1, 2$, with $a_1 = \sum_{i=1}^{l-1} ip_i + (l+1)[p_l/2] - p_l$, $M_1 = \sum_{i=1}^{l-2} p_i\omega_i + (p_l - 2[p_l/2])\omega_{l-2}$, $a_2 = -\sum_{i=2}^{l}(l+1-i)p_i + p_1 - (l+1)[p_1/2]$, $M_2 = \sum_{i=1}^{l-2} p_{i+2}\omega_i + (p_1 - 2[p_1/2])\omega_1$. It is easy to see that $a_1 \leq 0 \Leftrightarrow \Lambda = 3\omega_2$ for $l = 2$; $= \omega_l$; $= \omega_1 + \omega_l$ and that the noneffectivity kernel of the $(Z_G(T)/T)'$-module $R(M_1)$ is infinite $\Leftrightarrow \Lambda = p_{l-1}\omega_{l-1} + p_l\omega_l$ when $l \geq 3$ and p_l is even. Similarly, $a_2 \geq 0 \Leftrightarrow \Lambda = 3\omega_1$ for $l = 2$; $= \omega_1$; $= \omega_1 + \omega_l$ and the noneffectivity kernel of the $(Z_G(T)/T)'$-module $R(M_2)$ is infinite $\Leftrightarrow \Lambda = p_1\omega_1 + p_2\omega_2$ when $l \geq 3$ and p_1 is even. Therefore if Λ is different from the weights $3\omega_1$ for $l = 2$; ω_1; $\omega_1 + \omega_l$; $p_1\omega_1 + p_2\omega_2$ for $l \geq 3$ and even p_1 and the corresponding contragradient weights, then the torus T corresponding to α_1 satisfies 2a) and 2c). Now let us consider those exceptional weights (excluding ω_1 and ω_l, see the statement of the theorem).

Notice that if two weights are contragradient to each other then it suffices to consider only one of them: if we found a torus with required properties for Λ, then applying an outer automorphism we can also find one for Λ^*.

Let $l \geq 3$; $\Lambda = p_1\omega_1 + p_2\omega_2$; $p_1 = 2S$, where $s \in \mathbb{N}$ and T be the torus corresponding to α_1. By Lemma 5.1 and Theorem 4.1, the $Z_G(T)/T$-module $R(p_1\omega_1)^T$ contains, when $p_1 \neq 0$, submodules $\chi^{s(1-l)} \otimes 1$ and $\chi^{p_1} \otimes R(p_1\omega_1)$ and the $Z_G(T)/T$-module $R(p_2\omega_2)^T$ contains, when $p_2 \neq 0$, submodules $\chi^{p_2(1-l)} \otimes 1$ and $\chi^{2p_2} \otimes R(p_2\omega_2)$ (when $l = 3$ we set $\omega_2 = 0$ for $(Z_G(T)/T)'$. By Theorem 4.1, the $Z_G(T)/T$-module $R(\Lambda)^T$ contains, when $p_1p_2 \neq 0$, submodules $W_1 = \chi^{p_1+2p_2} \otimes R(p_1\omega_1 + p_2\omega_2)$ and $W_2 = \chi^{(1-l)(s+p_2)} \otimes 1$. They have the properties mentioned in 2c) and therefore when $p_1p_2 \neq 0$ the torus T satisfies all of the conditions of the theorem. This is also true when $p_2 = 0$ and $p_1 \neq 0$ (we can set $W_1 = \chi^{p_1} \otimes R(p_1\omega_1)$, $W_2 = \chi^{s(1-l)} \otimes 1$) and when $p_1 = 0$, $p_2 \neq 0$ and $l \geq 4$ (we can set $W_1 = \chi^{2p_2} \otimes R(p_2\omega_2)$, $W_2 = \chi^{p_2(1-l)} \otimes 1$).

Thus it remains to consider $p_2\omega_2$ when $l = 3$; $3\omega_1$ when $l = 2$ and $\omega_1 + \omega_l$. We keep the notation from §5.

Let $l = 3$ and $\Lambda = p_2\omega_2$. Let T be the torus with Lie algebra $\operatorname{Ker}\alpha_1 \cap \operatorname{Ker}\alpha_3$. It is easy to see that $Z_G(T)/T = \mathrm{SL}_2 \times \mathrm{SL}_2$, $(Z_G(T)/T, R(\omega_2)^T) = R(\omega_1 + \omega_2)$ and therefore $(Z_G(T)/T, R(\Lambda)^T) \supset R(p_2(\omega_1 + \omega_2))$. Thus T is a desired torus.

Let $l = 2$ and $\Lambda = 3\omega_1$. Let T be the torus with Lie algebra $\operatorname{Ker}\alpha_1$. Clearly $Z_G(T)/T = \mathrm{SL}_2$ and $(Z_G(T)/T, R(\Lambda)^T) = R(2\omega_1)$. Hence T is a desired torus.

Let $\Lambda = \omega_1 + \omega_l$. Consider the torus with Lie algebra $\mathfrak{t} = \bigcap_{i\geq 2} \operatorname{Ker}\alpha_i$. Then $\mathfrak{z}_\mathfrak{g}(\mathfrak{t}) = \mathfrak{t} \oplus \mathfrak{a}$, where $\mathfrak{a} = A_{l-1}$ and is generated by all X_{α_i} and $X_{-\alpha_i}$, $i \geq 2$. Because $R(\Lambda)$ is the adjoint $\mathfrak{g}$-module, we have $(\mathfrak{a}, R(\Lambda)^\mathfrak{t}) = R(\omega_1 + \omega_{l-1}) \oplus 1$ and T is a desired torus. □

REMARK. It is easy to see that when $G = \mathrm{SL}_{l+1}$ and $\Lambda = \omega_1, \omega_l$, there exists no torus T with the properties specified in the theorem.

3. Proof of the main theorem: the second case. If $p \geq 1$, let $\mathfrak{T}_i$ be a finite set of one-dimensional tori in G_i whose existence is guaranteed by Theorem 8.1 (applied to $G = G_i$). For each i and $T \in \mathfrak{T}_i$ consider the constant b defined for $P = (Z_G(T)/T)'$ as was indicated in the beginning of part 3 of §7. We assume that c is not less than the maximum of these constants taken over all i and all $T \in \mathfrak{T}_i$.

Let $\Lambda \in \mathfrak{X}^0 - \mathfrak{X}^0_c$. Several cases can occur:

1) Suppose that in (7.1) $p \geq 1$ and $G_i \neq \mathrm{SL}_{r_i+1}$ for some i.

Let T_i denote a one-dimensional torus in G whose existence is guaranteed by Theorem 8.1 (applied to $G = G_i$). Formula (7.3) remains valid. Let V, V_j, and U_l have the same meaning as in (7.4); then (7.4) is also valid. By Theorem 8.1, $Z_{G_i}(T_i)/T_i$ is a semisimple group and the $Z_{G_i}(T_i)/T_i$-module $U_i^{T_i}$ contains a nonzero irreducible submodule W whose noneffectivity kernel is finite. Then $Z_{G_i}(T_i)/T_i$ is semisimple (see (7.3)) and, in view of (7.4), the $Z_G(T_i)/T_i$-module V^{T_i} contains the nonzero irreducible submodule $(\bigotimes_{j=1}^s V_j) \otimes (\bigotimes_{t=1}^{i-1} U_t) \otimes W \otimes (\bigotimes_{r=i+1}^p U_r)$ whose noneffectivity kernel is finite and whose dimension is bigger than or equal to $\prod_{j=1}^s \dim V_j = \prod_{j=1}^s (n_j + 1) \geq c$ (because $\Lambda \in \mathfrak{X}^0 - \mathfrak{X}^0_c$). Therefore the generic stabilizer for this submodule is contained in the noneffectivity kernel. It follows from Lemma 7.1 that the noneffectivity kernel of $(Z_G(T_i)/T_i, V^{T_i})$ is finite and the corresponding generic stabilizer is contained in this kernel. It now follows from Theorem 3.3 that all conditions of Theorem 3.1 are satisfied (when $T = T_i$). For the torus $T = T_i$ we shall use the notation adopted in the beginning of §2. Because G_i is simple, $T_i \subset G_i$ and $\Lambda_i \neq 0$ we have $\dim(T_i, U_i^{\pm}) \geq 1$. It then follows from (7.4) that $\dim(T_i, V^{\pm}) \geq \prod_{j=1}^s \dim V_j \geq c$. From this and from the remark after Theorem 3.2 we deduce that if c is bigger than the maximum of the numbers $20+6d+\max\{\dim(T, \mathfrak{g}^+), \dim(T, \mathfrak{g}^-)\}$ taken over all i and $T \in \mathfrak{T}$ (henceforth this condition will be automatically assumed) then for the groups under consideration $\{\Lambda \in \mathfrak{X}^0 - \mathfrak{X}^0_c \mid \operatorname{hd} k[R(\Lambda)]^G \leq d\} = \varnothing$.

2) Suppose that in (7.1) $p \geq 1$, $G_j = \mathrm{SL}_{r_j+1}$ for all j and $\Lambda_i \neq \omega_{i1}, \omega_{ir_i}$

for some i. We keep the same notation as in 1). If $Z_{G_i}(T_i)/T_i$ is semisimple then repeating the argument from 1) we have $\operatorname{hd} k[R(\Lambda)]^G > d$. Suppose now that the center of $Z_{G_i}(T_i)/T_i$ is one-dimensional. Then the center of $Z_G(T_i)/T_i$ is also one-dimensional. By Theorem 8.1, $(Z_{G_i}(T_i)/T_i, U_i^{T_i})$ contains two nonzero submodules W_1 and W_2 such that the noneffectivity kernel of the action of $(Z_{G_i}(T_i)/T_i)'$ on W_1 is finite, and the weights of the center of $Z_{G_i}(T_i)/T_i$ on W_1 and W_2 are different from zero and have different signs. Then $(Z_G(T_i)/T_i, V^{T_i})$ contains the submodules $(\bigotimes_{j=1}^s V_j) \otimes (\bigotimes_{t=1}^{i-1} U_t) \otimes W_j \otimes (\bigotimes_{r=i+1}^p U_r)$, $j = 1, 2$, the weights of the center of $Z_G(T_i)/T_i$ on them are different from zero and have different signs, and $(Z_G(T_i)/T_i)'$ acts on the first submodule with finite noneffectivity kernel. As it was in 1), the dimensions of those modules are bigger than or equal to c, therefore the generic stabilizer of the first submodule with respect to $(Z_G(T_i)/T_i)'$ is contained in the noneffectivity kernel.

Henceforth we shall assume that c is bigger than the maximum of the numbers $\dim(Z_G(T)/T)'$ taken over all i and all $T \in \mathfrak{T}$.

Because the dimension of a locally transitive module does not exceed the dimension of the group, the action of $(Z_G(T_i)/T_i)'$ on $(\bigotimes_{j=1}^s V_j) \otimes (\bigotimes_{t=1}^{i-1} U_t) \otimes W_2 \otimes (\bigotimes_{r=i+1}^p U_r)$ is not locally transitive. It now follows from Lemma 7.1, Propositions 3.3 and 3.2, and Theorem 3.3 that (for $T = T_i$) all conditions of Theorem 3.1 are satisfied.

Arguing now as in the concluding part of 1) we see that in this case $\operatorname{hd} k[R(\Lambda)]^G > d$.

3) Suppose that in (7.1) $p \geq 0$, $G_i = \mathrm{SL}_{r_i+1}$, and Λ_i is either ω_i or ω_{ir_i} for all i.

Reshuffling, if necessary, H_i and applying an outer automorphism (the group of outer automorphisms of any reductive group P acts naturally on the set of equivalence classes of P-modules and, obviously, the classes from the same orbit have isomorphic algebras of invariants) we may (and will) assume that

$$n_1 \geq \cdots \geq n_s; \qquad \Lambda_i = \omega_i \quad \forall i. \tag{8.1}$$

We still keep the notation V, V_j, U_l from (7.4). Consider the following one-dimensional tori: $T_j = \{\lambda_j(t) = \operatorname{diag}(t, t^{-1}) \in H_j = \mathrm{SL}_2 \mid t \in k^*\}$; $\{\mu_l(t) = \operatorname{diag}(1, t, t^{-1}, t^2, t^{-2}, \dots) \in G_l = \mathrm{SL}_{r_l+1} \mid t \in k^*\}$, if r_l is even, $\{\mu_l(t) = \operatorname{diag}(t, t^{-1}, t^2, t^{-2}, \dots) \in G_l = \mathrm{SL}_{r_l+1} \mid t \in k^*\}$, if r_l is odd, and $\{\nu_l(t) = \operatorname{diag}(t, t^{-1}, t^3, t^{-3}, \dots) \in G_l = \mathrm{SL}_{r_l+1} \mid t \in k^*\}$, if r_l is odd. Let B_l be the maximal torus of G_l consisting of diagonal matrices. We fix a B_l-weighted basis $u_{l_0}, \dots, u_{lr_l}$ of U_l in which $(\operatorname{diag}(a_0, \dots, a_{r_l}))u_{li} = a_i u_{li}$ and a T_j-weighted basis $v_{j0}, \dots, v_{jn_j}$ of V_j in which $\lambda(t)v_{ji} = t^{n_j-2i}v_{ji}$

(see part 5 of §2). We also fix isomorphisms $\operatorname{Hom}(T_i, k^*) \to \mathbb{Z}$ under which 1 corresponds to the characters $\lambda_i(t) \mapsto t$.

We may have the following possibilities.

3a) Let $s \geq 2$ and suppose that n_i is even for some $i \geq 2$.

Because $Z_{H_i}(T_i)/T_i = \{e\}$, we see, in view of (7.3), that $Z_G(T_i)/T_i$ is a semisimple group isomorphic to $\prod_{j=1,j\neq i}^{s} H_j \times \prod_{t=1}^{p} G_t$. It follows from the assumption that $\dim V_i^{T_i} = 1$ and therefore the $Z_G(T_i)/T_i$-module $V^{T_i} = (\bigotimes_{j=1}^{i-1} V_j) \otimes V_i^{T_i} \otimes (\bigotimes_{t=i+1}^{s} V_t) \otimes (\bigotimes_{r=1}^{p} U_r)$ is isomorphic to $(\bigotimes_{j=1}^{i-1} V_j) \otimes (\bigotimes_{t=i+1}^{s} V_t) \otimes (\bigotimes_{r=1}^{p} U_r)$. Its dimension is not less than $n_1 + 1$. Because $\Lambda \in \mathfrak{X}^0 - \mathfrak{X}_c^0$, we have, in view of (8.1), that $(n_1+1)^s \geq c$.

We shall now impose a new restriction on c. For each $j = 1, \dots, s$ consider the constant $b = b_j$ defined for $P = Z_G(T_j)/T_j$ as was indicated in the beginning of part 3 of §7. We now assume that $c \geq \max_j b_j^s$.

We have, similarly to 1), that all of the conditions of Theorem 3.1 are satisfied (for $T = T_i$). Moreover, using the same argument as the one at the end of 1) and replacing G_i with H_i, Λ_i with M_i, and $U_i^{\pm}$ and $V_i^{\pm}$, we see that $\dim(T_i, V^{\pm}) \geq n_1 + 1 \geq c^{1/s}$ in addition, if

$$c > \max_i(20 + 6d + \max\{\dim(T_i, \mathfrak{g}^+), \dim(T_i, \mathfrak{g}^-)\})^s$$

(henceforth this restriction will be assumed automatically), then in this case $\operatorname{hd} k[R(\Lambda)]^G > d$.

3b) Suppose that $s \geq 3$ and all h_i are odd.

Consider the one-dimensional torus $T = \{\lambda_2(t) \times \lambda_3(t) \mid t \in k^*\} \subset T_2 \times T_3$; of $H_2 \times H_3$. Because $H_2 \times H_3$ is viewed as a subgroup of G, T is also a one-dimensional torus of G. We fix the isomorphism $\operatorname{Hom}(T, k^*) \to \mathbb{Z}$ under which 1 corresponds to the character $\lambda_2(t) \times \lambda_3(t) \mapsto t$. Clearly, $Z_G(T) = H_1 \times T_2 \times T_3 \times H_4 \times \cdots \times H_s \times \prod_{j=1}^{p} G_j$ and, in view of (8.1) $V^T = V_1 \otimes \langle\{v_{2,(n_2+n_3)/2-l} \otimes v_{3,l} \mid 0 \leq l \leq n_3\}\rangle \otimes V_4 \otimes \cdots \otimes V_s \otimes (\bigotimes_{j=1}^{p} U_j)$. The torus T is a direct summand of $T_2 \times T_3$: for a complement we can take T_3. Thus the center of $Z_G(T)/T$ is one-dimensional and the $Z_G(T)/T$-module V^T contains irreducible modules $V_1 \otimes \langle v_{2,(n_2-n_3)/2} \otimes v_{3,n_3}\rangle \otimes V_4 \otimes \cdots \otimes V_s \otimes (\bigotimes_{j=1}^{p} U_j)$ and $V_1 \otimes \langle v_{2,(n_2+n_3)/2} \otimes v_{3,0}\rangle \otimes V_4 \otimes \cdots \otimes V_s \otimes (\bigotimes_{j=1}^{p} U_j)$ of dimension not less than $n_1 + 1$. The center of $Z_G(T)/T$ has weight $-n_3$ on the former and n_3 on the latter. Clearly, $(Z_G(T)/T)'$ is isomorphic to $H_1 \times H_4 \times \cdots \times H_s \times \prod_{j=1}^{p} G_j$ and each of the above submodules is isomorphic, as a $(Z_G(T)/T)'$-module, to $V_1 \otimes V_4 \otimes \cdots \otimes V_s \otimes (\bigotimes_{j=1}^{p} U_j)$.

We are about to impose yet another restriction on c. For each $1 \leq i < j \leq s$ consider the constant $b = b_{ij}$ defined for $P = G/(H_i \times H_j)$ as was indicated in the beginning of part 3 of §7. We now assume that $c \geq \max_{i,j} b_{ij}^s$.

Similar to 3a), we have $(n_1+1)^s \geq c$. It now follows from Lemma

7.1, Propositions 3.3 and 3.2 and Theorem 3.3 that all of the conditions of Theorem 3.1 are satisfied.

The vectors $v_{2,0} \otimes v_{3,0}$ and $v_{2,n_2} \otimes v_{3,n_3}$ are T-weighted with respective weights $n_2 + n_3 > 0$ and $-n_2 - n_3 < 0$. It follows from this and from (7.4) that $\dim(T, V^{\pm}) \geq n_1 + 1$. Arguing as at the end of 3a) and assuming that c is bigger than the maximum of the numbers $(20 + 6d + \max\{\dim(T_{ij}, \mathfrak{g}^+), \dim(T_{ij}, \mathfrak{g}^-)\})^s$ taken over all i and j where $1 \leq i < j \leq s$ and $T_{ij} = \{\lambda_i(t) \times \lambda_j(t) | t \in k^*\}$ we see that in this case $\operatorname{hd} k[R(\Lambda)]^G > d$.

3c) Suppose that $s = 2$, n_2 is odd, and n_1 is even.

For $P = H_2 \times \prod_{j=1}^p G_j$ consider the constant b defined as in the beginning of part 3 of §7.

First we assume that $n_2 < b$. Let

$$\gamma(t) = \lambda_1(t) \times \lambda_2(t^2) \times \prod_{j=1}^{p} \mu_j(t^2)$$

and consider the one-dimensional torus $T = \{\gamma(t) | t \in k^*\}$ of G. It is clear that $Z_G(T) = T_1 \times T_2 \times \prod_{j=1}^p B_j$ and that the torus $T_2 \times \prod_{j=1}^p B_j$ is complementary to T in $Z_G(T)$. Any vector $v_{1,m_1} \otimes v_{2,m_2} \otimes (\bigotimes_{j=1}^p u_{jl})$ is an eigenvector for $\gamma(t)$ with eigenvalue $t^{n_1 - 2m_1 + 2n_2 - 4m_2 + f}$ where f is an even integer (depending on the vector) and $|f| \leq r_1 + \cdots + r_p + p$. But all such vectors form a T-weighted basis of V. Because of the restriction imposed on n_2 we have $|2n_2 - 4m_2 + f| < 2b + \sum_{j=1}^p r_j + p$. Similar to 3a) we also have $(n_1 + 1)^2 \geq c$. Because n_1 is even and $0 \leq m_1 \leq n_1$, we see that if $c \geq (2b + p + 1 + \sum_{j=1}^p r_j)^2$ (henceforth this inequality will be automatically assumed) then for any $m_2, l_1, \ldots, l_p$ there exists (a unique) m_1 such that the aforementioned vector is contained in V^T. This means that the sets of weights of the torus $T_2 \times \prod_{j=1}^p B_j$ on V^T and on $V_2 \otimes (\bigotimes_{j=1}^p U_j)$ are the same. But the action of $T_2 \times \prod_{j=1}^p B_j$ on $V_2 \otimes (\bigotimes_{j=1}^p U_j)$ is induced by an action of the semisimple group $H_2 \times \prod_{j=1}^p G_j$ in which this torus is a maximal one. The noneffectivity kernel of this action is finite. By the corollary of Proposition 2 of Chapter 3, the system of weights of the torus $T_2 \times \prod_{j=1}^p B_j$ is not contained in any closed half-space of $\operatorname{Hom}(T_2 \times \prod_{j=1}^p B_j, k^*) \otimes_{\mathbb{Z}} \mathbb{Q}$. On the other hand, for any torus and any finite-dimensional representation of it the generic stabilizer coincides with the noneffectivity kernel which, in turn, is the intersection of the kernels of all of the weights: this easily follows from the existence of weighted bases. It now follows from these remarks and from Proposition 3.1 that all conditions of Theorem 3.1 are satisfied. We remark that, keeping the same notation, for any $m_2, l, \ldots, l_p$ there exist at least $(n_1 - 2b - p - \sum_{j=1}^p r_j)/2 \geq (C^{1/2} - 2b - 1 - p - \sum_{j=1}^p r_j)/2$ values of m_1 for which $n_1 - 2m_1 + 2n_2 - 4m_2 + f > 0$ and at least the same number of values of m_1 for which $n_1 - 2m_1 + 2n_2 - 4m_2 + f < 0$. It follows from the remark after Theorem 3.2 that if

$$c > \left(41 + 12d + 2b + p + \sum_{j=1}^{p} r_j + 2\max\{\dim(T, \mathfrak{g}^+), \dim(T, \mathfrak{g}^-)\}\right)^2$$

(henceforth this inequality will be automatically assumed) then in our case $\operatorname{hd} k[R(\Lambda)]^G > d$.

We now assume that $n_2 > b$. Consider the torus T_1. Because n_1 is even, $\dim V_1^{T_1} = 1$. Similar to 3a), $Z_G(T_1)/T_1$ is isomorphic to $H_2 \times \prod_{j=1}^p G_j$ and the $Z_G(T_1)/T_1$-module V^T is isomorphic to $V_2 \otimes (\bigotimes_{j=1}^p U_j)$. Its dimension is not less than $n_2 + 1 > b$, and therefore (when $T = T_1$) all conditions of Theorem 3.1 are satisfied. Because $\dim(T_1, V_1^{\pm}) = (n_1)/2$ and, obviously, $(T_1, V^{\pm}) = V^{\pm} = V_1^{\pm} \otimes V_2 \otimes (\bigotimes_{j=1}^p U_j)$, we have $\dim(T_1, V^{\pm}) \geq n_1(n_2+1)/2$ (one should also take into account that $n_2 \geq 3$). Arguing as at the end of 3a) and using the restrictions on c imposed there we have $\operatorname{hd} k[R(\Lambda)]^G > d$.

3d) Let $S = 2$ and suppose that n_2 is odd and n_1 is even.

For $P = \prod_{j=1}^p G_j$ we consider the constant b defined as was indicated in the beginning of part 3 of §7 (if $p = 0$ we set $P = \{1\}$ and consider $\bigotimes_{j=1}^p U_j$ as the trivial one-dimensional module).

First we assume that $n_2 < 2b$. Let $\eta(t) = \lambda_1(t) \times \lambda_2(t) \times \prod_{j=1}^p \mu_j(t^2)$ and consider the one-dimensional torus $T = \{\eta(t) \mid t \in k^*\}$ of G. Then $Z_G(T) = T_1 \times T_2 \times B_1 \times \cdots \times B_p$ and $T_2 \times \prod_{j=1}^p B_j$ is a torus complementary to T in $Z_G(T)$. The vector $v_{1,m_1} \otimes v_{2,m_2} \otimes u_{1,l_1} \otimes \cdots \otimes u_{pl_p}$ is an eigenvector of $\eta(t)$ with eigenvalue $t^{n_1 - 2m_1 + n_2 - 2m_2 + f}$, where f is an integer (depending on the vector) and $|f| \leq \sum_{j=1}^p r_j + p$. We have $|n_2 - 2m_2 + f| < 2b + p + \sum_{j=1}^p r_j$. Because both n_1 and n_2 are odd, an argument similar to the one in 3c) shows that if $c \geq (2b + p + 1 + \sum_{j=1}^p r_j)^2$ then all conditions of Theorem 3.1 are satisfied, and that if

$$c > \left(41 + 12d + 2b + \sum_{j=1}^{p} r_j + p + 2\max\{\dim(T, \mathfrak{g}^+), \dim(T, \mathfrak{g}^-)\}\right)^2$$

(henceforth this condition will be assumed) then $\operatorname{hd} k[R(\Lambda)]^G > d$.

Suppose now that $n_2 \geq 2b$. We set $T = \{\lambda_1(t) \times \lambda_2(t) \mid t \in k^*\}$. Similar to 3b) we have

$$Z_G(T) = T_1 \times T_2 \times \prod_{j=1}^{p} G_j;$$

$$V^T = \langle \{v_{1,(n_1+n_2)/2-l} \otimes v_{2,l} \mid 0 \leq l \leq n_2\} \rangle \otimes \left(\bigotimes_{j=1}^{p} U_j\right),$$

T_2 is a torus complementary to T in the torus $T_1 \times T_2$; the center of the group $Z_G(T)/T$ is one-dimensional, and $(Z_G(T)/T)'$ is isomorphic to $\prod_{j=1}^p G_j$. It now follows that the $Z_G(T)/T$-module V^T is the direct sum of the submodules $L_1 = \langle\{v_{1,(n_1+n_2)/2-l} \otimes v_{2,l} | 0 \le l \le (n_2-1)/2\}\rangle \otimes (\bigotimes_{j=1}^p U_j)$ and $L_2 = \langle\{v_{1,(n_1+n_2)/2-l} \otimes v_{2,l} | (n_2+1)/2 \le l \le n_2\}\rangle \otimes (\bigotimes_{j=1}^p U_j)$ and that all weights of T_2 (and therefore of the center of $Z_G(T)/T$) are positive on L_1, negative on L_2, and that the $(Z_G(T)/T)'$-modules L_1 and L_2 are isomorphic to the direct sum of $(n_2+1)/2$ copies of $\prod_{j=1}^p U_j$. Therefore $\dim L_1 = \dim L_2 \ge (n_2+1)/2 > b$. Because the generic stabilizer of a locally transitive representation of a connected semisimple group is at least one dimensional (see [154]), it follows from the definition of b, Lemma 7.1, Propositons 3.3 and 3.2 and Theorem 3.3 that all conditions of Theorem 3.1 are satisfied. We now fix the isomorphism $\mathrm{Hom}(T, k^*) \to \mathbb{Z}$ under which 1 corresponds to the character $\lambda_1(t) \times \lambda_2(t) \mapsto t$. The vectors $v_{1,l} \otimes v_{2,m}$ are T-weighted of positive weights for $0 \le l \le (n_1-1)/2$ and $0 \le m \le (n_2-1)/2$. Because T acts trivially on $\bigotimes_{j=1}^p U_j$ we have $\dim(T, V^{\pm}) \ge (n_1+1)(n_2+1)/4 \ge c/4$. It now follows from the remark after Theorem 3.2 that if $c > 80 + 24d + 4\max\{\dim(T, \mathfrak{g}^+), \dim(T, \mathfrak{g}^-)\}$ (henceforth this inequality will be automatically assumed) then $\mathrm{hd}\, k[R(\Lambda)]^G > d$.

It remains to consider the case $s = 1$. Because $\mathrm{rk}\, G \ge 2$ we then automatically have $p \ge 1$.

3e) Let $s = 1$ and suppose that n_1 is even.

Let $\varphi(t) = \lambda_1(t) \times \prod_{j=1}^p \mu_j(t^2)$ and $T = \{\varphi(t) | t \in k^*\}$. The following argument is similar to one from the first part of 3c). We have $Z_G(T) = T_1 \times \prod_{j=1}^p B_j$, the torus $\prod_{j=1}^p B_j$ is a torus complementary to T in $Z_G(T)$, each vector of the form $v_{1,m_1} \otimes u_{1,l_1} \otimes \cdots \otimes u_{pl_p}$ is an eigenvector for $\varphi(t)$ with eigenvalue $t^{n_1-2m_1+f}$, where f is an even integer (depending on the vector) and $|f| \le p + \sum_{j=1}^p r_j$. Because n_1 is even and $n_1+1 \ge c$ we see that if $c > 1 + p + \sum_{j=1}^p r_j$ (henceforth this inequality will be automatically assumed) then for any $l_1, \dots, l_p$ there exists (a unique) m_1 such that the aforementioned vector is contained in V^T. Therefore the sets of weights of torus $\prod_{j=1}^p B_j$ on V^T and $\prod_{j=1}^p U_j$ are the same. It follows that the conditions of Theorem 3.1 are satisfied. Furthermore, for any $l_1, \dots, l_p$ there exist at least $(n_1 - 1 - p - \sum_{j=1}^p r_j)/2 \ge (c - 2 - p - \sum_{j=1}^p r_j)/2$ values of m_1 for which $n_1 - 2m_1 + f > 0$ and at least the same number of values of m_1 for which $n_1 - 2m_1 + f < 0$. Therefore if $c > 42 + 12d + p + \sum_{j=1}^p r_j + 2\max\{\dim(T, \mathfrak{g}^+), \dim(T, \mathfrak{g}^-)\}$ (henceforth this inequality will be automatically assumed) then in this case $\mathrm{hd}\, k[R(\Lambda)]^G > d$.

3f) Let $s = 1$ and suppose that n_1 is odd and r_i is even for some i.

Let $\psi(t) = \lambda_i(t) \times \prod_{j=1}^{i-1} \mu_j(t^2) \times \nu_i(t) \times \prod_{j=i+1}^p \mu_j(t^2)$ and $T = \{\psi(t) | t \in k^*\}$. Then $Z_G(T) = T \times \prod_{j=1}^p B$, the torus $\prod_{j=1}^p B$ is complementary to

T in $Z_G(T)$, the vector $v_{1,m_1} \otimes u_{1,l_1} \otimes \cdots \otimes u_{p,l_p}$ is an eigenvector of $\psi(t)$ with eigenvalue $t^{n_1-2m_1+f}$, where f is an odd integer (depending on the vector) with $|f| \leq p-1+\sum_{j=1}^{p} r_j$. Because n_1 is odd we see, similar to 3e), that under the restrictions on c from 3e) (with torus T defined as was just mentioned) $\operatorname{hd} k[R(\Lambda)]^G > d$. We now assume that those restrictions on c are satisfied.

To complete the proof of the main theorem it remains to consider the case when $s = 1$, n_1 is odd, and all r_i are even (and $\Lambda_i = \omega_{i,1}$ for all i). This will be done in the next section.

§9. Proof of the main theorem: the third case

1. We keep the notation of §§7 and 8. The group G is of the form (7.1), where $s = 1$, $p \geq 1$, $G_l = \mathrm{SL}_{r_l+1}$, r_l is even for all l. We now consider the weights Λ of the form (7.2), where n_1 is odd, $n_1 + 1 \geq c$, $\Lambda_l = w_{l1}$, for all l and we want to show that for any given constant $d \in \mathbb{N}$ we can choose c so that $\operatorname{hd} k[R(\Lambda)]^G > d$.

We need some additional notation: $r_i = 2k_i$ for all i, $T = T_1 \times \prod_{j=1}^{p} B_j$ a maximal torus in G.

2. The following proposition explains why the remaining case cannot be dealt with by an appropriate choice of a one-dimensional torus of G, the technique that worked in previous cases.

PROPOSITION 9.1. *Let $p = 1$, n_1 be any odd integer, r_1 be any even integer, and $v \in V$ a nonzero vector whose orbit Gv is closed. Then G_v is a finite group.*

PROOF. Suppose that $\dim G_v \geq 1$. Because Gv is closed, G_v is reductive and therefore we can find a one-dimensional torus S of G_v. Let $\varphi\colon k^* \to S$ be a fixed isomorphism. Replacing v with gv for a suitable $g \in G$ we may assume that $\varphi(t) = \lambda_1(t^a) \times \operatorname{diag}(t^{b_0}, \ldots, t^{b_{r_1}}) \in H_1 \times G_1 = G$ for all $t \in k^*$, where $b_0 + \cdots + b_{r_1} = 0$. Vectors $v_{1,l} \otimes u_{1,j} \in V_1 \otimes U_1 = V$ form an S-weighted basis on V with $\varphi(t)(v_{1,l} \otimes u_{1,j}) = t^{(n_1-2l)a+b_j} u_{1,l} \otimes u_{1,j}$. Therefore V^S is the linear hull of all $v_{1,l} \otimes u_{1,j}$, such that $(n_1 - 2l)a + b_j = 0$.

We shall now show that for some j the vector $v_{1,l} \otimes v_{1,j}$ is not contained in V^S for any l. Assuming the opposite, we see that, because n_1 is odd, $b_j = af_j$ for any $j = 0, \ldots, r_1$, where f_j is an odd integer. Therefore $0 = \sum_{j=0}^{r_1} b_j = a \sum_{j=0}^{r_1} f_j$, and because r_1 is even, we have $a = 0$. Therefore V^S is the linear hull of those $v_{1,l} \otimes u_{1,j}$ for which $b_j = 0$. It is clear that $b_j = 0$ cannot hold for all j (however there is some j for which $b_j = 0$ because $v \in V^T$ and therefore $V^T \neq 0$). Let, say, $b_0 \neq 0$. In G consider the one-dimensional torus $\{\delta(t) = \lambda_1(t) \times \operatorname{diag}(t^{-r_1}, t, t, \ldots, t) \in H_1 \times G_1 = G \mid t \in k^*\}$. If $j > 0$ then $\delta(t)(v_{1,l} \otimes u_{1,j}) = t(v_{1,l} \otimes u_{1,j})$. Therefore

$\delta(t)$ acts on V^S as the homothety with coefficient t. Because $v \in V^S$, this contradicts the closedness of the orbit Gv.

Therefore for some j the vector $v_{1,l} \otimes u_{1,j}$ is not contained in V^S for any l; let, say, this happens for $j = 0$. Then, as before, $\delta(t)$ acts on V^S as the homothety with coefficient t, which contradicts the closedness of Gv. □

3. We will later need the following auxilliary result.

LEMMA 9.1. *There exists a set of integers a_{lj}, where $1 \le l \le p$, $0 \le j \le 2k_l = r_l$, such that $a_{l,0} = 0$, $a_{l,2j-1} = -a_{l,2j}$ for all l and all $j = 1, \dots, k_l$ and for any positive integer q bigger than a certain (depending on the set) integer constant the following conditions are satisfied (below, $e^{2\pi i/q}$ is denoted by ε): a) $\varepsilon^{a_{lj}} = \varepsilon^{a_{lt}} \Leftrightarrow j = t$ for all l; b) $\varepsilon^{a_{1,j_1} + \dots + a_{p,j_p}} = \varepsilon^{a_{1,l_1} + \dots + a_{p,l_p}} \Leftrightarrow j_1 = l_1, \dots, j_p = l_p$.*

PROOF. We induct on p. If $p = 1$ we set $a_{1,0} = 0$, $a_{1,2j-1} = -a_{1,2j} = j$ for $j = 1, \dots, k_1$. Then it is easy to check directly that Condition a) (coinciding in this case with b)) is satisfied whenever $q > 2k_1 = r_1$.

Suppose now that the numbers a_{lj}, $1 \le l \le p-1$ have already been constructed. We will construct additional numbers $a_{p,0}, \dots, a_{p,r_p}$ so as to satisfy all of the requirements. Let M be the maximum of the numbers $\sum_{t=1}^{p-1} a_{tj_t}$ taken over all $(p-1)$-tuples $(j_1, \dots, j_{p-1})$. It follows from the properties of a_{lj} that the minimum of those numbers is equal to $-M$. We set $a_{p,0} = 0$ and $a_{p,2j-1} = -a_{p,2j} = 3Mj$ for $j = 1, \dots, k_p$. Assume now that q is bigger than the constant mentioned in the statement of the lemma for the set $a_{l,j}$, $1 \le l \le p-1$, and also bigger than $2M(1+3k_p)$. We shall show that Conditions a) and b) are both satisfied.

Let $3Mj - 3Ml \equiv 0 \mod q$ for $-k_p \le j, l \le k_p$. Because $|3Mj - 3Ml| \le 6Mk_p \le q$, we have $j = l$ so a) is satisfied. Let now $(\sum_{t=1}^{p-1} a_{t,j_t} + 3Mj_p) - (\sum_{t=1}^{p-1} a_{t,l_t} + 3Ml_p) \equiv 0 \mod q$ for $-k_p \le j_p, l_p \le k_p$. It is easy to see that the absolute value of the left-hand side does not exceed $2M(1+3k_p) < q$ and therefore it is equal to zero. Hence $M \ge |\sum_{t=1}^{p-1} a_{tj_t}| = |\sum_{t=1}^{p-1} a_{tl_t} + 3M(l_p - j_p)| \ge 3M|l_p - j_p| - |\sum_{t=1}^{p-1} a_{tl_t}| \ge 3M|l_p - j_p| - M$. It now follows that $l_p = j_p$ and therefore, by the induction hypothesis, $l_1 = j_1, \dots, l_{p-1} = j_{p-1}$. This yields b). □

4. Proof of the main theorem: the third case.

1) We fix a set $\{a_{lj}\}$ from Lemma 9.1. Let M be the maximum of the numbers $\sum_{t=1}^{p} a_{tj_t}$ taken over all tuples $(j_1, \dots, j_p)$. It is clear that the minimum of those numbers is equal to $-M$. We now fix an odd integer $q \in \mathbb{N}$ bigger than both M and 10, for which Conditions a) and b) of Lemma 9.1 are satisfied. As in the lemma, we set $\varepsilon = e^{2\pi i/q}$.

It follows from the properties of a_{lj} that T contains the element $g = \lambda_1(\varepsilon) \times \prod_{t=1}^{p} \operatorname{diag}\{\varepsilon^{a_{t,0}}, \dots, \varepsilon^{a_{t,r_t}}\}$. Let A be the cyclic subgroup of T generated by g. It follows from Condition a) of Lemma 9.1 that $Z_G(A) = T_1 \times \prod_{j=1}^{p} B_j = T$.

2) For any $w = v_{1,l} \otimes u_{1,j_1} \otimes \cdots \otimes u_{p,j_p}$ we have

$$gw = \varepsilon^{n_1 - 2l + a_{1j_1} + \cdots + a_{p,j_p}} w. \tag{9.1}$$

Therefore V^A is the linear hull of those w for which

$$n_1 - 2l + \sum_{t=1}^{p} a_{tj_t} \equiv 0 \mod q. \tag{9.2}$$

We now assume that $c > M + 2q$. Then $n_1 \geq M + 2q$.

For any fixed values of $j_1, \dots, j_p$ we consider all numbers of the form $n_1 - 2l + \sum_{t=1}^{p} a_{t,j_t}$, where $l = 0, 1, \dots, n_1$. We want to prove the existence of two values $l = l_+$ and $l = l_-$ of l, where $0 \leq l_+, l_- \leq n_1$ such that $n_1 - 2l_+ > 0$, $n_1 - 2l_- < 0$ and (9.2) holds. Suppose that $\sum_{t=1}^{p} a_{tj_t}$ is odd. Because the numbers $-\sum_{t=1}^{p} a_{t,j_t} \pm 2q$ are odd and their absolute values do not exceed $M + 2q \leq n_1$ we can find, taking into account that n_1 is odd, two numbers $l_+, l_- \in \mathbb{Z}$, $0 \leq l_+, l_- \leq n_1$, such that $n_1 - 2l_\pm = -\sum_{t=1}^{p} a_{t,j_t} \pm 2q$. Because, by assumption, $q > M$, it is clear that l_+ and l_- are the desired numbers. Suppose now that $\sum_{t=1}^{p} a_{t,j_t}$ is even. Because q is odd, the numbers $-\sum_{t=1}^{p} a_{t,j_t} \pm q$ are also odd. Their absolute values do not exceed $M + q < n$, and therefore, because n_1 is odd, we can find l_+ and l_-, $0 \leq l_+, l_- \leq n_1$, such that $n_1 - 2l_\pm = -\sum_{t=1}^{p} a_{t,j_t} \pm q$. Because $q > M$, l_+ and l_- are the desired numbers.

3) We shall now deduce from this that the action of $Z_G(A) = T$ on $V^A \neq 0$ is stable.

Let $X = \operatorname{Hom}(T, k^*) \otimes_{\mathbb{Z}} Q$, $X_1 = \operatorname{Hom}(T_1, k^*) \otimes_{\mathbb{Z}} \mathbb{Q}$ and $X_2 = \operatorname{Hom}(\prod_{j=1}^{p} B_j, k^*) \otimes_{\mathbb{Z}} \mathbb{Q}$. Then $X = X_1 \oplus X_2$ and X_1, being one-dimensional, has only two open half-spaces. We endow X with scalar product $(\ ,\ )$ with respect to which X_1 and X_2 are orthogonal. Let $\chi \in X$, $\chi \neq 0$. We want to find two weights λ_+ and λ_- of T on V^A for which $(\chi, \lambda_+) > 0$ and $(\chi, \lambda_-) < 0$. If we can do that then the stability of the action would follow from Proposition 3.1. First we prove the existence of λ_+; for λ_- the argument is identical. Let $\chi = \chi_1 + \chi_2$, $\chi_i \in X_i$. By the corollary to Proposition 2 of Chapter 3, there exists a set of indices $(j_1, \dots, j_p)$ such that $(\beta, \chi_2) \geq 0$ whenever $\beta \in X_2$ is a $\prod_{j=1}^{p} B_j$-weight of the vector $u_{1,j_1} \otimes \cdots \otimes u_{p,j_p}$ and $(\beta, \chi_2) = 0$ only when $\chi_2 = 0$. As was shown above there exist l_+ and l_-, $0 \leq l_+, l_- \leq n_1$, such that $v_{1,l_\pm} \otimes u_{1,j_1} \otimes \cdots \otimes u_{p,j_p} \in V^A$, $n_1 - 2l_+ > 0$ and $n_1 - 2l_- < 0$. The vectors v_{1l_+} and v_{1l_-} are T_1-weighted. Let α_+ and, respectively, $\alpha_- \in X_1$ be their T_1-weights. Then α_+ and α_- are contained in different open half-spaces of X_1. Choose the one contained in the same half-space as χ_1 (if $\chi_1 = 0$ we choose any) and add it to β. It is easy to see that this sum can be taken for λ_+.

4) Thus the action of $Z_G(A)$ on V^A is indeed stable. It follows from [110] and the definition of stability that the orbit Gv of a point v in general position in V^A is closed in V. Now we are going to investigate the group G_v. We will show that it is finite and find an upper bound on its order. As it will turn out this bound depends only on q and $r_1, \dots, r_p$ (but not on n_1 !). From this we will deduce easily the desired inequality $\operatorname{hd} k[R(\Lambda)]^G > d$ for a sufficiently large c.

5) Let $\pi\colon G_v \to H_1 = \mathrm{SL}_2$ be the natural projection. What is the structure of $\operatorname{Ker}\pi$? Let $h = \lambda_1(1) \times \prod_{j=1}^p h_j$, where $h_j \in G_j = \mathrm{SL}_{r_j+1}$ is an element of $\operatorname{Ker}\pi$. We can write v in the form

$$v = \sum z_{j_1\cdots j_p} \otimes u_{1,j_1} \otimes \cdots \otimes u_{p,j_p}, \tag{9.3}$$

where $z_{j_1\cdots j_p}$ is a linear combination of those v_{1l} for which (9.2) holds (the summation is over all $j_1 = 0, \dots, r_1; \dots; j_p = 0, \dots, r_p$). As was shown before, $z_{j_1\cdots j_p} \neq 0$ for all $j_1, \dots, j_p$. Moreover, because the numbers a_{ij} satisfy, for any $l \in \mathbb{N}$, $0 \le l \le n$, Condition b), there will be at most one set of indices $(j_1, \dots, j_p)$ such that v_{1l} has a nonzero coefficient in the decomposition of $z_{j_1\cdots j_p}$ with respect to the basis $v_{10}, \dots, v_{1n_1}$; in particular all vectors $z_{j_1\cdots j_p}$ are linearly independent. Because $h \in \operatorname{Ker}\pi$, we deduce from (9.3) that $u_{1,j_1} \otimes \cdots \otimes u_{p,j_p} = h_1 u_{1,j_1} \otimes \cdots \otimes h_p u_{p,j_p}$ for all $j_1, \dots, j_p$. This means that the element $h_1 \times \cdots \times h_p \in \prod_{j=1}^p G_j$ is contained in the noneffectivity kernel of the action of $\prod_{j=1}^p G$ on $\bigotimes_{j=1}^p u_j$, which is finite and contained in $Z_{G_1 \times \cdots \times G_p}$. Thus $\operatorname{Ker}\pi$ is a finite subgroup of Z_{G_v} of order at most $\prod_{j=1}^p (r_j + 1)$.

6) Because Gv is closed, G_v is reductive. Therefore $\pi(G_v)$ is a reductive subgroup of H_1. Because $H = \mathrm{SL}_2$, we deduce from the description of reductive subgroups of SL_2 (see part 8 of §1, Chapter 6) that either $\pi(G_v)$ is finite, or $\pi(G_v)^0$ is a one-dimensional torus, or $\pi(G_v) = H_1$.

We shall now show that only the first case is possible, and therefore, in view of the finiteness of $\operatorname{Ker}\pi$, the group G_v is also finite.

7) Suppose that $\pi(G_v) = H_1$. Then $\pi^{-1}(T_1)^0$ is a one-dimensional torus. Let $f \in \pi^{-1}(T_1)^0$ and $s \in A \subset G_v$. Because $\pi(s) \in T_1$ and T_1 is commutative, we have $fsf^{-1}s^{-1} \in \operatorname{Ker}\pi$. Furthermore, because $|\operatorname{Ker}\pi| < \infty$ and the variety $\{fsf^{-1}s^{-1} \mid f \in \pi^{-1}(T_1)^0\}$ is irreducible and contains the identity of G_v, we see that $\pi^{-1}(T_1) \subset Z_{G_v}(A) \subset Z_G(A) = T$. Let $\psi^*\colon k^* \to \pi^{-1}(T_1)^0$

be a fixed isomorphism. Then $\psi(t) = \lambda_1(t^a) \times \prod_{j=1}^{p} \operatorname{diag}(t^{b_{j,0}}, \ldots, t^{b_{j,r_j}}) \in T_1 \times \prod_{j=1}^{p} B_j = T$ for some $a, b_{l,j} \in \mathbb{Z}$. Because $\pi(\pi^{-1}(T_1)^0) = T_1$, we have $a \neq 0$. Moreover, for any $w = v_{1,l} \otimes u_{1,j_1} \otimes \cdots \otimes u_{p,j_p}$ we have

$$\psi(t)w = t^{(n_1-2l)a+b_{1,j_1}+\cdots+b_{p,j_p}}w. \tag{9.4}$$

Being a point in general position of V^A, the vector v is a linear combination with nonzero coefficients of those w for which (9.2) holds. Because $\pi^{-1}(T_1)^0 \subset G_v$, it follows from (9.4) that these w satisfy the condition

$$(n_1 - 2l)a + b_{1,j_1} + \cdots + b_{p,j_p} = 0. \tag{9.5}$$

As was shown before, for any set $(j_1, \ldots, j_p)$ we can find l, $0 \leq l \leq n_1$, such that (9.2) holds; for each $(j_1, \ldots, j_p)$ we now fix such an l and denote it by $l_{j_1 \cdots j_p}$. We then deduce from (9.5) that

$$a\sum(n_1 - 2l_{j_1\cdots j_p}) + \sum(b_{1,j_1} + \cdots + b_{p,j_p}) = 0 \tag{9.6}$$

(the summation is over all $j_1 = 0, \ldots, r_1; \ldots; j_p = 0, \ldots, r_p$). The first sum in (9.6) contains $\prod_{t=1}^{p}(r_t + 1)$ summands of the form $n_1 - 2l_{j_1\cdots j_p}$ and, because n_1 is odd and $r_1, \ldots, r_p$ are even, it is different from zero. On the other hand, because all possible vectors $u = u_{1,j_1} \otimes \cdots \otimes u_{pj_p}$ form a basis of $U_1 \otimes \cdots \otimes U_p$ consisting of eigenvectors of the element $\sigma(t) = \prod_{j=1}^{p} \operatorname{diag}(t^{b_{j,0}}, \ldots, t^{b_{j,r_j}}) \in \prod_{j=1}^{p} G_j = \prod_{j=1}^{p} \mathrm{SL}_{r_j+1}$, and $\sigma(t)u = t^{b_{1,j_1}+\cdots+b_{p,j_p}}u$ the other sum in (9.6) is equal to zero. It now follows from (9.6) that $a = 0$, a contradiction.

8) We now suppose that $\pi(G_v)^0$ is a one-dimensional torus. The element $\pi(g)$ either centralizes this torus or induces (by conjugation) its only nontrivial automorphism. Therefore $\pi(g^2) = \operatorname{diag}(\varepsilon^2, \varepsilon^{-2})$ centralizes $\pi(G_v)^0$. Because q is odd, $\varepsilon^2 \neq \varepsilon^{-2}$ and therefore T_1 is the centralizer of $\pi(g^2)$ in H_1. Because $\pi(G_v)^0$ and T_1 are one-dimensional tori we have $\pi(G_v)^0 = T_1$. From this point on we use the same argument as in the case $\pi(G_v) = H_1$ and arrive at a contradiction.

9) Thus we have proved that G_v is finite. We shall now find an upper bound on its order. To do this we need similar information about $Z_{G_v}(A)$.

Because $\lambda_1(\varepsilon) \in \pi(Z_{G_v}(A)) \subset T_1$ and $|G_v| < \infty$, we see that $\pi(Z_{G_v}(A))$ is a finite cyclic group generated by an element of the form $\lambda_1(\theta) = \operatorname{diag}(\theta, \theta^{-1})$, where $\theta = e^{2\pi i/qm}$, $m \in \mathbb{Z}^+$. Thus in $Z_{G_v}(A)$ we can find an element of the form $z = \lambda_1(\theta) \times \prod_{j=1}^{p} \operatorname{diag}(\theta^{b_{j,0}}, \ldots, \theta^{b_{j,r_j}}) \in T_1 \times \prod_{j=1}^{p} B_j = T$ for some $b_{l,j} \in \mathbb{Z}$. For any $w = v_{1,l} \otimes u_{1,j_1} \otimes \cdots \otimes u_{p,j_p}$ we have

$$zw = \theta^{n_1-2l+b_{1,j_1}+\cdots+b_{p,j_p}}w. \tag{9.7}$$

Because $zv = v$, we can argue as in 7) and deduce from (9.7) that if (9.2) holds for w then

$$n_1 - 2l + b_{1,j_1} + \cdots + b_{p,j_p} \equiv 0 \mod qm. \tag{9.8}$$

We now impose yet another condition on c: we assume that $c > 3q$. Then $n_1 \geq 3q$ and, because both n_1 and q are odd, there exist l' and $l'' \in \mathbb{N}$, $l', l'' \leq n_1$, such that $n_1 - 2l' = q$ and $n_1 - 2l'' = 3q$. Therefore if we set $j_1 = \cdots = j_p = 0$ and $l = l'$ or $l = l''$ then (in view of $a_{l,0} = 0$) (9.2) would hold. Therefore (9.8) holds, i.e., $q + \sum_{j=1}^p b_{j,0} \equiv 0 \mod qm$ and $3q + \sum_{j=1}^p b_{j,0} \equiv 0 \mod qm$. Hence $2q \equiv 0 \mod qm$ and therefore either $m = 1$ or $m = 2$. Thus $\pi(Z_{G_v}(A))$ is a cyclic group of order $\leq 2q$.

Now we consider $\pi(G_v)$. This is a finite subgroup of $H_1 = \mathrm{SL}_2$, i.e., a binary polyhedral group (see [48]) containing the cyclic group of order q generated by $\pi(g)$. Because, by assumption, $q > 10$ and the binary tetrahedral, octahedral, and icosahedral groups do not contain elements of order > 10, we see that $\pi(G_v)$ is either a cyclic group or a binary dihedral group. We now examine each possibility.

If $\pi(G_v)$ is a cyclic group then, because $\operatorname{Ker}\pi \subset Z_{G_v}$, we see that G_v is a central extension of the abelian group $\operatorname{Ker}\pi$ by the cyclic group $\pi(G_v)$. Therefore G_v itself is abelian (see [24]). Therefore $G_v = Z_{G_v}(A)$. Because $|\pi(Z_{G_v}(A))| \leq 2q$, and $|\operatorname{Ker}\pi| \leq \prod_{j=1}^p (r_i + 1)$ we have in this case $|G_v| \leq 2q \prod_{j=1}^p (r_j + 1)$.

Suppose now that $\pi(G_v)$ is a binary dihedral group, i.e., $\pi(G_v)$ is conjugate in $H_1 = \mathrm{SL}_2$ to the group D_n for some n (see part 5 of §2).

Notice that any element of D_n of the form $a_n^l b$ has order 4 and therefore all elements of order > 4 are of the form a_n^l. On the other hand, it is easy to see that if a_n^l has order > 4 then the centralizer of a_n^l in D_n is the cyclic group generated by a_n. Thus the centralizer of any element of order > 4 in D_n is a cyclic group of order $2n$.

Because $\pi(Gv)$ contains the element $\pi(g)$ of order q and, by assumption, $q > 4$ we see that the centralizer of $\pi(g)$ in $\pi(Gv)$ is a cyclic group B of order $2n$. Thus $\pi^{-1}(B)$ is a central extension of the abelian group $\operatorname{Ker}\pi$ by the cyclic group B. Therefore $\pi^{-1}(B)$ is an abelian group (see [24]). Because $A \subset \pi^{-1}(B)$, it means that $\pi^{-1}(B) \subset Z_{G_v}(A)$. It follows easily that $B = \pi(Z_{G_v}(A))$. Therefore $|B| = 2n = |\pi(Z_{G_v}(A))| \leq 2q$ so that $n \leq q$. Hence $|\pi(G_v)| = |D_n| = 4n \leq 4q$. Because G_v is an extension of $\operatorname{Ker}\pi$ by $\pi(G_v)$, we see that in this case $|G| \leq 4q \prod_{j=1}^p (r_j + 1)$.

10) Thus we have proved that regardless of the value of n_1 the group G_v is a finite group of order less than or equal to $4q \prod_{j=1}^p (r_j + 1)$.

Up to isomorphism there are only finitely many groups of order $\leq 4q \prod_{j=1}^p (r_j + 1)$. By Theorem 1.3, for each one of them there are, up to isomorphism, only finitely many modules (without nonzero fixed vectors) whose algebras of invariants are of dimension $\leq d$. Let K be the maximum of dimensions of those modules taken over all aforementioned finite groups; K depends only on $q_1 G$, and d.

Because there are n_1+1 odd integers between $-n_1$ and n_1, and at most $[n_1/q]+1$ of them are divisible by q, we deduce from (9.1) and (9.2) that there are at least $n_1-[n_1/q]$ different values of l, $0 \leq l \leq n_1$, for which the vectors $v_{1,l} \otimes u_{1,0} \otimes \cdots \otimes u_{p,0} \in V$ are eigenvectors of g with eigenvalue different from 1. All these vectors are linearly independent. Because $g \in G_v$, it means that $\dim V - \dim V^{G_v} \geq n_1 - [n_1/q] \geq n_1(q-1)/q$. Consider the slice representation (G_v, N_v). It follows from (0.2) and from the obtained inequality that $\dim N_v - \dim N_v^{G_v} \geq (\dim V - \dim V^{G_v}) - (\dim \mathfrak{g} - \dim \mathfrak{g}^{G_v}) \geq n_1(q-1)/q+1-\dim G$. We now assume that $c \geq (K+\dim G-1)q/(q-1)+2$. It follows from $n_1+1 \geq c$ that $\dim N_v - \dim N_v^{G_v} > K$. Because $N_v^{G_v}$ has a G_v-invariant complement in N_v, we deduce from the definition of K and from Theorem 1.2 that $\operatorname{hd} k[R(\Lambda)]^G \geq \operatorname{hd} k[N_v]^{G_v} \geq \operatorname{hd} k[N_v/N_v^{G_v}]^{G_v} > d$.

This completes the proof of the main theorem.

§10. Examples

1. In the conclusion we will show with examples how to use the method developed in this chapter to obtain an explicit classification of modules with prescribed homological dimension of their algebras of invariants.

2. Algebras of invariants of small homological dimensions for $G = \mathrm{SL}_2$ (the algebras of invariants of binary forms). First we consider irreducible G-modules $V = S_n$, $n \geq 1$ (i.e., invariants of a binary form of degree n).

Minimal systems of generators of algebras $k[S_n]^G$ for $n \leq 6$ were already determined in the classical papers (see [48, 76]): when $n = 1, 2, 3, 4$ they are free (and are of dimensions $0, 1, 1, 2$, respectively), and when $n = 5, 6$ they are hypersurfaces (of dimensions 3 and 4). According to Shioda [143], the six-dimensional algebra $k[S_8]^G$ has a minimal system of homogeneous generators consisting of 9 elements and a minimal resolution of length 3 with Betti numbers $b_0 = b_3 = 1$, $b_1 = b_2 = 5_G$. Thus $\operatorname{hd} k[S_8]^G = 3$ (unlike the previous cases, $k[S_8]^G$ is not a complete intersection).

THEOREM 10.1. *The algebra of invariants of a binary form of degree $n \geq 1$ has homological dimension $d \leq 10$ exactly in the following cases:*

d	0	1	2	3	4	5	6	7	8	9	10
n	1, 2, 3, 4	5, 6	none	8	none	none	none	none	none	none	none

PROOF. It follows from the remarks above and from Propositions 2.4 and 2.5 that if $\operatorname{hd} k[S_n]^G \leq 10$, $n \geq 1$, then $n \leq 12$, $n \neq 9, 11$. According to what was said above, we only have to consider the cases $n = 7, 10, 12$.

When $n = 7, 10, 12$ the Poincaré series of $k[S_n]^G$ were determined in [65] and they are, respectively,

$$\begin{gathered} 1+x^4+4x^8+10x^{12}+4x^{14}+18x^{16}+13x^{18}+35x^{20}+\cdots \\ 1+x^2+2x^4+6x^6+12x^8+5x^9+24x^{10}+13x^{11}+48x^{12}+33x^{13}+\cdots \\ 1+x^2+x^3+2x^4+3x^5+7x^6+10x^7+18x^8+26x^9+\cdots \end{gathered} \tag{10.1}$$

(dots indicate higher order terms). It is not difficult to deduce, from (10.1) and from the algorithm in part 3 of §1, Chapter 2 for the construction of a minimal homogeneous system of generators of the algebra of invariants, that $\operatorname{ed} k[S_n]^G \geq 23, 42, 38$ when, respectively, $n = 7, 10, 12$. Because $\dim k[S_n]^G = 5, 8, 10$ when $n = 7, 10, 12$, in view of (0.1), we have, respectively, $\operatorname{hd} k[S_n]^G \geq 18, 3428$. □

REMARK. This theorem shows that $\operatorname{hd} k[S_n]^G$ is not a monotone function of n.

We shall now consider reducible G-modules V. Those with free algebras of invariants have already been determined in [2, 94, 135]: $2S_1$, $S_1 \oplus S_2$, $2S_2$, and $3S_1$. They were already studied in [76, 132]. We shall make the next step and determine all V for which the algebra of invariants is a hypersurface.

Let $V = \bigoplus_{j=1}^p W_j$, where $W_j = S_{n_j}$, $n_j \geq 1$ for all j; $p \geq 2$, and $k[V]^G$ is a hypersurface. We consider W_j as a submodule of V. By Theorem 10.1 and Proposition 1.4, among the numbers n_j at most one can be equal to 5 or 6.

Suppose that one of the n_j (say, n_1) is equal to 6. Let $v \in W_1 = S_6$, $v = x^6 - y^6$. By Lemma 2.1, the orbit Gv is closed in V and $G_v = D_3$. The element $a_3 \in D_3$ acts nontrivially on S_n for $1 \leq n \leq 4$. The adjoint G-module is S_2. The multiplicity of the eigenvalue $e^{2\pi i/3}$ of a_3 is equal to 1 on S_2 and 2 on S_6, so that, by (0.2), (G_v, N_v) is nontrivial. This contradicts part e) of Proposition 1.4. Therefore $n_j \leq 5$ for all j.

Suppose now that one of the n_j (say n_1) is equal to 5. Let $v \in W_1 = S_5$, $v = x^5 - y^5$. By Lemma 2.1, the orbit Gv is closed in V and $G_v = T_5$. The group T_5 acts nontrivially on S_n for $1 \leq n \leq 4$. The element $a_5^2 \in T_5$ has eigenvalue $e^{2\pi i/5}$ on S_5, and does not have this eigenvalue on S_2. Therefore (G_v, N_{v, W_1}) is nontrivial. As above, this contradicts Proposition 1.4. Hence $n_j \leq 4$ for all j.

Suppose now that one of the n_j (say n_1) is equal to 2. Let $v \in W_1 = S_2$, $v = xy$. Then Gv is closed in V and $G_v = T$ (see §3, Chapter 6). By the corollary of Theorem 1.2 and Proposition 2.1, we have $1 \geq \operatorname{hd} k[\bigoplus_{j=2}^p W_j]^T \geq (\sum_{j=2}^p [(n_j+1)/2] - 1)^2$. The G-modules V satisfying this inequality are exactly $S_2 \oplus S_3$; $S_2 \oplus S_4$; $2S_2$; $S_1 \oplus S_2$; $3S_2$; $S_1 \oplus 2S_2$; $2S_1 \oplus S_2$. Of the modules in this list $2S_2$ and $S_1 \oplus S_2$ have, as was mentioned above, free algebras of invariants. In the remaining cases the algebras of invariants were described by the classics (see [76, 132, 64]), and they all turned out to be hypersurfaces.

Suppose now that n_j is either 1 or 3 or 4 for all j and one of the n_j (say, n_1) is equal to 3. Let $v \in W_1 = S_3$, $v = x^3 - y^3$. By Lemma 2.1, Gv is closed in V and $G_v = T_3$. It follows easily from (0.2) that (G_v, N_{v,W_1}) is the trivial one-dimensional representation. By Proposition 1.4, at most one of the representations (Gv, W_j), $j \geq 2$ is nontrivial. Because the representations (T_3, S_n), $n = 1, 3, 4$ are all nontrivial, we have $p = 2$. Moreover, by Proposition 1.4, the image (T_3, S_{n_2}) must be generated by those g for which $\operatorname{rk}(g - \mathrm{Id}) = 2$. Therefore n_2 is either 1 or 3, and V is one of the modules $S_1 \oplus S_3$ and $2S_3$. Their algebras of invariants were also described in classical papers (see [76, 64]): $k[S_1 \oplus S_3]^G$ is a hypersurface and $k[2S_3]^G$ has a minimal system of homogeneous generators consisting of 7 elements, whence $\operatorname{hd} k[2S_3]^G = 2$.

Suppose now that n_j is equal to either 1 or 4 for any j and one of the n_j (say, n_1) is equal to 4. By Lemma 2.1 there exists a point $v \in W_1 = S_4$ such that Gv is closed and $G_v = D_2$. It follows easily from (0.2) that (D_2, N_{v,W_1}) is the trivial two-dimensional representation. Because (D_2, S_n) is nontrivial for $n = 1$ and $n = 4$ we deduce from Proposition 1.4 that $p = 2$. Thus V is either $S_1 \oplus S_4$ or $2S_4$. Again these cases can be found in the classical literature (see [76, 132, 64]): their algebras of invariants are also hypersurfaces.

It remains to consider the case when $n_j = 1$, i.e., the case of invariants of a system of p vectors in 2-space. By classical invariant theory (see [11]), $\operatorname{ed} k[V]^G = p(p-1)/2$. Because $\dim V/G = 2p - 3$, we have $\operatorname{hd} k[V]^G = (p-2)(p-3)/2$, which is equal to 1 only when $p = 4$. Thus we have proved

THEOREM 10.2. *The following is a complete list of* SL_2*-modules* V *whose algebras of invariants are hypersurfaces*: S_5; S_6; $S_2 \oplus S_3$; $S_2 \oplus S_4$; $2S_1 \oplus S_2$; $S_1 \oplus 2S_2$; $3S_2$; $S_1 \oplus S_3$; $S_1 \oplus S_4$; $2S_4$; $4S_1$.

REMARK. One can proceed in the same manner and classify representations with $\operatorname{hd} k[V]^{\mathrm{SL}_2} = d = 2, 3, \ldots$ (with the growth of d the only problems that appear are of a technical nature, see part 2 of §0). For example, when $d = 2$ the list consists of only one module $2S_3$, and when $d = 3$ there are two modules in the list: S_8 and $5S_1$.

3. Algebras of invariants of exceptional simple algebraic groups. We shall find a lower bound on the homological dimension of algebras of invariants of irreducible representations of such groups. A general conclusion based on our investigations is that algebras of invariants of the representations different from the adjoint and simplest are quite complicated; in particular they are never hypersurfaces.

We assume that the groups under consideration are connected and simply connected.

a) Invariants of $G = E_8$.

We investigate this case in detail because the arguments used here underscore the role of the assertions from §§3 and 4 in the problems of explicit classification and explain the necessity of the detailed information obtained in the lemmas of §§5 and 6. Moreover we will use those arguments in the remaining cases as well (the difference will be only of a technical nature and there we will only indicate the key relations and answers).

Let T and C be the same as in Lemma 6.1. Then $Z_G(T)/T$ is the group of type E_7.

By [52, 26], if $M \neq \omega_1, \omega_7$ then the generic stabilizer for $(E_7, R(M))$ is contained in the noneffectivity kernel. Let V be an irreducible G-module. It follows from Lemma 6.1 and Theorem 4.1 that if the highest weight of V is different from 0 and ω_8 then $(Z_G(T)/T, V^T) \supset R(M)$, where $M \neq 0, \omega_1, \omega_7$, and therefore the generic stabilizer for $(Z_G(T)/T, V^T)$ is contained in the noneffectivity kernel. Because $Z_G(T)/T$ is connected and simple, it follows from Theorem 3.3 that for such V all conditions of Theorem 3.1 are satisfied. Henceforth we will assume that $V = R(\Lambda)$, $\Lambda \neq 0, \omega_8$.

By Theorem 3.1, if $v \in V^T$ is a point in general position, then $(G_v)^0$ is a reductive group with T as a maximal torus. Let Λ be any of the weights ω_i, $i = 1, 2, 5, 7$; $2\omega_i$, $i = 3, 4, 6, 8$; $\omega_3 + \omega_i$, $i = 4, 6, 8$; $\omega_4 + \omega_i$, $i = 6, 8$; $\omega_6 + \omega_8$.

We shall now show that $(G_v)^0 = T$. Suppose this is not the case and $\dim G_v = 3$. Then both C and $(G_v)^0$ are three-dimensional simple subgroups containing T. Lemma 6.1 shows that C has nontrivial even-dimensional representations and therefore (see §3) $C \cong \mathrm{SL}_2$. Because $\Lambda \neq \omega_8$, we have, by [52], that the generic stabilizer for (G, V) is finite. But then it follows from Proposition 3.4 that $(G_v)^0 \cong \mathrm{SL}_2$. Hence $C \cong (G_v)^0$. Therefore by Lemma 3.2 we can replace v with a suitable point from $Z_G(T)v$ and assume that $(G_v)^0 = C$. But then it follows from Lemma 6.1, formula (0.2) and Theorem 4.1 that for the chosen Λ the module $((G_v)^0, N_v)$ has a nontrivial irreducible submodule of odd dimension. This contradicts Proposition 3.4. Therefore for the Λ listed above we have $(G_v)^0 = T$.

Because $Z_G(T)/T$ is simple, it follows from Proposition 4.1 that if Λ majorizes one of those weights, i.e., $\Lambda \neq \omega_i$, $i = 3, 4, 6, 8$ then all conditions of the second part of Theorem 3.2 are satisfied. If Λ is one of the weights ω_3, ω_4, or ω_6 then, as was shown before, all conditions of the first part of that theorem are satisfied. Thus if we want to apply this theorem we only need to calculate $q = q_{(C, V)} - q_{(C, \mathfrak{g})}$.

It follows from Lemma 6.1 that $q_{(C, \mathfrak{g})} = 57$. On the other hand, because, by Lemma 6.1, $R(\omega_i)^C \neq 0$ for all i we deduce from Theorem 4.1 that $q_{(C, R(\Lambda))} \geq q_{(C, R(M))}$ when Λ majorizes M. Moreover, we deduce from

Lemma 6.1 and Theorem 4.1 that the C-module $V = R(\Lambda)$ contains the following submodules: $86184(S_6 \oplus S_3)$ for $\Lambda = 2\omega_3$; $23760S_{10} \oplus 365750S_5$ for $\Lambda = 2\omega_4$; $27664(S_6 \oplus S_3)$ for $\Lambda = 2\omega_6$; $S_4 \oplus 133(S_2 \oplus S_1) \oplus 56S_3$ for $\Lambda = 2\omega_8$; $86184S_8 \oplus 8645S_5 \oplus 365750S_3$ for $\Lambda = \omega_3 + \omega_4$; $86184(S_6 \oplus S_3)$ for $\Lambda = 2\omega_6$; $86184(S_5 \oplus S_4 \oplus S_3) \oplus 8645(S_2 \oplus S_1)$ for $\Lambda = \omega_3 + \omega_8$; $27664S_8 \oplus 23760S_5 \oplus 365750S_3$ for $\Lambda = \omega_4 + \omega_6$; $23760(S_7 \oplus S_6 \oplus S_5) \oplus 27664(S_2 \oplus S_1)$ for $\Lambda = \omega_4 + \omega_8$; $27664(S_5 \oplus S_4 \oplus S_3) \oplus 1539(S_2 \oplus S_1)$ for $\Lambda = \omega_6 + \omega_8$. From this (and from Lemma 6.1) we obtain the following bounds: $q_{(C, R(\omega_1))} = 1101$; $q_{(C, R(\omega_2))} \geq 8645$; $q_{(C, R(\omega_3))} \geq 172368$; $q_{(C, R(\omega_4))} \geq 71280$; $q_{(C, R(\omega_5))} \geq 731500$; $q_{(C, R(\omega_6))} \geq 55328$; $q_{(C, R(\omega_7))} \geq 9294$; $q_{(C, R(2\omega_3))} \geq 430920$; $q_{(C, R(2\omega_4))} \geq 1216050$; $q_{(C, R(2\omega_6))} \geq 138320$; $q_{(C, R(2\omega_8))} \geq 380$; $q_{(C, R(\omega_3+\omega_4))} \geq 1102171$; $q_{(C, R(\omega_3+\omega_6))} \geq 430920$; $q_{(C, R(\omega_3+\omega_8))} \geq 620578$; $q_{(C, R(\omega_4+\omega_6))} \geq 913436$; $q_{(C, R(\omega_4+\omega_8))} \geq 292928$; $q_{(C, R(\omega_6+\omega_8))} \geq 196726$.

Thus if $\Lambda \neq \omega_i$, $i = 3, 4, 6, 8$, then $q \geq 323$ and, by Theorem 3.2, $\operatorname{hd} k[V]^G \geq 51681$. If Λ is one of the weights ω_3, ω_4, or ω_6, then $q \geq 55271$, and by the same theorem, $\operatorname{hd} k[V]^G \geq 9208$. Because the algebra of invariants of $R(\omega_8)$ is free (see [101]), we have a proof of

THEOREM 10.3. *The homological dimension of the algebra of invariants of a nontrivial irreducible E_8-module V is equal to* 0 *when V is the adjoint module and* ≥ 9208 *otherwise.*

b) Invariants of $G = E_7$.

Let T and C be the same as in Lemma 6.1. Then $Z_G(T)/T$ is of type D_6. Utilizing again the argument for E_8 and using the tables from [52, 26] we see that if $\Lambda \neq \omega_1, \omega_7$ and $V = R(\Lambda)$ then for (G, V) all conditions from the second part of Theorem 3.2 are satisfied (moreover the generic stabilizer for $Z_G(T)/T$ in V^T is contained in the noneffectivity kernel). It now follows from Lemma 6.1 and Theorem 4.1 that $q_{(C, R(\omega_1))} = 33$; $q_{(C, R(\omega_2))} = 264$; $q_{(C, R(\omega_3))} = 2738$; $q_{(C, R(\omega_4))} = 127325$; $q_{(C, R(\omega_5))} = 9064$; $q_{(C, R(\omega_6))} = 450$; $q_{(C, R(2\omega_1))} \geq 198$; $q_{(C, R(2\omega_7))} \geq 44$; $q_{(C, R(\omega_1+\omega_7))} \geq 121$.

This information is sufficient for our purposes, but to find better bounds we will improve the last three inequalities. To this end we use the following isomorphisms of E_7-modules (see [15]): $S^2R(\omega_7) = R(2\omega_7) \oplus R(\omega_1)$; $S^2R(\omega_1) = R(2\omega_1) \oplus R(\omega_6) \oplus 1$; $R(\omega_1) \otimes R(\omega_7) = R(\omega_1 + \omega_7) \oplus R(\omega_2) \oplus R(\omega_7)$. The structure of the C-module $R(\omega_i)$ is described in Lemma 6.1. Thus from the first isomorphism we have another isomorphism of C-modules: $S^2(32S_0 \oplus 12S_1) = (C, R(2\omega_7)) \oplus (66S_0 \oplus 32S_1 \oplus S_2)$. The left-hand side of this relation is of the form $32S^2S_0 \oplus 32 \cdot 12(S_0 \otimes S_1) \oplus \binom{32}{2}(S_0 \otimes S_0) \oplus \binom{12}{2}(S_1 \otimes S_1) \oplus 12S^2S_1 = 594S_0 \oplus 384S_1 \oplus 78S_2$, whence $(C, R(2\omega_7)) = 528S_0 \oplus 352S_1 \oplus 77S_2$. Similarly, from the other two isomorphisms we have

$(C, R(2\omega_1)) = 2134S_0 \oplus 1760S_1 \oplus 528S_2 \oplus 32S_3 \oplus S_4$, $(C, R(\omega_1 + \omega_7)) = 2112S_0 \oplus 1584S_1 \oplus 384S_2 \oplus 12S_3$. Therefore $q_{(C, R(2\omega_1))} = 2353$, $q_{(C, R(2\omega_7))} = 429$, $q_{(C, R(\omega_1+\omega_7))} = 1992$.

Therefore, for $\Lambda \neq \omega_1, \omega_7$ we have $q \geq 231$, and, by Theorem 3.2, $\operatorname{hd} k[V]^G \geq 26335$. Because the algebras of invariants of $R(\omega_1)$ and $R(\omega_7)$ are free (see [94]) we have a proof of

THEOREM 10.4. *The homological dimension of the algebra of invariants of a nontrivial irreducible E_7-module V is equal to 0 if V is the simplest or adjoint module, and ≥ 26335 otherwise.*

c) Invariants of $G = E_6$.

Let T and C be defined as in Lemma 6.1. Then the group $Z_G(T)/T$ is of type A_5. As before, if $\Lambda \neq \omega_1, \omega_2, \omega_6$ and $V = R(\Lambda)$ then for (G, V) all conditions of the second part of Theorem 3.2 are satisfied (and the generic stabilizer for $(Z_G(T)/T, V^T)$ is contained in the noneffectivity kernel). It remains to find a bound for q. With the aid of Lemma 6.1 and Theorem 4.1 we have $q_{(C, R(2\omega_1))} = q_{(C, R(2\omega_6))} \geq 21$; $q_{(C, R(\omega_2))} \geq 21$; $q_{(C, R(2\omega_2))} \geq 112$; $q_{(C, R(\omega_3))} = q_{(C, R(\omega_5))} = 105$; $q_{(C, R(\omega_4))} = 965$; $q_{(C, R(\omega_1+\omega_6))} \geq 21$; $q_{(C, R(\omega_1+\omega_2))} = q_{(C, R(\omega_2+\omega_6))} \geq 67$.

Next we use the following isomorphisms of E_6-modules (see [15]): $S^2R(\omega_1) = R(2\omega_1) \oplus R(\omega_6)$; $R(\omega_1) \otimes R(\omega_6) = R(\omega_1 + \omega_6) \oplus R(\omega_2) \oplus 1$; $R(\omega_1) \otimes R(\omega_2) = R(\omega_1 + \omega_2) \oplus R(\omega_5) \oplus R(\omega_1)$. Similar to the E_7 case, we now deduce from Lemma 6.1 that $(C, R(2\omega_1)) = 120S_0 \oplus 84S_1 \oplus 21S_2$; $(C, R(\omega_1+\omega_6)) = 225S_0 + 160S_1 \oplus 35S_2$; $(C, R(\omega_1+\omega_2)) = 504S_0 \oplus 420S_1 \oplus 120S_2 \oplus 6S_3$. Therefore $q_{(C, R(2\omega_1))} = q_{(C, R(2\omega_6))} = 105$, $q_{(C, R(\omega_1+\omega_6))} = 195$, $q_{(C, R(\omega_1+\omega_2))} = q_{(C, R(\omega_2+\omega_6))} = 552$. Combining this with the bounds obtained before, we see that when $\Lambda \neq \omega_1, \omega_2, \omega_6$ we have $q \geq 84$ and therefore $\operatorname{hd} k[V]^G \geq 3403$.

Because $k[R(\omega_i)]^G$, $i = 1, 2, 6$ is free (see [94]) we have a proof of

THEOREM 10.5. *The homological dimension of the algebra of invariants for a nontrivial irreducible E_6-module V is equal to 0 if V is the simplest, its contragradient or adjoint module, and ≥ 3403 otherwise.*

d) Invariants of $G = F_4$.

Again T and C are defined as in Lemma 6.1. As before, we have that when $\Lambda \neq \omega_1, \omega_4$ then for (G, V), where $V = R(\Lambda)$, all conditions of the second part of Theorem 3.2 are satisfied (and the generic stabilizer for $Z_G(T)/T$ on V^T is contained in the noneffectivity kernel). From Lemma 6.1 and Theorem 4.1 we have that $q_{(C, R(\omega_1))} = 15$; $q_{(C, R(\omega_2))} = 434$; $q_{(C, R(\omega_3))} = 84$; $q_{(C, R(2\omega_1))} \geq 72$; $q_{(C, R(2\omega_4))} \geq 20$; $q_{(C, R(\omega_1+\omega_4))} \geq 40$. We also have the isomorphisms of F_4-modules (see [15]): $S^2R(\omega_4) = R(2\omega_4) \oplus R(\omega_4) \oplus 1$;

$S^2R(\omega_1) = R(2\omega_1) \oplus R(2\omega_4) \oplus 1$; $R(\omega_1) \otimes R(\omega_4) = R(\omega_1 + \omega_4) \oplus R(\omega_3) \oplus R(\omega_4)$. Therefore $(C, R(2\omega_4)) = 105S_0 \oplus 78S_1 \oplus 21S_2$, $(C, R(2\omega_1)) = 217S_0 \oplus 230S_1 \oplus 105S_2 \oplus 14S_3 \oplus S_4$; $(C, R(\omega_1 + \omega_4)) = 273S_0 \oplus 252S_1 \oplus 84S_2 \oplus 6S_3$, so that $q_{(C, R(2\omega_4))} = 99$, $q_{(C, R(2\omega_1))} = 365$, $q_{(C, R(\omega_1+\omega_4))} = 348$. It now follows from the bounds obtained before that for $\Lambda \neq \omega_1, \omega_4$ we have $q \geq 69$ and therefore $\operatorname{hd} k[V]^G \geq 2278$.

Because $k[R(\omega_i)]^G$, $i = 1, 4$ is free (see [94]) we have a proof of

THEOREM 10.6. *The homological dimension of the algebra of invariants for a nontrivial irreducible F_4-module V is equal to 0 when V is the simplest or adjoint module, and ≥ 2278 otherwise.*

e) Invariants of $G = G_2$.

Again we define T and C as in Lemma 6.1. Then $Z_G(T)/T$ is of type A_1. By [26, 27], of all SL_2-modules only $R(n\omega_1)$, $n = 1, 2, 3, 4$, have their generic stabilizers not contained in the their noneffectivity kernels. We need the following isomorphism of G_2-modules (see [15]): $S^2R(\omega_1) = R(2\omega_1) \oplus 1$ from which we deduce that (in the notation of part 2 §6) the weights of $(G, R(2\omega_1))$ are of the form $\pm 2\varepsilon_i$, $\pm\varepsilon_i$, 0, $\varepsilon_i - \varepsilon_j$ where $\pm 2\varepsilon_i$ and $\varepsilon_i - \varepsilon_j$ are simple, $\pm\varepsilon_i$ have multiplicity 2, and 0 has multiplicity 3. Because $R(2\omega_1)^T$ is the linear hull of the weight subspaces corresponding to $\pm 2\varepsilon_2$, $\pm\varepsilon_2$ and 0, the $Z_G(T)/T$-module $R(2\omega_1)^T$ is isomorphic to $R(4\omega_1) \oplus R(2\omega_1) \oplus 1$. Hence, in view of Lemma 6.1, if $\Lambda \neq \omega_1\omega_2$ and $V = R(\Lambda)$ then all conditions of Theorem 3.1 are satisfied for V (and the generic stabilizer for $(Z_G(T)/T, V^T)$ is contained in the noneffectivity kernel). The same argument as in the E_8 case shows that for such Λ all conditions of the second part of Theorem 3.2 are also satisfied. Using the G_2-module isomorphisms (see [15]) $R(\omega_1) \otimes R(\omega_2) = R(\omega_1 + \omega_2) \oplus R(2\omega_1) \oplus R(\omega_1)$ and $S^2R(\omega_2) = R(2\omega_2) \oplus R(2\omega_1) \oplus 1$ we have $(C, R(2\omega_1)) = 6S_0 \oplus 6S_1 \oplus 3S_2$; $(C, R(2\omega_2)) = 6S_0 \oplus 10S_1 \oplus 10S_2 \oplus 4S_3 \oplus S_4$ and $(C, R(\omega_1 + \omega_2)) = 8S_0 \oplus 12S_1 \oplus 8S_2 \oplus 2S_3$. It now follows from Lemma 6.1 that $q_{(C, R(\omega_2))} = 5$, $q_{(C, R(2\omega_1))} = 9$, $q_{(C, R(2\omega_2))} = 29$, $q_{(C, R(\omega_1+\omega_2))} = 24$. Hence if $\Lambda \neq \omega_1, \omega_2$ then $q \geq 4$ and therefore $\operatorname{hd} k[V]^G \geq 3$.

Because $k[R(\omega_i)]^G$, $i = 1, 2$, is free (see [94]) we have a proof of

THEOREM 10.7. *The homological dimension of the algebra of invariants of a nonzero irreducible G_2-module V is equal to 0 when V is the simplest or adjoint module, and ≥ 3 otherwise.*

CHAPTER 5

Representations with Free Modules of Covariants

In this chapter we assume that the characteristic of the field k is zero. We also keep the notation and conventions from part 5 of §0, Chapter 4.

§1. Connections with equidimensionality: finiteness theorems

1. If G is a reductive group and V is a nonzero finite-dimensional G-module then the graded $k[V]^G$-module $k[V]$ is called the module of covariants of the G-module V (or of the associated with V representation $\rho: G \to \mathrm{GL}(V)$). We call V (or ρ) equidimensional if $\pi_{V,G}$ is an equidimensional morphism. It is clear that the equidimensionality of the G-module V is equivalent to that of the G^0-module V.

Henceforth we will assume that G is a nontrivial connected simply connected semisimple group.

THEOREM 1. *The following properties of the module V are equivalent*: a) *its module of covariants is free,* b) *V is equidimensional and $k[V]^G$ is a free algebra.*

PROOF. a) $\Rightarrow$ b). Because the module of covariants is free, the morphism $\pi_{V,G}$ is flat, and therefore (because the varieties V and V/G are irreducible) equidimensional (see [112]). Because V is nonsingular and $\pi_{V,G}$ is flat, the variety V/G is also nonsingular (see [112]). But then, by Corollary 2 of Proposition 1.1, the algebra $k[V]^G$ is free.

b) $\Rightarrow$ a). Because $k[V]^G$ is free, the variety V/G is nonsingular. Because V is a Cohen–Macaulay variety, the equidimensionality of $\pi_{V,G}$ implies that $\pi_{V,G}$ is a flat morphism (see [51]). Thus $k[V]$ is a flat $k[V]^G$-module. But because it is graded, it is also a free $k[V]^G$-module. □

It now follows easily that for any trivial G-module L the module of covariants of V is free if and only if the module of covariants of $V \oplus L$ is free. The same is true for equidimensionality (see part 5 of §3, Chapter 3). Because of this, from now on we will assume that $V^G = \{0\}$.

Theorem 1, along with Theorem 6 of Chapter 3, yields the following finiteness theorem.

THEOREM 2. *There exist only finitely many, up to isomorphism and addition of a trivial direct summand, finite-dimensional G-modules with free modules of covariants.*

2. In view of Theorem 1, the classification problem of representations with free module of covariants reduces to identifying equidimensional modules in the set of all G-modules with free algebra $k[V]^G$ (which, according to Chapter 3, can be explicitly determined in principle). Later we will obtain several methods for solution of this problem by considering a more general problem of classification of all equidimensional G-modules. As an application, we will describe all irreducible representations of simple groups with free modules of covariants.

3. LEMMA 1. *Let* $f_1, \dots, f_m \in k[\mathbb{A}^n]$ *be homogeneous polynomials of positive degrees and* $\tau: \mathbb{A}^n \to \mathbb{A}^m$, $\tau = (f_1, \dots, f_m)$. *Then* $\dim \tau^{-1}(\tau(0)) \geq \dim \tau^{-1}(\tau(x))$ *for all* $x \in \mathbb{A}^n$.

PROOF. Let $(t_0 : \cdots : t_n)$ be homogeneous coordinates on $\mathbb{P}^n$ and $\Gamma = \{p \in \mathbb{P}^n \mid t_0(p) = 0\}$. We assume that $\mathbb{A}^n = \mathbb{P}^n - \Gamma$ and $\{x_i = t_i/t_0 \mid i = 1, \dots, n\}$ are coordinates on $\mathbb{A}^n$. If $a = (a_1, \dots, a_m) \in \mathbb{A}^m$, then $\tau^{-1}(a)$ is given by the system of equations $\{f_i(x_1, \dots, x_n) = a_i \mid 1 \leq i \leq m\}$. Because each f_i is homogeneous, the closure $\overline{\tau^{-1}(a)}$ in $\mathbb{P}^n$ of this fiber is contained in the variety given by the system $\{f_i(t_1, \dots, t_n) = a_i t_0^{\deg f_i} \mid i = 1, \dots, m\}$. Because $\overline{\tau^{-1}(a)} - \tau^{-1}(a) = \overline{\tau^{-1}(a)} \cap \Gamma$, we have that $\dim (\overline{\tau^{-1}(a)} - \tau^{-1}(a)) = \dim \overline{\tau^{-1}(a)} - 1 = \dim \tau^{-1}(a) - 1$. But $\overline{\tau^{-1}(a)} \cap \Gamma$ is contained in the subvariety Z of Γ given by the system $\{f_i(t_1, \dots, t_n) = 0 \mid i = 1, \dots, m\}$ (we identify Γ with $\mathbb{P}^{n-1}$ endowed with homogeneous coordinates t_i, $i = 1, \dots, n$). Therefore Z is the base of the cone $\tau^{-1}(\tau(0))$ so that $\dim Z = \dim \tau^{-1}(\tau(0)) - 1$. Therefore $\dim \tau^{-1}(\tau(0)) - 1 \geq \dim \tau^{-1}(a) - 1$. $\square$

In the notation of part 1) we set $m_G = \max_{v \in V} \dim Gv$ and $\pi_{V,G}^{-1}(\pi_{V,G}(0)) = \mathfrak{N}$.

THEOREM 3. 1) $\dim \mathfrak{N} \geq \dim \pi_{V,G}^{-1}(a) \geq m_G$ *for all* $a \in V/G$. 2) V *is equidimensional if and only if* $\dim \mathfrak{N} = m_G$. 3) *If each irreducible component of* $\mathfrak{N}$ *contains an open orbit then* V *is equidimensional.*

PROOF. Because G is semisimple, we have $m_G = \min_{a \in V/G} \dim \pi_{V,G}^{-1}(a)$ (see [30]). Assertions 1) and 2) follow now from Lemma 1. If condition 3) is satisfied then $\dim \mathfrak{N} \leq m_G$ so that the equidimensionality of V follows from 1). $\square$

COROLLARY. *If* $\mathfrak{N}$ *contains only finitely many orbits then* V *is equidimensional.*

PROOF. In this case condition 3) of Theorem 3 is satisfied because any orbit is open in its own closure. □

4. THEOREM 4. *The following properties of* V *are equivalent*: a) *its module of covariants is free*, b) *every minimal system of homogeneous generators of* $k[V]^G$ *is a regular* $k[V]$*-sequence.*

PROOF. Because G is semisimple, we have $\dim V/G = \dim V - m_G$ (see [30]). Suppose a) holds and let $f_1, \dots, f_s$ be a minimal system of homogeneous generators of $k[V]^G$. By Theorem 1, $s = \dim V/G$ and, by Theorems 1 and 3, $s = \operatorname{codim} \mathfrak{N}$. On the other hand, $\mathfrak{N} = \{v \in V \mid f_i(v) = 0 \text{ for all } i\}$. Because $k[V]$ is Cohen–Macaulay, we have b) (see [112]). Conversely, assume b) and let $h_1, \dots, h_t$ be a minimal system of homogeneous generators of $k[V]^G$. Then $\mathfrak{N} = \{v \in V \mid h_i(v) = 0 \text{ for all } i\}$ and, because $k[V]$ is Cohen–Macaulay, we have $t = \operatorname{codim} \mathfrak{N}$. But the elements of a regular sequence are algebraically independent (see [112], so that $k[V]^G$ is free, $t = \dim V/G$, and therefore $\dim \mathfrak{N} = m_G$. The assertion follows now from Theorem 3. □

PROPOSITION 1. *If* $\dim V/G \leq 2$ *then the module of covariants is free.*

PROOF. This is obvious when $\dim V/G = 0$. If $\dim V/G = 1$ then $V/G \cong \mathbb{A}^1$ (see [38]) and the assertion follows from Theorem 4. If $\dim V/G = 2$ then $V/G \cong \mathbb{A}^2$ (see [98]). Let f and g be homogeneous generators of $k[V]^G$. Because G is connected and semisimple and $k[V]$ is factorial, it follows that any irreducible polynomial dividing an invariant is an invariant itself (see [29]). It now follows from the algorithm for finding a minimal system of homogeneous generators of $k[V]^G$ (see part 3 of §1, Chapter 2) that f and g are irreducible. Therefore the desired assertion follows from Theorem 4 and the factoriality of $k[V]$. □

5. We shall now calculate $\dim \mathfrak{N}$.

Let T be a maximal torus in G and $\mathfrak{X} = \operatorname{Hom}(T, k^*) \otimes_{\mathbb{Z}} \mathbb{Q}$ (we view $\operatorname{Hom}(T, k^*)$ as a lattice in $\mathfrak{X}$). Let Σ be the system weights of T in V. For any $f \in \mathfrak{X}^*$ we set $\Sigma(f) = \{\Lambda \in \Sigma \mid f(\Lambda) > 0\}$ and $V(f) = \bigoplus_{\Lambda \in \Sigma(f)} V_\Lambda \subset V$. Let Δ be the root system of G with respect to T, $\{\alpha_1, \dots, \alpha_r\}$ be a fixed system of simple roots in Δ, and Δ_+ be the corresponding system of positive roots in Δ. The set $\{d\alpha \mid \alpha \in \Delta\} \subset \mathfrak{t}^*$ is the root system of $\mathfrak{g}$ with respect to $\mathfrak{t}$. The element $X_{d\alpha} \in \mathfrak{g}$ (see [9, 10]) will be denoted X_α. The one-dimensional unipotent subgroup of G corresponding to root α will be denoted U_α (see [50]). Let $\{\omega_1, \dots, \omega_r\}$ be the system of fundamental weights. Let $\pi_V \colon V \to V_\Lambda$ denote the projection parallel to $\bigoplus_{M \in \Sigma - \Lambda} V_M$. For any $v \in V$ we set $v_\Lambda = \pi_\Lambda(v)$. Also let W denote the Weyl group.

6. THEOREM 5. $\mathfrak{N} = \bigcup_{f \in \mathfrak{X}^*} GV(f)$.

PROOF. Because each fiber of $\pi_{V,G}$ contains a unique closed orbit, we have $\mathfrak{N} = \{v \in V \mid \overline{Gv} \ni 0\}$. It is clear that if $0 \in \overline{Tv}$ then $0 \in \overline{Gv}$. Conversely, suppose that $v \in \mathfrak{N}$. By the Hilbert–Mumford theorem (see [114, 21]), there exists a one-parameter group $\lambda: k^* \to G$ such that $\overline{\{\lambda(t)v \mid t \in k^*\}} \ni 0$. Because maximal tori in G are conjugate to each other, there exists $g \in G$ such that $g\lambda(t)g^{-1} \in T$ for all t. Therefore $0 \in \overline{Tgv}$. On the other hand (see [16, 99] and also part 4 of §1, Chapter 6), $\{v \in V \mid \overline{Tv} \ni 0\} = \bigcup_{f\in\mathfrak{X}^*} V(f)$. □

REMARKS. 1) Because $|\Sigma| < \infty$, it follows that $\bigcup_{f\in\mathfrak{X}^*} V(f)$ is the union of only finitely many subspaces.

2) If f and $f' \in \mathfrak{X}^*$ are in the same orbit of the Weyl group, it follows that $V(f)$ and $V(f')$ are in the same orbit of $N_G(T)$.

3) Because $|\Sigma| < \infty$, it follows that for any $f \in \mathfrak{X}^*$ there exists $f' \in \mathfrak{X}^*$, such that $f'(\Lambda) \neq 0$ for all $\Lambda \neq 0$, $\Lambda \in \Sigma$, and $V(f) \subset V(f')$. Therefore, $\mathfrak{N}$ is the union of the G-orbits of only those $V(f)$ for which $f(\Lambda) \neq 0$ for all $\Lambda \neq 0$, $\Lambda \in \Sigma$.

4) If $V(f) \neq \varnothing$ then $G_{V(f)}$ contains the parabolic subgroup corresponding to $\{\alpha \in \Delta \mid f(\alpha) \geq 0\}$. Therefore $GV(f)$ is closed in V (see [151]).

COROLLARY. 1) $\dim \mathfrak{N} \geq (\dim V - \dim V_0)/2 + 1$;
2) $\dim \mathfrak{N} \geq \dim V(f) + 1$ *for all* f.

PROOF. If $\mathfrak{N} = V(f)$ for some $f \in \mathfrak{X}^*$ then $V(f)$ is a nonzero G-invariant subspace of V on which the action of T is not stable. In view of Proposition 3.1 of Chapter 4 and the corollary of Proposition 2 of Chapter 3, this is impossible. Whence 2). Let now $f \in \mathfrak{X}^*$ and $f(\Lambda) \neq 0$ for all $\Lambda \neq 0$, $\Lambda \in \Sigma$. Because $V = \bigoplus_{\Lambda\in\Sigma} V_\Lambda$ one of the numbers $\dim V(f)$ or $\dim V(-f)$ is bigger than or equal to $(\dim V - \dim V_0)/2$. Therefore 1) follows from 2). □

7. LEMMA 2. *Suppose V is irreducible and let $\Lambda \in \Sigma$ be a weight of V different from the highest and lowest weights. Then* $\dim V_\Lambda \leq \sum_{\alpha\in\Delta_+} \dim X_{-\alpha} V_\Lambda$ *and* $\dim V_\Lambda \leq \sum_{\alpha\in\Delta_+} \dim X_\alpha V_\Lambda$.

PROOF. Consider the map $\pi: V_\Lambda \to \bigoplus_{\alpha\in\Delta_+} V_{\Lambda-\alpha}$, $\pi(v) = \sum_{\alpha\in\Delta_+} X_{-\alpha} v$ for all $v \in V_\Lambda$. It is clear that $\bigoplus_{\alpha\in\Delta_+} X_{-\alpha} V_\Lambda \supset \pi(V_\Lambda)$; hence $\dim \pi(V_\Lambda) \leq \sum_{\alpha\in\Delta_+} \dim X_{-\alpha} V_\Lambda$. If the first inequality in the assertion of the lemma is not true then $\dim V_\Lambda > \dim \pi(V_\Lambda)$, so that $\operatorname{Ker} \pi \neq \{0\}$. Because V is irreducible, any nonzero vector from $\operatorname{Ker} \pi$ is lowest, which contradicts the definition of Λ. A similar argument proves the other inequality. □

PROPOSITION 2. $\dim V_0 \leq (1/3) \dim V$.

PROOF. Because $\dim V$ and $\dim V_0$ are additive functions with respect to direct sums of G-modules, it suffices to prove the inequality in the case when V is irreducible. To this end we apply Lemma 2 when $\Lambda = 0$: V contains the subspace

$$\left(\bigoplus_{\alpha\in\Delta_+} X_{-\alpha}V_0\right) \oplus V_0 \oplus \left(\bigoplus_{\alpha\in\Delta_+} X_{\alpha}V_0\right),$$

whose dimension is

$$\sum_{\alpha\in\Delta_+} \dim X_{-\alpha}V_0 + \dim V_0 + \sum_{\alpha\in\Delta_+} \dim X_{\alpha}V_0 \geq 3\dim V_0. \quad \square$$

8. PROPOSITION 3. *If V is equidimensional then:*

1) $\dim V \leq 2m_G + \dim V_0 - 2$;
2) $m_G \geq \dim V(f) + 1$ *for all* f;
3) $\dim V \leq 3m_G - 3 \leq 3\dim G - 3$.

PROOF. By Theorem 3, inequalities 1) and 2) follow, respectively, from inequalities 1) and 2) of the corollary of Theorem 5. Inequalities 3) follow from inequality 1), Proposition 2, and the definition of m_G. $\square$

Because there are only finitely many nonisomorphic G-modules of dimension not exceeding a given constant (this follows from Weyl's dimension formula (see [10])), we deduce from inequality 3) of Proposition 3 the following finiteness theorem.

9. THEOREM 6. *There exist only finitely many, up to isomorphism and addition of trivial direct summands, finite-dimensional equidimensional G-modules.*

This theorem along with Theorem 1 yields another proof of Theorem 2 (notice that the explicit bounds from Proposition 3 give more precise information than do the bounds from Chapter 3).

In determining equidimensional modules in practice it is more convenient to use, in addition to the bounds from Proposition 3, the following two propositions.

10. PROPOSITION 4. *Suppose that V is irreducible and its noneffectivity kernel is finite. Let us cross out several (≥ 1) vertices from its Dynkin diagram (along with their numerical labels); the remaining part of the diagram will be the Dynkin diagram of the irreducible module V' of the corresponding semisimple regular subgroup H of G (see* [15]*). Then this module is an H-submodule of V and is contained entirely in $V(f)$ for some $f \in \mathfrak{X}^*$, $f \neq 0$.*

PROOF. Let Λ_0 be the highest weight of V and $v \in V_{\Lambda_0}$, $v \neq 0$. Let $\alpha_{l_1}, \dots, \alpha_{l_d}$ be the simple roots corresponding to the remaining vertices and Z be their linear hull in $\mathfrak{X}$. By definition, $\mathfrak{h}$ is generated by those X_α for which α is a linear combination of $\alpha_{l_1}, \dots, \alpha_{l_d}$. It now follows that, first, $V' = \langle Hv \rangle$ is an irreducible H-module whose Dynkin diagram is the remaining part of the Dynkin diagram of V, and, second, $V' \subset \bigoplus_{\Lambda\in\Sigma\cap(\Lambda_0+Z)} V_\Lambda$

(see [50, §31, part 2]). Notice that $0 \notin \Lambda_0 + Z$. Otherwise $\Lambda_0 = \sum_{j=1}^d p_j\alpha_{l_j}$ for some $p_j \in \mathbb{Q}$. We now switch to the notation of part 1 of §2, Chapter 3. Because Λ_0 is the highest weight, we have $\Lambda_0 = \sum_{i=1}^r n_i\omega_i$, where $n_i \in \mathbb{N}$ for all i. Because the noneffectivity kernel of V is finite, it follows that for any $i = 1, \dots, s$ there exists j, $r_{i-1} + 1 \le j \le r_i$, such that $n_j \neq 0$. On the other hand ω_j is a linear combination of $\alpha_{r_{i-1}+1}, \dots, \alpha_{r_i}$ with positive rational coefficients (see [15, 9]). Therefore $\Lambda_0 = \sum_{i=1}^r q_i\alpha_i$, $q_i \in \mathbb{Q}$, $q_i > 0$ for all i. Because $\alpha_{l1}, \dots, \alpha_{l_d}$ are not all the roots, this contradicts (in view of the equality $\sum_{i=1}^r q_i\alpha_i = \sum_{j=1}^d p_j\alpha_{l_j}$) the linear independence of $\alpha_l, \dots, \alpha_r$.

Finally, because $0 \notin \Lambda_0 + Z$, there exists $f \in \mathfrak{X}^*$ such that $\Lambda_0 + Z \subset \{x \in \mathfrak{X}^* \mid f(x) > 0\}$. Therefore $V' \subset V(f)$. □

COROLLARY. *If V is equidimensional then* $\dim V' \le m_G - 1 \le \dim G - 1$.

This follows from inequality 2) of Proposition 3.

PROPOSITION 5. *Suppose that for some $f \in \mathfrak{X}^*$ there exists a set of roots $\beta_1, \dots, \beta_d \in \Delta$ and a set of weights $\Lambda_1, \dots, \Lambda_d \in \Sigma - \Sigma(f)$ such that $\Lambda_i - \beta_i \in \Sigma(f)$ for all i and $\Lambda_j - \beta_i \notin \Sigma(f)$ for all i and $j < i$. Then* $\dim GV(f) \ge \dim V(f) + d$.

PROOF. Examining the morphism $G \times V(f) \to GV(f)$, $(g, v) \mapsto gv$, we see that $V(f)$ contains a nonempty open subset Ω such that each point $v \in \Omega$ is a nonsingular point of $GV(f)$ and $\Theta_{v, GV(f)} = V(f) + \mathfrak{g}v$. It is known (see [15]) that if $\alpha \in \Delta$ and $\Lambda \in \Sigma$ then $X_\alpha V_\Lambda$ is a nonzero subspace of $V_{\alpha+\Lambda}$ when $\alpha + \Lambda \in \Sigma$ and $X_\alpha V_\Lambda = 0$ when $\alpha + \Lambda \notin \Sigma$. On the other hand, because Ω is open, $\pi_\Lambda(\Omega)$ contains a nonempty open subset of V_Λ for each $\Lambda \in \Sigma(f)$. It now follows that Ω contains a point x such that for any $\Lambda \in \Sigma(f)$ and $\alpha \in \Delta$ the condition $\alpha + \Delta \in \Sigma$ implies $X_\alpha x_\Lambda \neq 0$. Because $V = \bigoplus_{M \in \Sigma} V_M$, we have $X_\alpha x_\Lambda = \pi_{\alpha+\Lambda}(X_\alpha x)$. It now follows from the hypothesis that $\pi_{\Lambda_i}(X_{\beta_i}x) \neq 0$ for all i, $\pi_{\Lambda_j}(X_{\beta_i}x) = 0$ for all i, and $j < i$. Because $\Lambda_i \in \Sigma - \Sigma(f)$ for all i, it means that the images of $X_{\beta_i}x$, $i = 1, \dots, d$, in $(\mathfrak{g}x + V(f))/V(f)$ are linearly independent. Therefore our assertion follows from the fact that $\dim GV(f) = \dim(V(f) + \mathfrak{g}x)$. □

11. Finally we mention several other properties of V related to the equidimensionality and the freeness of the module of covariants.

We call a linear subspace L of V a Cartan subspace and the group $W_{(L)} = N(L)|Z(L)$, where $N(L) = G_L (= \{g \in G \mid gL = L\})$, $Z(L) = \{g \in G \mid gu = u \text{ for all } u \in L\}$ its Weyl group if: a) $W_{(L)}$ is finite, b) restriction of functions gives rise to an isomorphism between $k[V]^G$ and $k[L]^{W_{(L)}}$ (notice that this definition differs from the one in [72]).

A linear subvariety S in V (which is not necessarily a subspace) will be called a section if S intersects each fiber of $\pi_{V,G}$ at a unique point.

THEOREM 7. *If V has a Cartan subspace L or a section S then its module of covariants is free.*

PROOF. 1) Suppose L exists. Then, in view of [25], the algebra $k[V]^G$ is free. On the other hand, by a), $\dim L = \dim L/W_{(L)}$ and, by b), $\dim L/W_{(L)} = \dim V/G = \dim V - m_G$. Therefore, by Theorem 3, $\dim L + \dim \mathfrak{N} \geq \dim V$. If the inequality were strict, we would have, in view of $L \cap \mathfrak{N} \neq \varnothing$, that $\dim L \cap \mathfrak{N} \geq \dim L + \dim \mathfrak{N} - \dim V > 0$ (see [51]) so that there would be a point $v \in L \cap \mathfrak{N}$, $v \neq 0$. Because $|W_{(L)}| < \infty$ we would have $\overline{W_{(L)}v} = W_{(L)}v \ni 0$. Therefore we could find a homogeneous element $f \in k[L]^{W_{(L)}}$ such that $f(v) = 1$ (invariants separate closed orbits). But this contradicts b). Therefore $\dim \mathfrak{N} = \dim V - \dim L$ and, by Theorem 3, the module V is equidimensional. Our assertion now follows from Theorem 1.

2) Suppose S exists. Then the restriction of $\pi_{V,G}$ to S is a bijective morphism. Because $\operatorname{char} k = 0$ and V/G is normal (see [16]), this restriction is an isomorphism between S and V/G. Because $S \cong \mathbb{A}^d$, the algebra $k[V]^G$ is free. The argument from 1) now shows that V is equidimensional. □

If $\dim V/G = 0$ then, obviously, V has both a Cartan subspace and a section ($L = S = \{0\}$).

PROPOSITION 6. *If $\dim V/G = 1$ then V has a Cartan subspace.*

PROOF. By [38], $k[V]^G = k[f]$, where f is a homogeneous polynomial of degree $d > 0$. Because G is semisimple, there exists a point v in V such that $v \neq \mathfrak{N}$ and $\overline{Gv} = \pi_{V,G}^{-1}(\pi_{V,G}(v)) = \{u \in V \mid f(u) = f(v)\}$ (see [30, 29]). We shall now show that conditions a) and b) from the definition of a Cartan subspace are satisfied for $L = \langle v \rangle$. Because $\dim\langle v \rangle = 1$, we have $gv = \chi(v)v$ for all $g \in G_{\langle v \rangle}$, where $\chi \in \operatorname{Hom}(G_{\langle v \rangle}, k^*)$. It is clear that $W_{(\langle v \rangle)} = G_{\langle v \rangle}/G_v = \chi(G_{\langle v \rangle})$. But $f(v) = f(gv) = \chi(g)^d f(v)$, therefore (because $f(v) \neq 0$) $\chi(G_{\langle v \rangle}) \subset \mu_d = \{\varepsilon \in k^* \mid \varepsilon^d = 1\}$. Hence a) is satisfied. To prove b) we have to show that $\chi(G_{\langle v \rangle}) = \mu_d$. Let $\varepsilon \in \mu_d$. Then $f(\varepsilon v) = \varepsilon^d f(v) = f(v)$, so that (by the definition of v) $\varepsilon v = gv$ for some $g \in G$. Because $\langle \varepsilon v \rangle = \langle v \rangle$ we now have $g \in G_{\langle v \rangle}$. □

The argument in the proof of the following lemma is a generalization of the method for finding the degree of a basis invariant from [38].

LEMMA 3. *Suppose that the algebra $k[V]^G$ is free and let $f_1, \ldots, f_m$ be a minimal system of homogeneous generators of it. Suppose also that the orbit Gv of some point $v \in \mathfrak{N}$ is open in $\mathfrak{N}$. Then the existence of a section of V containing v is equivalent to the linear independence of $(df_1)_v, \ldots, (df_m)_v$.*

PROOF. Let S be a section and $v \in S$. Then $\pi_{V,G|S}$ is an isomorphism between S and V/G. But $V/G \cong \mathbb{A}^m$ and $\pi_{V,G}$ can be identified with $\pi = (f_1, \ldots, f_m)\colon V \to \mathbb{A}^m$. Therefore $f_1|_S, \ldots, f_m|_S$ have linearly independent differentials at v. But then $(df_1)_v, \dots, (df_m)_v$ are, a fortiori, linearly independent.

Conversely, suppose that $(df_1)_v, \dots, (df_m)_v$ are linearly independent. We may (and will) assume that $G \subset \mathrm{GL}(V)$. It follows from the hypothesis that $\overline{Gv}$ is an irreducible component of $\mathfrak{N}$ and therefore, by the theorem on the dimension of a fiber (see [51]), $\dim \overline{Gv} \geq \dim V - \dim V/G = m_G$. From $m_G \geq \dim Gv = \dim \overline{Gv}$, we have $\dim \overline{Gv} = m_G$. Because Gv is open in $\overline{Gv}$, v is a nonsingular point of $\overline{Gv}$. Therefore $\dim \Theta_{v,\overline{Gv}} = m_G$.

Consider the one-parameter subgroup $\eta\colon k^* \to \mathrm{GL}(V)$, $\eta(t) = t \cdot \mathrm{Id}$. Because $\eta(t)\mathfrak{N} = \mathfrak{N}$, it is true that $\eta(t)\overline{Gv} = \overline{Gv}$. Therefore $\eta(t)Gv = G\eta(t)v$ is also an open orbit in $\overline{Gv}$ and $Gv = \eta(t)Gv$. Hence $G_{\langle v\rangle}/G_v \cong k^*$. Because each unipotent element from $G_{\langle v\rangle}$ must be contained in G_v, it follows that there exists a one-parameter subgroup $\lambda\colon k^* \to G_{\langle v\rangle}$, such that $\lambda(t)v = t^p v$ for all $t \in k^*$, where $p \in \mathbb{Z}, p \neq 0$.

It is clear that $\sigma\colon k^* \to \mathrm{GL}(V)$, $\sigma(t) = \lambda(t^{-1})\eta(t^p)$ for all $t \in k^*$ is a one-parameter subgroup. Let $T = \{\sigma(t) \mid t \in k^*\}$. Then $Tv = v$ and T acts naturally on $\Theta_{v,V}$. But $\sigma(t)\overline{Gv} = \overline{Gv}$ and therefore $\Theta_{v,\overline{Gv}}$ is a T-invariant subspace of $\Theta_{v,V}$. Because T is a torus, we have $\Theta_{v,V} = \Theta_{v,\overline{Gv}} \oplus S$ for some T-invariant subspace S. We shall show that S is a section of V (both $\Theta_{v,\overline{Gv}}$ and S are considered to be naturally embedded in V as linear varieties).

As $\mathfrak{N} = \{u \in V \mid f_i(u) = 0 \text{ for all } i\}$, we have $\Theta_{v,\overline{Gv}} \subset \bigcap_{i=1}^m \mathrm{Ker}(df_i)_v$. It follows from the linear independence of $(df_1)_v, \dots, (df_m)_v$ that $\dim \bigcap_{i=1}^m \mathrm{Ker}(df_i)_v = \dim \Theta_{v,V} - m = \dim V - \dim V/G = m_G$. Let x_i be the restriction of $(df_i)_v$ to S. Because $\dim \Theta_{v,\overline{Gv}} = m_G$, it follows from those equalities that $x_1, \dots, x_m$ is a basis of S^*.

On the other hand, because the f_i are homogeneous and G-invariant we have $(\sigma(t)f_i)(u) = f_i(\lambda(t)\eta(t^{-p})u) = t^{-pd_i}f_i(\lambda(t)u) = t^{-pd_i}f_i(u)$ for all i, for all $t \in k^*$, $u \in V$, where $d_i = \deg f_i$. Hence $\sigma(t)(df_i)_v = t^{-pd_i}(df_i)_v$ and therefore $\sigma(t)x_i = t^{-pd_i}x_i$ for all i, for all $t \in k^*$.

For each $i = 1, \dots, m$, let h_i be a polynomial in m variables with coefficients in k such that $h_i(x_1, \dots, x_m) = f_i|_S$. As we already showed, $t^{-pd_i}h_i(x_1, \dots, x_m) = h_i(t^{-pd_1}x_1, \dots, t^{-pd_m}x_m)$. Because $p \neq 0$, it means that h_i is a quasihomogeneous polynomial of type $(d_i; d_1, \dots, d_m)$. Therefore $\pi = \pi_{V,G}|_S = (h_1(x_1, \dots, x_m), \dots, h_m(x_1, \dots, x_m))$ is a quasihomogeneous morphism $\mathbb{A}^m \to \mathbb{A}^m$ of type $(d_1, \dots, d_m; d_1, \dots, d_m)$. Because $d_i > 0$ for all i and $\mathrm{rk}(d\pi)_v = m$, it follows from the general properties of such morphisms that π is an isomorphism (see [145]). □

REMARK. Suppose that the hypothesis of Lemma 3 is satisfied. Then $\dim \Theta_{v,\overline{Gv}} = m_G$ and, as in the proof of Lemma 3, we can construct λ, σ, T, and S (this part of the proof is not based on the linear independence of differentials). Consider a T-weighted basis $e_1, \dots, e_m$ of S and let $\sigma(t)e_i = t^{a_i}e_i$ for all i, for all $t \in k^*$. Then the linear independence of $(df_1)_v, \dots, (df_m)_v$ is equivalent to saying that (after a suitable reordering) $a_i = pd_i$ for all i. In one direction this was proved in Lemma 3. Conversely, if $a_i = pd_i$ for all i then (because of the equality $\sigma(t)f_i = t^{-pd_i}f_i$ for all i, for all $t \in k^*$ proved in Lemma 3) $\pi = \pi_{V,G}|_S$ is a quasihomogeneous morphism $\mathbb{A}^m \to \mathbb{A}^m$ of type $(d_1, \dots, d_m; d_1, \dots, d_m)$. Its fiber over zero is $\mathfrak{N} \cap S$. If $w \in \mathfrak{N} \cap S$ and $w \neq v$ then $Tw \in \mathfrak{N} \cap S$. Because $d_i > 0$ for all i and $p \neq 0$, we have $\dim \overline{Tw} = 1$ and $v \in \overline{Tw}$. On the other hand, $\Theta_{v,\overline{Tw}} \subset \Theta_{v,\overline{Gv}} \cap S = v$, a contradiction. Therefore $\mathfrak{N} \cap S = v$. From the general theory for such morphisms (see [145]) we know that π is an isomorphism and therefore $\operatorname{rk}(d\pi)_v = \operatorname{rk}((df_1|_S)_v, \dots, (df_m|_S)_v) = m$.

We note that (in the notation of part 5 of §0, Chapter 4) $(T, \Theta_{v,V}) = (T, V)$ and $(T, \Theta_{v,\overline{Gv}}) = (T, \mathfrak{g}v) = (T, \mathfrak{g}/\mathfrak{g}_v)$ (the last representation is induced by Ad). It follows easily that the system $P_1 = \{t^{b_1}, \dots, t^{b_l}\}$ of weights λ on $\mathfrak{g}/\mathfrak{g}_v$ is contained in $t^{-p}P_2 = \{t^{c_1-p}, \dots, t^{c_s-p}\}$, where $P_2 = \{t^{c_1}, \dots, t^{c_s}\}$ is the system of weights λ on V (the weights are considered with their multiplicities), and that $t^{-p}P_2 - P_1 = \{t^{-a_i} \mid i = 1, \dots, m\}$.

PROPOSITION 7. *Suppose that* $\dim V/G = 1$ *and* $\mathfrak{N}$ *contains an open orbit* Gv. *Then* V *has a section.*

PROOF. We have $k[V]^G = k[f]$, where f is irreducible because G is connected and semisimple (see [29]). Therefore $(df)_v \neq 0$ and the assertion follows from Lemma 3. □

§2. Classification and equivalent characterizations: Igusa's condition

1. In this section, unless stated otherwise, we assume that G is simple and V is irreducible. The ordering of simple roots will be taken from [9] and their decompositions by the elements ε from [15].

As the first step in our classification of equidimensional modules V we shall find those V that satisfy the following necessary conditions: $\dim V \leq 3m_G - 3$ (inequality 3) from Proposition 3), $\dim V \leq 2m_G - 2$ if zero is not a weight of V (inequality 1) from Proposition 3), and $\dim V' \leq m_G - 1$ (the inequality from the corollary of Proposition 4).

This is a simple exercise of a computational nature. Indeed, all the numbers m_G are known (see [52, 53]), and zero is a weight of $V = R(\Lambda)$ if and only if Λ is a $\mathbb{Z}$-linear combination of simple roots (see, e.g., [15]). Hence zero is a weight of V if and only if V is congruent (in the sense of [22]) to 0. This gives, in view of [22], a criterion (in terms of the numerical labels of Λ) for zero weight to be a weight of V. Thus the problem is reduced to the examination of the tables of dimensions of irreducible G-modules (see [15]) and the systematic use of the following observation: if the ith numerical label of the G-module V is not less than the ith numerical label of the

G-module U for all i, then $\dim V \geq \dim U$ (this follows immediately from Weyl's dimension formula (see [10])).

2. EXAMPLE 1. G is of type G_2 and we have $3\dim G - 3 = 39$, $\dim R(\omega_1+\omega_2) = 64$, $\dim R(2\omega_2) = \dim R(3\omega_1) = 77$, $\dim R(2\omega_1) = 27$, $\dim R(\omega_2) = 14$, and $\dim R(\omega_1) = 7$ (see [15]). Moreover $m_G = \dim G$ if $V \neq R(w_1)$, $R(\omega_2)$ (see [52, 53]), and zero is a weight of any V (see [22]). It now follows that all inequalities mentioned in this section are true only for $R(\omega_1)$, $R(2\omega_1)$, and $R(\omega_2)$.

EXAMPLE 2. Let G be of type B_l, $l \geq 3$. Suppose $V = R(\Lambda)$ satisfies the above inequalities and $\Lambda = \sum_{i=1}^{l} a_i w_i$. By [15], $\dim G = l(2l+1)$, $\dim R(\omega_p) = (2l+1)!/p(2l+1-p)!$ for $1 \leq p \leq l-1$; $\dim R(\omega_l) = 2^l$; $\dim R(2\omega_2) = (l-1) \times (2l+1)!(2l+3)!/6(2l)!(2l+1)!$; $\dim R(\omega_1+\omega_2) = (2l-1)(2l+1)!(2l+3)!/3(2l)!(2l+2)!$; $\dim R(\omega_1+\omega_l) = l2^{l+1}$; $\dim R(\omega_2+\omega_l) = 2^{l-1}(2l-2)(2l+1)$; $\dim R(3\omega_1) = (2l+5)l(2l+1)/3$; $\dim R(2\omega_l) = (2l+1)!/l!(l+1)!$. It is easy to see that $\dim R(\omega_i) > \dim R(\omega_j)$ for $1 \leq j < i \leq l-1$. If $l \geq 4$, then, solving the inequality $\dim R(\omega_3) \leq 3\dim G - 3$, we obtain $l \leq 4$. Therefore if $l \geq 5$ then $a_3 = a_4 = \cdots = a_{l-1} = 0$. From the inequality $\dim R(\omega_l) \leq 3\dim G - 3$ we have $l \leq 8$ so that $a_l = 0$ for $l \geq 9$. It is easy to check that the numbers $\dim R(\Lambda)$ for $\Lambda = 2\omega_2$, $\omega_1+\omega_2$, $\omega_2+\omega_l$, $3\omega_1$ for all l, and $\omega_1+\omega_l$, $l \geq 4$, are bigger than $3\dim G - 3$. Hence $a_2 \leq 1$, $a_1a_2 = 0$, $a_2a_l = 0$, $a_1 \leq 2$ for all l and $a_1a_l = 0$ for $l \geq 4$. Therefore if $l \geq 9$ then V can only be either $R(\omega_1)$ or $R(2\omega_1)$ or $R(\omega_2)$. On the other hand, zero is a weight of each of these modules (see [22]), and an elementary calculation shows that the aforementioned inequalities are satisfied for all values of l. Furthermore, from the inequality $\dim R(2\omega_l) \leq 3\dim G - 3$, we have $l = 3$, therefore $a_l \leq 1$ for $l \geq 4$. Hence for $5 \leq l \leq 8$ we can only have one new module: $R(\omega_l)$. Zero is not a weight of this module (see [22]) and in this case $m_G = 7, 15, 31, 62, 105$, and 136 for $l = 3, 4, 5, 6, 7$, and 8, respectively, so that all of the inequalities are true. If $l = 4$ then (see [15]) $\dim R(\Lambda) = 27\cdot 22, 50\cdot 33, 90\cdot 22$, and $16\cdot 48$ for $\Lambda = \omega_1+\omega_3$, $\omega_2+\omega_3$, $2\omega_3$, and $\omega_3+\omega_4$, respectively. Because each of these numbers is bigger than $3\dim G-3$ we can only have one new module, $R(\omega_3)$, and for this module all of the inequalities are satisfied. Finally when $l = 3$ we have (see [15]) $\dim R(2\omega_1+\omega_3) = 28\cdot 6$, $\dim R(\omega_1+2\omega_3) = 21\cdot 9$, and $\dim R(3\omega_3) = 112$. Because each of these numbers is bigger than $3\dim G-3$ and $\dim R(\omega_1+\omega_3) = 48 > 2\dim G-2$ and zero is not a weight of $R(\omega_1+\omega_3)$, we should add to our list the module $R(2\omega_3)$ only (all of the inequalities are true in this case).

3. The remaining types of groups can be dealt with similarly (these cases differ from each other only by the amount of computations). Because of this we only give the final results. The modules $V = R(\Lambda)$ satisfying the conditions mentioned in the beginning of this section are exactly the modules from the following list (and the modules that can be obtained from them by outer automorphisms): G is of type $A_l (l \geq 1)$, $\Lambda = \omega_1, 2\omega_1$ for $l \geq 1$; $\omega_1 + \omega_l, \omega_2$ for $l \geq 2$; ω_3 for $5 \leq l \leq g$; ω_4 for $l = 7, 8$; $3\omega_1$ for $1 \leq l \leq 4$; $\omega_1 + \omega_3$ for $l = 4$; $2\omega_2$ for $l = 3$; $4\omega_1$ for $l = 1$. G is of type $B_l (l \geq 3)$, $\Lambda = \omega_1, 2\omega_1, \omega_2$ for $l \geq 3$; ω_l for $3 \leq l \leq 8$; ω_3 for $l = 4$; $2\omega_3$ for $l = 3$. G is of type $C_l (l \geq 2)$, $\Lambda = \omega_1, 2\omega_1, \omega_2$ for $l \geq 2$; ω_l for $2 \leq l \leq 4$; ω_3 for $l = 4$; $2\omega_2, \omega_1 + \omega_2$ for $l = 2$. G is of type $D_l (l \geq 4)$, $\Lambda = \omega_1, 2\omega_1, \omega_2$ for $l \geq 4$; ω_l for $5 \leq l \leq g$. G is of type E_6, $\Lambda = \omega_1, \omega_2$. G is of type E_7, $\Lambda = \omega_1, \omega_7$. G is of type E_8, $\Lambda = \omega_8$. G is of type F_4, $\Lambda = \omega_1, \omega_4$. G is of type G_2, $\Lambda = \omega_1, 2\omega_1, \omega_2$.

4. In the proofs of the forthcoming results we will use Proposition 5 and Inequality 2) from Proposition 3.

PROPOSITION 8. *There exists an element* $f \in \mathfrak{X}^*$ *such that* $m_G < \dim V(f)$ *for each G-module* $V = R(\Lambda)$ *from the following list*: 1) G *is of type* A_4, $\Lambda = \omega_1 + \omega_3$; 2) G *is of type* A_8, $\Lambda = \omega_4$; 3) G *is of type* D_g, $\Lambda = \omega_g$; 4) G *is of type* B_4, $\Lambda = \omega_3$.

PROOF. 1) Because $\Lambda = 2\varepsilon_1 + \varepsilon_2 + \varepsilon_3$ and W consists of all possible permutations of ε_i (see [15]), $W(\Lambda)$ contains 30 weights of multiplicity 1. Because $\Lambda - w(-\varepsilon_5)$ is a sum of simple roots for any $w \in W$ it is true that $-\varepsilon_5 \in \Sigma$ (see [10]). We also have $|W(-\varepsilon_5)| = 5$. Computing the multiplicity (it can be easily done with the aid of [113]) we see that $\dim V_{-\varepsilon_5} = 3$. Therefore $\Sigma = W(\Lambda) \cup W(-\varepsilon_5)$ (because $\dim V = 45 = 30 + 3 \cdot 5$) Now we define $f \in \mathfrak{X}^*$ by the conditions $f(\varepsilon_i) = -1$ for all $i = 1, 2, 3, 4$, $f(\varepsilon_5) = 4$. Obviously $|\Sigma(f) \cap W(\Lambda)| = 18$, $|\Sigma(f) \cap W(-\varepsilon_5)| = 4$, so that $\dim V(f) = 18 + 3 \cdot 4 = 30 > m_G = 24$ (see [52]).

2) Because $\Lambda = \varepsilon_1 + \varepsilon_2 + \varepsilon_3 + \varepsilon_4$, it is true that $|W(\Lambda)| = 126$. Since $\dim V = 126$, we see that $\Sigma = W(\Lambda)$ (all weights are of multiplicity 1). We define an element $f \in \mathfrak{X}^*$ by the conditions $f(\varepsilon_1) = f(\varepsilon_2) = 4$, $f(\varepsilon_i) = -8/7$ for all $i = 3, \dots, 9$. Then $|\Sigma(f)| = \dim V(f) = 91 > m_G = 80$ (see [52]).

3) All weights are of multiplicity 1 and have the form $(\pm\varepsilon_1 \pm \cdots \pm \varepsilon_9)/2$ with an even number of minuses (see [15]). Therefore, if we define $f \in \mathfrak{X}^*$ be the conditions $f(\varepsilon_i) = 1$ for all $i = 1, \dots, 9$, then $\dim V(f) = 163 > m_G = 153$ (see [52]).

4) Because $\Lambda = \varepsilon_1 + \varepsilon_2 + \varepsilon_3$ and W consists of all possible permutations of ε_i with an arbitrary change of sign, $W(\Lambda)$ consists of 32 weights of multiplicity 1. Because $\Lambda - w(\varepsilon_1)$ is a sum of simple roots for all $w \in W$, it is true that $\varepsilon_1 \in \Sigma$. Similarly $\varepsilon_1 + \varepsilon_2 \in \Sigma$. Computing the multiplicities (see [113]) we see that $\dim V_{\varepsilon_1} = 3$, $\dim V_{\varepsilon_1+\varepsilon_2} = 1$. Because $\dim V = 84$, $|W(\varepsilon_1)| = 8$, $|W(\varepsilon_1 + \varepsilon_2)| = 24$, and $0 \in \Sigma$ (see [22]), we have $\dim V_0 \leq 84 - 32 - 3 \cdot 8 - 24 = 4$. Therefore, if $f \in \mathfrak{X}^*$ has the property $f(\Lambda) \neq 0$ for all $\Lambda \in \Sigma - \{0\}$, then either $\dim V(f)$ or $\dim V(-f)$ is bigger than or equal to $(\dim V - \dim V_0)/2 \geq (84 - 4)/2 = 40 > m_G = 36$, (see [52]). $\square$

PROPOSITION 9. *The following G-modules $V = R(\Lambda)$ are not equidimensional*: 1) G *is of type* A_4, $\Lambda = 3\omega_1$; 2) G *is of type* A_3, $\Lambda = 3\omega_1$; 3) G *is of type* C_4, $\Lambda = \omega_3$; 4) G *is of type* B_3, $\Lambda = 2\omega_3$; 5) G *is of type* B_8, $\Lambda = \omega_8$; 6) G *is of type* G_2, $\Lambda = 2\omega_1$; 7) G *is of type* C_2, $\Lambda = \omega_1 + \omega_2$.

PROOF. 1) All weights are of multiplicity 1 and have the form $\sum_{i=1}^5 k_i\varepsilon_i$, where $\sum_{i=1}^5 k_i = 3$, $k_i \in \mathbb{N}$ for all i (see [15]). Let $f \in \mathfrak{X}^*$ be given by the conditions $f(\varepsilon_1) = -7$, $f(\varepsilon_2) = 6$, $f(\varepsilon_3) = -5$, $f(\varepsilon_4) = 4$, $f(\varepsilon_5) = 2$. Then $\dim V(f) = 19$ and the sets $\beta_1 = \varepsilon_1 - \varepsilon_2$, $\beta_2 = \varepsilon_3 - \varepsilon_4$, $\beta_3 = \varepsilon_3 - \varepsilon_5$, $\beta_4 = \varepsilon_1 - \varepsilon_4$, $\beta_5 = \varepsilon_1 - \varepsilon_5$, $\beta_6 = \varepsilon_1 - \varepsilon_3$, and $\Lambda_1 = 2\varepsilon_1 + \varepsilon_5$, $\Lambda_2 = 2\varepsilon_3 + \varepsilon_5$, $\Lambda_3 = \varepsilon_3 + 2\varepsilon_5$, $\Lambda_4 = \varepsilon_1 + \varepsilon_3 + \varepsilon_5$, $\Lambda_5 = \varepsilon_1 + 2\varepsilon_5$, $\Lambda_6 = \varepsilon_1 + \varepsilon_4 + \varepsilon_5$ satisfy the conditions from Proposition 5. Therefore $\dim GV(f) \geq 19 + 6 > 24 = m_G$ (see [52]).

2) All weights are of multiplicity 1 and have the form $\sum_{i=1}^4 k_i\varepsilon_i$, where $\sum_{i=1}^4 k_i = 3$, $k_i \in \mathbb{N}$ for all i (see [15]). Let $f \in \mathfrak{X}^*$ be given by the conditions $f(\varepsilon_1) = 5$, $f(\varepsilon_2) = 3$, $f(\varepsilon_3) = -1$, $f(\varepsilon_4) = -7$. Then $\dim V(f) = 11$ and the sets $\beta_1 = \varepsilon_4 - \varepsilon_1$, $\beta_2 = \varepsilon_4 - \varepsilon_2$, $\beta_3 = \varepsilon_3 - \varepsilon_1$, $\beta_4 = \varepsilon_4 - \varepsilon_3$, $\beta_5 = \varepsilon_2 - \varepsilon_1$ and $\Lambda_1 = \varepsilon_2 + 2\varepsilon_4$, $\Lambda_2 = 2\varepsilon_3 + \varepsilon_4$, $\Lambda_3 = 3\varepsilon_3$, $\Lambda_4 = \varepsilon_1 + \varepsilon_2 + \varepsilon_3$, $\Lambda_5 = 2\varepsilon_2 + \varepsilon_4$ satisfy the conditions from Proposition 5. Therefore $\dim GV(f) \geq 11 + 5 > 15 = m_G$ (see [52]).

3) Because $\Lambda = \varepsilon_1 + \varepsilon_2 + \varepsilon_3$ and W consists of all possible permutations of ε_i with an arbitrary change of sign, $W(\Lambda)$ consists of 32 weights of multiplicity 1. Because $(\lambda_3 \mid \varepsilon_1 + \varepsilon_2) = 2$, it is true that $\varepsilon_3 \in \Sigma$. It is easy to compute (see [113]) that $\dim V_{\varepsilon_3} = 2$ and $|W(\varepsilon_3)| = 8$. Because $\dim V = 48 = 32 + 2 \cdot 8$, we have $\sum = W(\Lambda) \cup W(\varepsilon_3)$. It follows that if $f \in \mathfrak{X}^*$ is defined by the conditions $f(\varepsilon_1) = 1$, $f(\varepsilon_2) = 2$, $f(\varepsilon_3) = 4$, $f(\varepsilon_4) = 7$, then $\dim V(f) = 24$, and the sets $\beta_1 = -2\varepsilon_4$, $\beta_2 = -\varepsilon_3 - \varepsilon_4$, $\beta_3 = -\varepsilon_2 - \varepsilon_4$, $\beta_4 = -\varepsilon_1 - \varepsilon_4$, $\beta_5 = \varepsilon_1 - \varepsilon_4$, $\beta_6 = -\varepsilon_2 - \varepsilon_3$, $\beta_7 = \varepsilon_2 - \varepsilon_4$, $\beta_8 = \varepsilon_3 - \varepsilon_4$, $\beta_9 = -2\varepsilon_3$, $\beta_{10} = \varepsilon_2 - \varepsilon_3$, $\beta_{11} = -\varepsilon_1 - \varepsilon_3$, $\beta_{12} = -\varepsilon_1 - \varepsilon_2$, $\beta_{13} = -2\varepsilon_1$ and $\Lambda_1 = -\varepsilon_1 + \varepsilon_2 + \varepsilon_4$, $\Lambda_2 = -\varepsilon_1 - \varepsilon_2 - \varepsilon_4$, $\Lambda_3 = \varepsilon_1 - \varepsilon_2 - \varepsilon_4$, $\Lambda_4 = -\varepsilon_1 + \varepsilon_2 - \varepsilon_4$, $\Lambda_5 = -\varepsilon_2 + \varepsilon_3 - \varepsilon_4$, $\Lambda_6 = \varepsilon_1 - \varepsilon_2 - \varepsilon_3$, $\Lambda_7 = -\varepsilon_1 + \varepsilon_3 - \varepsilon_4$, $\Lambda_8 = \varepsilon_2 + \varepsilon_3 - \varepsilon_4$, $\Lambda_9 = -\varepsilon_1 - \varepsilon_2 - \varepsilon_3$, $\Lambda_{10} = \varepsilon_1 + \varepsilon_2 - \varepsilon_3$, $\Lambda_{11} = -\varepsilon_1 + \varepsilon_2 - \varepsilon_3$, $\Lambda_{12} = -\varepsilon_2$, $\Lambda_{13} = -\varepsilon_1$ satisfy the conditions from Proposition 5. Therefore $\dim GV(f) \geq 24 + 13 > 36 = m_G$, (see [52]).

4) Because $\Lambda = \varepsilon_1 + \varepsilon_2 + \varepsilon_3$, $W(\Lambda)$ consists of 8 weights of multiplicity 1. Because $(\Lambda \mid \varepsilon_3) = 1$, it is true that $\varepsilon_1 + \varepsilon_2 \in \Sigma$, and because $\Lambda - w(\varepsilon_1)$ is a sum of positive roots for all $w \in W$, it is also true that $\varepsilon_1 \in \Sigma$. We have $|W(\varepsilon_1 + \varepsilon_2)| = 12$ and $|W(\varepsilon_1)| = 6$. From [113] we see that $\dim V_{\varepsilon_1} = 2$, $\dim V_0 = 3$, $\dim V_{\varepsilon_1 + \varepsilon_2 = 1}$. Because $\dim V = 35$, it means that $\sum = W(\Lambda) \cup W(\varepsilon_1 + \varepsilon_2) \cup W(\varepsilon_1) \cup \{0\}$. Let $f \in \mathfrak{X}^*$ be defined by the conditions $f(\varepsilon_1) = 1 - 1$, $f(\varepsilon_2) = -2$, $f(\varepsilon_3) = -4$. Then $\dim V(f) = 16$ and the sets $\beta_1 = \varepsilon_2 + \varepsilon_3$, $\beta_2 = \varepsilon_1 + \varepsilon_3$, $\beta_3 = -\varepsilon_1 + \varepsilon_3$, $\beta_4 = -\varepsilon_2 + \varepsilon_3$, $\beta_5 = \varepsilon_3$,

$\beta_6 = \varepsilon_1 + \varepsilon_2$, $\beta_7 = \varepsilon_2$, $\beta_8 = \varepsilon_1$ and $\Lambda_1 = \varepsilon_1 + \varepsilon_3$, $\Lambda_2 = \varepsilon_3$, $\Lambda_3 = -\varepsilon_2 + \varepsilon_3$, $\Lambda_4 = -\varepsilon_1 - \varepsilon_2 + \varepsilon_3$, $\Lambda_5 = -\varepsilon_1 + \varepsilon_2$, $\Lambda_6 = \varepsilon_2$, $\Lambda_7 = \varepsilon_1$, $\Lambda_8 = 0$ satisfy the conditions from Proposition 5. Therefore $\dim GV(f) \geq 16 + 8 > 21 = m_G$ (see [52]).

5) All weights are of multiplicity 1 and have the form $\pm\varepsilon_1 \pm \cdots \pm \varepsilon_8)/2$ where the choice of signs is arbitrary (see [15]). Therefore, if $f \in \mathfrak{X}^*$ is given by the conditions $f(\varepsilon_1) = 1$, $f(\varepsilon_2) = f(\varepsilon_3) = 2$; $f(\varepsilon_i) = 0$ for $i \geq 4$ then $\dim V(f) = 128$ and the sets $\beta_i = -\varepsilon_2 + \varepsilon_{i+3}$ for $1 \leq i \leq 5$, $\beta_i = -\varepsilon_1 + \varepsilon_{i-2}$ for $6 \leq i \leq 10$, and $\Lambda_i = (\varepsilon_1 + \varepsilon_{i+3} - \sum_{j \neq 1, i+3} \varepsilon_j)/2$ for $1 \leq i \leq 5$, $\Lambda_i = (\varepsilon_2 + \varepsilon_{i-2} - \sum_{j \neq 2, i-2} \varepsilon_j)/2$ for $6 \leq i \leq 10$ satisfy the conditions from Proposition 5. Therefore $\dim GV(f) \geq 128 + 10 > 136 = m_G$ (see [52]).

In the remaining cases $\operatorname{rk} G = 2$ and therefore the weights can be easily determined graphically as the points of the lattice $\Lambda + Q$ (where Q is the lattice of roots) inside $\operatorname{conv} W(\Lambda)$ (see [10] and Chapter 3); the multiplicities of the weights can be computed with the aid of [113].

6) The weights are $\pm 2\varepsilon_i$, $i = 1, 2, 3$ (of multiplicity 1), $\pm(\varepsilon_i - \varepsilon_j)$, $1 \leq i < j \leq 3$ (of multiplicity 1), $\pm\varepsilon_i$, $1 \leq i \leq 3$ (of multiplicity 2), and 0 (of multiplicity 3). It follows that if $f \in \mathfrak{X}^*$ is defined by the conditions $f(\varepsilon_1) = 1$, $f(\varepsilon_2) = 2$, $f(\varepsilon_3) = -3$, then $\dim V(f) = 12$ and the sets $\beta_1 = \varepsilon_3 - \varepsilon_2$, $\beta_2 = \varepsilon_3$, $\beta_3 = -\varepsilon_1 + \varepsilon_3$ and $\Lambda_1 = -\varepsilon_1 + \varepsilon_3$, $\Lambda_2 = \varepsilon_1 - \varepsilon_2$, $\Lambda_3 = 0$ satisfy the conditions from Proposition 5. Therefore $\dim GV(f) \geq 12 + 3 > 14 = m_G$ (see [52]).

7) The weights are $\pm 2\varepsilon_i \pm \varepsilon_j$; $i, j = 1, 2, i \neq j$ (the choice of signs is arbitrary and the multiplicities $= 1$), $\pm\varepsilon_i$, $i = 1, 2$ (of multiplicity 2). Therefore, if $f \in \mathfrak{X}^*$ is defined by the conditions $f(\varepsilon_1) = 1$ and $f(\varepsilon_2) = 3$, then $\dim V(f) = 8$ and the sets $\beta_1 = -2\varepsilon_2$, $\beta_2 = -\varepsilon_1 - \varepsilon_2$, $\beta_3 = -2\varepsilon_1$ and $\Lambda_1 = \varepsilon_1 - 2\varepsilon_2$, $\Lambda_2 = -\varepsilon_2$, $\Lambda_3 = -\varepsilon_1$ satisfy the conditions from Proposition 5. Hence $\dim GV(f) \geq 8 + 3 > 10 = m_G$ (see [52]). □

5. Finally we will show how to use the results of Chapter 4 for the study of equidimensionality.

PROPOSITION 10. *The following G-modules $V = R(\Lambda)$ are not equidimentional*: 1) *G is of type B_7*, $\Lambda = \omega_7$; 2) *G is of type A_9*, $\Lambda = \omega_3$.

PROOF. It follows from [108] that any slice representation (see part 5 of §0, Chapter 4) of an equidimensional representation is also equidimensional. Therefore it suffices to construct a nonequidimensional slice representation (G_v, N_v) of V. We choose v so that a maximal torus T of G_v is one-dimensional and $\dim N_v^{\pm} \geq 4$ (see the notation of part 1 of §2, Chapter 4); in this case (G_v, N_v) would not be equidimensional. Indeed $N^{\pm} \subset \pi^{-1}_{N_v, G_v}(\pi_{N_v, G_v}(0))$, so that $\dim \pi^{-1}_{N_v, G_v}(\pi_{N_v, G_v}(0)) \geq 4$. On the other hand $(G_v)^0$, being reductive, either coincides with T or is a three-dimensional simple group. In the former case the action of T on N_v is stable and nontrivial, so that $\min_{u\in N_v} \dim \pi^{-1}_{N_v, T}(\pi_{N_v, T} \mid (u)) = 1$. In the latter case, because (G_v^0) is simple, we have $\min_{u\in N_v} \dim \pi^{-1}_{N_v, (G_v)^0}(\pi_{N_v, (G_v)^0}(u)) = \max_{u\in N_v} \dim(G_v)^0 u \leq 3$ (see [29]).

1) Let T and C be chosen for G as in Lemma 5.2 of Chapter 4 with $\alpha = \omega_2$. Because the generic stabilizer for the representation $R(\omega_1)\otimes R(\omega_5)$ of a group of type $A_1 \oplus B_5$ is finite (see [53]), it follows from Lemma 5.2, the remark after Theorem 3.1, and Theorem 3.3 of Chapter 4 that if v is a point in general in V^T then $\overline{Gv} = Gv$ and T is a maximal torus in G_v. Because, by Lemma 5.2 of Chapter 4, $(C, V) = 64S_0 \oplus 32S_1$ and $(C, \mathfrak{g}) = 58S_0 \oplus 22S_1 \oplus S_2$, we have $\dim V^{\pm} = 32$, $\dim \mathfrak{g}^{\pm} = 23$. Hence (by formula (0.2) of Chapter 4), $\dim N_v^{\pm} = 9$, if $(G_v)^0 = T$ and $\dim N_v^{\pm} = 10$, if $\dim(G_v)^0 = 3$.

2) Consider the following two one-dimensional tori in $G = \mathrm{SL}_{10}$:

$$T = \{\mathrm{diag}(t, t, t, 1, 1, 1, 1, t^{-1}, t^{-1}, t^{-1}) | t \in k^*\}$$

and

$$Z = \{h(t) = \mathrm{diag}(t^{-4}, t^{-4}, t^{-4}, t^3, t^3, t^3, t^3, 1, 1, 1) \mid t \in k^*\}.$$

Then

$$Z_G(T) = T \times Z \times Z_G(T)',$$

where

$$Z_G(T)' = \{\begin{pmatrix} A & & \\ & B & \\ & & C \end{pmatrix} | A, C \in \mathrm{SL}_3, B \in \mathrm{SL}_4\}$$

(compare with part 2 of §5, Chapter 4) so that $Z_G(T)/T \cong Z \times Z_G(T)' = k^* \times \mathrm{SL}_3 \times \mathrm{SL}_4 \times \mathrm{SL}_3$. Let $\chi: Z \to k^*$, $\chi(h(t)) = t$, be a generator of $\mathrm{Hom}(Z, k^*)$. Arguing as in part 2 of §5, Chapter 4 we see that

$$(Z_G(T)/T, V^T) = (\chi^{-1} \otimes U_1) \oplus (\chi^9 \otimes U_2),$$

where U_1 (resp., U_2) is the $\mathrm{SL}_3 \times \mathrm{SL}_4 \times \mathrm{SL}_3$-module $R(\omega_1)\otimes R(\omega_1)\otimes R(\omega_1)$ (resp., $1 \otimes R(\omega_3) \otimes 1$). By [53], the generic stabilizer for U_1 is finite (and therefore U_1 is not locally transitive, see [154]). By the remark after Proposition 3.3 of Chapter 4, it follows that the generic stabilizer for $(Z_G(T)/T, V^T)$ is also finite. Because U_1 is not locally transitive, we can find a homogeneous $Z_G(T)'$-invariant of positive degree contained in $k[U_1]$; clearly it is a Z-semi-invariant of positive weight.

It follows from [106] that in $k[U_1 \oplus U_2]$ there exists a $Z_G(T)'$-invariant that is homogeneous in U_i of degree a_i, $i = 1, 2$ with $9a_2 > a_1$. It is a Z-semi-invariant of negative weight. It now follows from the remarks after Proposition 3.2 and Theorem 3.1, as well as from Theorem 3.3, Chapter 4, that if v is a generic point in V^T then $\overline{Gv} = Gv$ and T is a maximal torus in G_v.

Because $V = \Lambda^3 R(\omega_1)$, it is true that $\dim V^{\pm} = 40$. Because $\Delta = -\Delta$, we have $\dim \mathfrak{g}^{\pm} = (\dim G - \dim Z_G(T))/2 = 33$. Therefore $\dim N_v^{\pm} = 7$, if $(G_v)^0 = T$, and $\dim N_v^{\pm} = 8$, if $\dim (G_v)^0 = 3$. □

6. THEOREM 8 (A classification). *The set of irreducible equidimensional G-modules of any connected simply connected simple group G coincides with the set of irreducible G-modules with free modules of covariants and is given (up to an outer automorphism of G) by the following list:*

G is of type $A_l (l \geq 1)$, $\Lambda = \omega_1 ; 2\omega_1 ; \omega_1 + \omega_l (l \geq 2)$; $\omega_2 (l \geq 2)$; $\omega_3 (5 \leq l \leq 8)$; $\omega_4 (l = 7)$; $3\omega_1 (l = 1, 2)$; $2\omega_2 (l = 3)$; $4\omega_1 (l = 1)$.

G is of type $B_l (l \geq 3)$, $\Lambda = \omega_1 ; 2\omega_1 ; \omega_2 ; \omega_l (3 \leq l \leq 6)$.

G is of type $C_l (l \geq 2)$, $\Lambda = \omega_1 ; 2\omega_1 ; \omega_2 ; \omega_l (3 \leq l \leq 4)$; $2\omega_l (l = 2)$.

G is of type $D_l (l \geq 4)$, $\Lambda = \omega_1 ; 2\omega_1 ; \omega_2 ; \omega_l (5 \leq l \leq 8)$.

G is of type E_6, $\Lambda = \omega_1 ; \omega_2$.

G is of type E_7, $\Lambda = \omega_1 ; \omega_7$.

G is of type E_8, $\Lambda = \omega_8$.

G is of type F_4, $\Lambda = \omega_1 ; \omega_4$.

G is of type G_2, $\Lambda = \omega_1 ; \omega_2$.

PROOF. It follows from part 1, the results of part 3 and Propositions 8, 9, 10, and 3 that (up to an outer automorphism of G) any irreducible equidimensional G-module is contained in the list. On the other hand, computing $\dim V/G = \dim V - m_G$ using [52] and the classification of Θ-groups from [13] (those are reductive linear groups associated, via a construction from [13], with each automorphism of finite order of any semisimple Lie algebra), we see that any module V from this table satisfies at least one of the following conditions: a) $\dim V/G \leq 2$; b) V is defined by a Θ-group. Because, by [13], in case b) the algebra $k[V]^G$ is free and $\mathfrak{N}$ contains only finitely many orbits, it follows from Theorem 1, the corollary of Theorem 3, and Proposition 1 that all of the $R(\Lambda)$ from the table have free modules of covariants. □

7. As was pointed out in part 11 of the Introduction, our tables show that (when G is simple and V is irreducible) the equidimensionality of V is equivalent to any of the following properties: a) $k[V]^G$ is free; b) $\mathfrak{N}$ contains only finite many orbits; c) any fiber of $\pi_{V,G}$ contains only finitely many orbits; d) the generic stabilizer for V is not contained in the noneffectivity kernel.

We can also add the following properties: e) each irreducible component of $\mathfrak{N}$ contains an open orbit; f) each irreducible component of any fiber of $\pi_{V,G}$ contains an open orbit. Indeed it follows from e) that $\dim \mathfrak{N} \leq m_G$ therefore, by parts 1) and 2) of Theorem 3, the module V is equidimensional. Our assertion now follows from the fact that an orbit is open in its closure.

We should remark that for reducible modules the above properties are not necessarily equivalent any longer. For example, if $G = \mathrm{GL}_2$ and $V = mR(\omega_1)$ then for $m = 2$ we have $\dim V/G = 1$ so that (by Proposition 1) the module of covariants of V is free. However $\mathfrak{N} = \{(v_1, v_2) \in 2R(\omega_1) \mid \langle v_1\rangle = \langle v_2\rangle\}$ and therefore $\mathfrak{N}$ contains infinitely many orbits (the coefficient of proportionality of v_1 and v_2 is constant on an orbit). If $m = 3$ then $\dim V/G = 3$ and $\mathrm{hd}\, k[V]^G = 0$ (see the end of the proof of Theorem 10.2, Chapter 4) so that $k[V]^G$ is free. However $\mathfrak{N} \cong \{X \in \mathrm{Mat}_{2\times 3} \mid \mathrm{rk}\, X = 1\}$, and therefore $\dim \mathfrak{N} = 4 > m_G = 3$. Hence V is not equidimensional. □

Theorem 9. *Suppose that G is simple and V is irreducible. Then the following conditions are equivalent*: 1) *V is equidimensional,* 2) *V has a Cartan subspace,* 3) *V has a section.*

Proof. By Theorems 7 and 8, it suffices to show that the modules from the list in Theorem 8 have sections and Cartan subspaces.

If either $\dim V/G \le 1$ or V is defined by a Θ-group then, by [13] and Proposition 6, the module V has a Cartan subspace. The only V that does not have the mentioned properties is $R(\omega_6)$ when G is of type B_6. In this case, however, a Cartan subspace exists by [72].

If $\dim V/G \le 1$ then, by property e) and Proposition 7, the module V has a section. The same is true when V is defined by the Θ-group corresponding to an automorphism of order $d \le 2$: when $d = 1$ then V is the adjoint module, and when $d = 2$ then V is the module given by the isotropy representation of a symmetric space. In those cases the existence of sections was proved in [101] and, respectively, [102]. This leaves out the following modules: 1) $R(3\omega_1)$ when G is of type A_2, 2) $R(\omega_3)$ when G is of type A_8, 3) $R(\omega_6)$ when G is of type B_6 (the first two are defined by the Θ-groups corresponding to automorphisms of order 3 of the Lie algebras E_8 and, respectively, D_4). We shall examine each module separately, using Lemma 3 and the remarks following it (and using the notation from those remarks).

1) This is the SL_3-module of cubic forms in variables x_1, x_2, x_3. In this case $\dim V/G = 2$ and $d_1 = 4, d_2 = 6$ (see [48, 13]). Let $v = x_1^3 + x_2^2 x_3$. Then $\lim_{t\to 0} \mathrm{diag}(t, t^2, t^{-3})v = 0$ and therefore $Gv \subset \mathfrak{N}$. Because the Hessian $H(v)$ of the form v is equal to $-24x_1x_2^2$, it is true that $\dim G_v = 0$ (see the end of the proof of Lemma 2.1 in Chapter 4). Therefore Gv is open in $\mathfrak{N}$. Let $\lambda(t) = \mathrm{diag}(t, t^4, t^{-5})$. Then $\lambda(t)v = t^3 v$ for all t, so that $p = 3$. An elementary calculation shows that $P_1 = \{t^0, t^0, t^{-3}, t^3, t^6, t^{-6}, t^9, t^{-9}\}$, $P_2 = \{t^0, t^3, t^3, t^{-3}, t^6, t^{-6}, t^9, t^{-9}, t^{12}, t^{-15}\}$. Hence $t^{-p}P_2 - P_1 = \{t^{-12}, t^{-18}\}$ and therefore $a_1 = 12$ and $a_2 = 18$. Thus $a_i = pd_i$ for any i and therefore V has a section.

2) This is the SL_9-module of 3-vectors in 9-space U (choosing a basis $e_1, \dots, e_9$ in U we identify G with the matrix group SL_9). By [17], we

have $\mathfrak{N} = \overline{Gv}$ for $v = e_{149} + e_{168} + e_{239} + e_{258} + e_{267} + e_{348} + e_{357} + e_{456}$ (here $e_{ijk} = e_i \wedge e_j \wedge e_k$), and $\dim G_v = 0$. It is easy to see that $\lambda(t)v = t^6 v$ for all t where $\lambda(t) = (t^{74}, t^{56}, t^{38}, t^{20}, t^2, t^{-16}, t^{-34}, t^{-52}, t^{-88})$. Therefore $p = 6$. By [17], $\dim V/G = 4$ and $d_i = 6i + 6$, $i = 1, 2, 3, 4$. Therefore to prove that $t^{-p}P_2 - P_1 = \{t^{-pd_i} \mid i = 1, 2, 3, 4\}$ (which implies the existence of a section) it suffices to show that $\dim V_i > \dim \mathfrak{g}_i$ for all $i = 1, 2, 3, 4$, where $V_i = V_{t^{-pd_i+p}}$, and $\mathfrak{g}_i = \mathfrak{g}_{t^{-pd_i}}$ (the weights are considered with respect to λ). An elementary calculation shows that $\dim \mathfrak{g}_i = 5, 3, 1$, and 0 when, respectively, $i = 1, 2, 3, 4$ and that $e_{478}, e_{568}, e_{189}, e_{279}, e_{369}, e_{459} \in V_1$; $e_{678}, e_{479}, e_{569}, e_{389} \in V_2$; $e_{679}, e_{589} \in V_3$, and $e_{789} \in V_4$. Hence the aforementioned inequalities hold.

3) This is a spinor module of the group Spin_{13}. Its weights are simple and of the form $(\pm\varepsilon_1 \pm \cdots \pm \varepsilon_6)/2$; the roots of $\mathfrak{g}$ are of the form $\pm\varepsilon_i$, $\pm\varepsilon_i \pm \varepsilon_j$, $1 \le i \le 6$, $1 \le i < j \le 6$ (we use the notation of part 3 of §5, Chapter 4). By [72], $\dim V/G = 2$, $d_1 = 4$, $d_2 = 8$, and $\mathfrak{N} = \overline{Gv}$ for some point $v = \sum_{i=1}^{5} v_i$, where $v_i \neq 0$ for all i and $v_1 \in V_{(\varepsilon_1+\varepsilon_2+\varepsilon_3+\varepsilon_4+\varepsilon_5+\varepsilon_6)/2}$, $v_2 \in V_{(-\varepsilon_1-\varepsilon_2-\varepsilon_3-\varepsilon_4+\varepsilon_5+\varepsilon_6)/2}$, $v_3 \in V_{(-\varepsilon_1-\varepsilon_2+\varepsilon_3+\varepsilon_4-\varepsilon_5-\varepsilon_6)/2}$, $v_4 \in V_{(\varepsilon_1+\varepsilon_2+\varepsilon_3-\varepsilon_4-\varepsilon_5-\varepsilon_6)/2}$, $v_5 \in V_{(-\varepsilon_1+\varepsilon_2-\varepsilon_3+\varepsilon_4-\varepsilon_5+\varepsilon_6)/2}$. Hence if we choose $h \in \mathfrak{h}$ so that $\varepsilon_1(h) = -6$, $\varepsilon_2(h) = 4$, $\varepsilon_3(h) = 4$, $\varepsilon_4(h) = -2$, $\varepsilon_5(h) = 2$, $\varepsilon_6(h) = 0$ then $hv = v$. There exists a one-parameter subgroup $\lambda \colon k^* \to G$ such that $h = (d\lambda(t))/(dt)|_{t=1}$ (in the notation of [38], $\lambda(t) = s_1(t^{-3})s_2(t^2)s_3(t^2)s_4(t^{-1})s_5(t)$). We now have $\lambda(t)v = tv$ for all t so that $p = 1$. An elementary calculation shows that the spectrum of h on $\mathfrak{h}$ is $\{O^{(12)}; \pm 2^{(12)}; \pm 4^{(9)}; \pm 6^{(7)}; \pm 8^{(3)}; \pm 10^{(2)}\}$, and on V is $\{\pm 1^{(10)}; \pm 3^{(10)}; \pm 5^{(6)}; \pm 7^{(4)}; \pm 9^{(2)}\}$ (the multiplicities of the eigenvalues are indicated in parentheses). Therefore $t^{-p}P_2 - P_1 \supset \{t^{-4}, t^{-8}\}$ and, because V/G is two dimensional, this inclusion is in fact an equality. Hence $a_i = pd_i$ for all i. □

8. In the conclusion of this chapter we shall consider modules called *exceptional in the sense of Igusa* (see the definition in part 11 of the Introduction). The assumptions on G and V will be the same as in part 1 of §1. In fact we consider an a priori bigger class of modules. Those are the modules satisfying the following condition (henceforth called Igusa's condition):

$$\left\{\begin{array}{l}\text{the null-variety } \mathfrak{N} \text{ contains an open (not necessarily dense) orbit } \mathcal{O} \text{ with} \\ \dim(\overline{\mathcal{O}} - \mathcal{O}) \le \dim \mathcal{O} - 2.\end{array}\right.$$

Clearly the modules exceptional in the sense of Igusa satisfy this condition. We shall obtain a classification of all irreducible modules of simple groups satisfying Igusa's condition, and, in particular, we will show that each one of them is exceptional.

PROPOSITION 11. *Suppose that V satisfies Igusa's condition. Then $m_G < \dim G$.*

PROOF. Because $\overline{\mathcal{O}}$ is an irreducible component of $\mathfrak{N}$ and $m_G = \dim V - \dim V/G$, we deduce from the theorem on the dimension of a fiber (see [51]) that $\dim \mathcal{O} = m_G$. We may assume that G acts on V nontrivially, i.e., $m_G \neq 0$. Then $\overline{\mathcal{O}} \neq \mathcal{O}$ (because $0 \in \overline{\mathcal{O}}$). If we had $m_G = \dim G$ then the G-stabilizer of a point from $\mathcal{O}$ would be finite, and, by Matsushima's criterion, $\mathcal{O}$ would be affine. But then $\dim(\overline{\mathcal{O}} - \mathcal{O}) = \dim \overline{\mathcal{O}} - 1$ (see [29]), contrary to Igusa's condition. □

PROPOSITION 12. *If* $\dim V/G = 0$ *then* V *is exceptional in the sense of Igusa.*

PROOF. Because $m_G = \dim V$, it follows that $V = \mathfrak{N}$ contains a dense orbit $\mathcal{O}$. If one of the irreducible components Γ of $V - \mathcal{O}$ were a hypersurface in V then, because G is connected and semisimple, a generator of the ideal $\{f \in k[V] \mid f(v) = 0 \text{ for all } v \in \Gamma\} = \mathcal{T}(\Gamma)$ would be contained in $k[V]^G$, which is impossible because $k[V]^G = k$. □

PROPOSITION 13. *Suppose that* $\dim V/G = 1$ *and* $V - \mathfrak{N}$ *contains a point* v *such that* G_v *is reductive and* $\dim Gv = m_G$. *Then all of the fibers of* $\pi_{V,G}$ *different from* $\mathfrak{N}$ *are orbits and* G_u *is conjugate to* G_v *for any* $u \in V - \mathfrak{N}$.

PROOF. Because $v \in \mathfrak{N}$, it is true that $|Gv \cap \langle v \rangle| < \infty$. Consider the morphism $\varphi\colon G \times k^* \to V$, $\varphi((g, t)) = tgv$. Because $\dim Gv = \dim V - 1$, it follows that the image of φ is dense in V. Because the G-stabilizers of the points of this image are conjugate to G_v and as G_v is reductive, it follows from Theorem 3.3, Chapter 4 that the action of G on V is stable. Therefore $\overline{Gv} = Gv$. Because $\dim V/G = 1$, it is true that $k[V]^G = k[f]$, where f is a nonconstant homogeneous polynomial (see [38]). Because $k[V]^G$ separates closed invariant subsets, it follows from the dimension count that $Gv = \{w \in V \mid f(w) = f(v)\}$. Therefore $Gtv = \{w \in V \mid f(w) = f(tv)\}$ for all $t \in k^*$. Because $f(v) \neq 0$ and $f(tv) = t^{\deg f} f(v)$, it follows that all of the fibers of $\pi_{V,G}$ except $\mathfrak{N}$ can be obtained this way. □

COROLLARY. *Under the assumptions of Proposition* 13, V *is exceptional in the sense of Igusa if and only if* V *satisfies Igusa's condition.*

REMARK. The example $V = R(3\omega_1)$ for $G = \mathrm{SL}_2$ shows that the conditions of Proposition 13 can be satisfied when Igusa's condition is not (in this case $m_G = \dim G$ (see [52])). The example $V = 2R(\omega_1) \oplus R(\omega_3)$ for $G = \mathrm{SL}_6$ shows that in the statement of Proposition 13 we cannot do away with the requirement that G_v be reductive (in this case the action of G on V is not stable because the generic stabilizer is not reductive (see [52])).

9. THEOREM 10 (A classification). *Suppose that* G *is simple and* V *is irreducible. Then the following are equivalent*: 1) V *satisfies Igusa's condition*; 2) V *is exceptional in the sense of Igusa*; 3) $\dim V \leq \dim G$; 4) V *is an equidimensional G-module* $R(\Lambda)$ (*see Theorem* 8) *different* (*up to an outer* *automorphism of G) from the following modules*: G *is of type* A_l $(l \geq 1)$, $\Lambda = \omega_3$ $(l = 8)$; ω_4 $(l = 7)$; $3\omega_1$ $(l = 1, 2)$; $2\omega_2$ $(l = 3)$; $4\omega_1$ $(l = 1)$. G *is of type* B_l $(l \geq 3)$, $\Lambda = 2\omega_1$. G *is of type* C_l $(l \geq 2)$, $\Lambda = \omega_l$ $(l = 4)$; $2\omega_l$ $(l = 2)$. G *is of type* D_l, $\Lambda = 2\omega_1$; ω_l $(l = 8)$.

PROOF. By [3], $m_G < \dim G \Leftrightarrow \dim V \leq \dim G$. Because of Proposition 11 this means that 1) $\Rightarrow$ 3). Comparing the list (see [13]) of all V for which $\dim V \leq \dim G$ with the list from Theorem 8, we have 3) $\Leftrightarrow$ 4). It remains to show that the modules from 4) are all exceptional. Excluding those that are adjoint (their exceptionality was proved by Kostant [101]) and those whose exceptionality was proved by Igusa himself in [87–89] (many of them are locally transitive (see [52]) so their exceptionality follows a priori from Proposition 12) we only have the following list:

a) $G = \mathrm{SL}_{l+1}$, $V = R(2\omega_1)$. In this case $V = S^2R(\omega_1)$ and therefore $G_v \cong (\mathrm{SO}_m \times \mathrm{SL}_{l+1-m})U_{m(l+1-m)}$ if $\mathrm{rk}\, v = m$, $v \in V$ (here $U_{m(l+1-m)}$ is the commutative unipotent $m(l+1-m)$-dimensional radical). Thus the conditions of Proposition 13 are satisfied, and, by the corollary of this proposition, we only have to check Igusa's condition. We have $\mathfrak{N} = \overline{\mathscr{O}}_l \supset \overline{\mathscr{O}}_{l-1} \supset \cdots \supset \{0\}$, where $\mathscr{O}_m = \{v \in V \mid \mathrm{rk}\, v = m \leq l\}$ and the dimension count yields $\dim \mathfrak{N} = \dim \mathscr{O}_{l-1} + 2$.

b) $G = \mathrm{SL}_{l+1}$, $V = R(\omega_3)$, $l = 5, 6, 7$. The conditions of Proposition 13 are again satisfied (see [52]). In this case $V = \Lambda^3 R(\omega_1)$ and classifications of 3-vectors in 6-, 7-, and 8-spaces were obtained, respectively, by Reichel, Schouten, and Gurevich (see [20]). It follows from their results that $\mathfrak{N}$ contains only finitely many orbits. Thus the verification of Igusa's condition reduces to the calculation of the dimensions of the stabilizers for a finite number of certain 3-vectors. Essentially this has already been done in [17] (although it is not difficult to do it directly). The results are as follows: if $l = 5$, then $\mathfrak{N}$ contains 4 orbits of dimensions $0, 10, 15, 19$; if $l = 6$, then $\mathfrak{N}$ contains 9 orbits of dimensions $0, 13, 20, 21, 25, 26, 28, 31, 34$; if $l = 7$, then $\mathfrak{N}$ contains 22 orbits of dimensions $0, 16, 25, 28, 31, 32, 35, 38, 40, 41, 41, 42, 43, 44, 46, 47, 48, 48, 50, 52, 53, 55$. Therefore Igusa's condition is satisfied.

c) G is of type B_6 and $V = R(\omega_6)$. The exceptionality was proved in [72].

d) G is of type D_7 and $V = R(\omega_7)$. The conditions of Proposition 13 are again satisfied (see [52]). There are finitely many orbits in $\mathfrak{N}$ and their dimensions are as follows (see [38]): $0, 22, 35, 43, 44, 50, 54, 59, 63$. Therefore Igusa's condition is satisfied. □

CHAPTER 6

A Classification of Normal Affine Quasihomogeneous Varieties of SL_2

The assumptions on the field k are the same as in Chapter 3.

§1. Some general results and the beginning of classification

1. We keep the definitions and notation from part 1 of §2, Chapter 1. When speaking about quasihomogeneous varieties of the group G, an isomorphism will always be understood as an equivariant isomorphism. If two quasihomogeneous varieties are isomorphic then their principal orbits are also isomorphic. Conversely, if the principal orbits $\mathscr{O}_X$ and $\mathscr{O}_Y$ of two quasihomogeneous varieties X and Y are isomorphic, then, although X and Y are generally not isomorphic, there exists a birational G-isomorphism $\varphi: X \to Y$; it is induced by the isomorphism $\mathscr{O}_X \to \mathscr{O}_Y$. Any birational G-isomorphism can be obtained this way (because the indeterminacy set of a rational G-map is G-invariant). Therefore, if φ is an isomorphism in the category of algebraic varieties, then φ is also an isomorphism in the category of quasihomogeneous varieties. Because of this, the classification problem for quasihomogeneous varieties reduces to the question whether a birational G-isomorphism is a morphism: this is the problem of extending a rational map to the boundary (and determining the obstructions to such extensions).

2. Let V be a finite-dimensional G-module and v a vector from V. Then $\overline{Gv}$ is an affine quasihomogeneous variety of the group G. In fact, all such varieties can be obtained this way: this follows immediately from the embedding theorem, according to which any affine variety X endowed with a regular action of G can be embedded equivariantly into a finite-dimensional G-module (proof: because $k[X]$ is an affine k-algebra with a rational action of G (see §1, Chapter 1), it is true that $k[X]$ contains a finite-dimensional G-submodule W generating $k[X]$ as algebra; if $f_1, \dots, f_n$ is a basis of it then $f = (f_1, \dots, f_n): X \to W^* := U$ is a desired embedding). This observation is crucial in applications of representation theory methods to the classification problem in the affine case; we will come back to it several times later. Notice that for normal projective varieties one can use the theorem on equivariant embedding in $\mathbb{P}^n$ (see [114]); however, as we will show in part

15 of §1, Chapter 7, a nonnormal variety may have no such embeddings. Henceforth in this chapter, unless stated otherwise, we only consider affine quasihomogeneous varieties X (so we may assume that G is linear (see [127])).

3. The classification problem for such varieties can be reformulated in purely algebraic terms: it is equivalent to the classification, up to right translations, of left-invariant finitely-generated subalgebras of $k[G]$. Indeed if $v \in \mathcal{O}_X$ and $\varphi_v \colon G \to X$, $\varphi_v(g) = gv$, then $\varphi_v^*(k[X])$ is a subalgebra of that type, and when v is changed the subalgebra is modified by a right translation. Conversely, if A is such a subalgebra, $X = \operatorname{Specm} A$ and $\varphi \colon G \to X$ is induced by the embedding $A \hookrightarrow k[G]$ then, because the image of the G-morphism φ is dense, X is quasihomogeneous with respect to G and $\varphi = \varphi_v$, where $v = \varphi(e)$.

These two formulations of the same classification problem, one geometric, in terms of affine quasihomogeneous varieties, the other algebraic, in terms of left-invariant subalgebras, are mutually beneficial. In some cases, it is more convenient to work with subalgebras; this leads to a classification of varieties (see below the example of S-varieties). In other cases, the geometric tools work better; this yields a description of subalgebras.

4. Example. S-varieties (see [16]). By definition, they are varieties X for which G_v, $v \in \mathcal{O}_x$, contains a maximal unipotent subgroup of G (clearly, we may assume that G is reductive; in [16] G was assumed to be connected).

Classical examples: the quadric $\sum_{i=1}^n x_i^2 = 0$ in $\mathbb{A}^n$ and the Grassmannian cone (here G is, respectively, the orthogonal and unimodular groups).

Suppose that G is connected. We keep the notation from parts 1 and 2 of §4, Chapter 4. If v is chosen so that $G_v \supset U$, then $\varphi_v^*(k[X])$ is an invariant subalgebra of $S = k[G/U]$. It is of the form $\bigoplus_{\Lambda \in \mathfrak{Z}} S(\Lambda)$ where $\mathfrak{Z}$ is a finitely-generated subsemigroup with zero of $\mathfrak{X}$. In this case we write $X = X(\mathfrak{Z})$. This construction gives rise to a bijection between the set of isomorphism classes of S-varieties and the set of all finitely-generated subsemigroups with zero of $\mathfrak{X}$. We call $\mathfrak{Z}$ the signature of $X(\mathfrak{Z})$. If $\Lambda_1, \dots, \Lambda_n$ is a (not necessarily minimal) system of generators of $\mathfrak{Z}$, v_i is a highest weight vector in $R(\Lambda_i)^*$ and $w = v_1 + \cdots + v_n \in \bigoplus_{i=1}^n R(\Lambda_i)^*$, then $gv \mapsto gw$ gives rise to a G-isomorphism between X and $\overline{Gw}$. Any equivariant linear embedding of X can be obtained this way. Let K_X be the polyhedral cone spanned by $\mathfrak{Z}$ in $\operatorname{Hom}(B, k^*) \otimes_{\mathbb{Z}} \mathbb{Q}$. If F is a face of K_X, then $\bigoplus_{\Lambda \in \mathfrak{Z} - (\mathfrak{Z} \cap F)} S(\Lambda)$ is an invariant prime ideal of $k[X]$ whose set of zeros is the S-variety defined by the subalgebra $\bigoplus_{\Lambda \in \mathfrak{Z} \cap F} S(\Lambda)$. If $\mathcal{O}_F$ is its principal orbit then the map $F \mapsto \mathcal{O}_F$ is an isomorphism between the partially-ordered set of faces of K and the partially-ordered set of orbits of G in X: $F_1 \subset F_2 \Leftrightarrow \mathcal{O}_{F_1} \subset \overline{\mathcal{O}}_{F_2}$. In particular, X is homogeneous if and only if K_X has no proper faces (in the case when $U = \{e\}$, i.e., G is a torus, this yields the following criterion: let V be a finite-dimensional G-module, and $v_i \in V_{\Lambda_i}$, $v_i \neq 0$, $\Lambda_i \in \mathrm{Hom}(G, k^*)$, $i = 1, \cdots, n$ then $\overline{Gv} \neq Gv$ (resp., $\overline{Gv} \ni 0$) if and only if $\Lambda_1, \dots, \Lambda_n$ are contained in some closed (resp., open) half-space of $\mathrm{Hom}(G, k^*) \otimes_{\mathbb{Z}} \mathbb{Q})$.

Later we will need the following lemma (it is not contained in [16]); among other things it establishes an interesting connection between the structure of the singularities of X and Weyl's formula for the dimensions of simple G-modules: with its aid one can calculate the Poincaré series of the local ring of a fixed point of X.

LEMMA 1. *Suppose X contains a fixed point o. Let* $\mathfrak{m} = \mathfrak{m}_{o,X} \cap k[X]$, $\mathfrak{Z}_0 = \mathfrak{Z} - \{0\}$ *and* $d\mathfrak{Z}_0 = \{M_1 + \cdots + M_d \mid M_i \in \mathfrak{Z}_0$ *for all* $i\}$ *for all* $d \in \mathbb{Z}^+$. *Then*: 1) $\mathfrak{m}^d = \bigoplus_{\Lambda \in d\mathfrak{Z}_0} S(\Lambda)$; 2) $\mathfrak{Z}_0$ *is a finitely-generated semigroup with a unique minimal system of generators* $\Lambda_1, \dots, \Lambda_m$ *(in the sense that no smaller subset of it is a system of generators)*; 3) $\Theta_{o,X} = \bigoplus_{i=1}^m R(\Lambda_i)^*$; 4) $\dim \Theta_{o,X} = \sum_{i=1}^m \dim R(\Lambda_i)$; 5) *the set* $d\mathfrak{Z}_0 - (d+1)\mathfrak{Z}_0$ *is finite for any* d *and when* $d = 1$ *coincides with* $\{\Lambda_1, \dots, \Lambda_m\}$, 6) *the Poincaré series of the local ring* $\mathcal{O}_{o,X}$ *is of the form* $P(\mathcal{O}_{o,X}, t) = 1 + \sum_{d=1}^{\infty} (\sum_{\Lambda \in d\mathfrak{Z}_0 - (d+1)\mathfrak{Z}_0} \dim R(\Lambda)) t^d$; 7) *if* $m = 1$ *(which is equivalent to the fact that X is normal and contains only two orbits: o and $\mathcal{O}$, (see [16])) then* $P(\mathcal{O}_{o,X}, t) = \sum_{d=0}^{\infty} \dim R(d\Lambda_1) t^d$.

PROOF. Because $k[X] = \bigoplus_{\Lambda \in \mathfrak{Z}} S(\Lambda)$ and $\mathfrak{m}$ is an invariant ideal with $k[X]/\mathfrak{m} = k = S(0)$, we have $\mathfrak{m} = \bigoplus_{\Lambda \in \mathfrak{Z}_0} S(\Lambda)$. It now follows from the definition of $\mathfrak{m}^d$ and from Property b) of algebra S (see part 2 of §4, Chapter 4) that 1) holds and that $\mathfrak{Z}_0$ is a semigroup. We choose a minimal system of generators $\Lambda_1, \dots, \Lambda_m$ of the semigroup (with zero) $\mathfrak{Z}$. Then $\Lambda_i \in \mathfrak{Z}_0$ for each i. Because $\mathfrak{Z}_0$ is a semigroup, it is true that $\Lambda_1, \dots, \Lambda_m$ is a system of generators of it. Notice that $\Lambda_i \notin 2\mathfrak{Z}_0$ for each i, otherwise if, say, $\Lambda_1 = M_1 + M_2$, $M_1, M_2 \in \mathfrak{Z}_0$ then, expressing M_i in terms of $\Lambda_1, \dots, \Lambda_{m_1}$, we would have $\Lambda_1 = n_1\Lambda_1 + \cdots + n_m\Lambda_m$, $n_i \in \mathbb{N}$, $n_1 + \cdots + n_m \geq 2$; because $0 \notin \mathfrak{Z}_0$ we see that $n_1 = 0$, contrary to the minimality assumption. Therefore $\mathfrak{Z}_0 - 2\mathfrak{Z}_0 = \{\Lambda_1, \dots, \Lambda_m\}$. It now follows from 1) that $\Theta_{o,X} \cong (\mathfrak{m}/\mathfrak{m}^2)^* = \bigoplus_{\Lambda \in \mathfrak{Z}_0 - 2\mathfrak{Z}_0} R(\Lambda)^* = \bigoplus_{i=1}^m R(\Lambda_i)^*$. Because the structure of the G-module $\Theta_{o,X}$ does not depend on the choice of generators, we now have 2), 3), and (because $\dim R(\Lambda) = \dim R(\Lambda)^*$) 4). By 1), $\mathfrak{m}^d/\mathfrak{m}^{d+1} = \bigoplus_{\Lambda \in d\mathfrak{Z}_0 - (d+1)\mathfrak{Z}_0} R(\Lambda)$ therefore, in view of $\dim \mathfrak{m}^d/\mathfrak{m}^{d+1} < \infty$ we obtain 5) and 6). Finally if $m = 1$ then, obviously, $d\mathfrak{Z}_0 - (d+1)\mathfrak{Z}_0 = d\Lambda_1$, which yields 7). □

It is easy to see that if G is not connected then each irreducible component of X is an S-variety of the group G^0 and the (finite) group $N_G(B)/B$ acts transitively on the set of those components. If U is a maximal unipotent subgroup of G^0 then the relation between S-varieties and left-invariant finitely-generated subalgebras of $k[G/U]$ still holds.

5. Let H be a subgroup of G. There is a general construction that assigns a quasihomogeneous variety of G to any quasihomogeneous variety Y of H (the varieties are not assumed to be affine): we just take the homogeneous bundle G/H with fiber Y. It can be defined formally as the geometric quotient $G \times_H Y$ of $G \times Y$ (it exists by [138]) by the action of H given by $H: (g, y) \overset{h}{\mapsto} (gh^{-1}, hy)$; the action of G is induced by the action of G on $G \times Y$ via left translations on the first factor (compare with part 3 of §1, Chapter 1). There is a natural bijection between G-orbits in $G \times_H Y$ and H-orbits in Y. Indeed, the image of (e, Y) in $G \times_H Y$ is obviously H-invariant and H-isomorphic to Y, and the intersection of any G-orbit in $G \times_H Y$ with this image is an H-orbit. It follows, in particular, that $G \times_H Y$ is quasihomogeneous. The generic stabilizer for G in $G \times_H Y$ coincides with the generic stabilizer for H in Y. The natural morphism $G \times_H Y \to G/H$ yields the structure of a locally-trivial bundle (both in the étale and $\mathbb{C}$-topology (see [138])). Therefore the smoothness (and normality) of $G \times_H Y$ is equivalent to that of Y. If $s \in N_G(H)$ and Z is the variety with the action of H obtained by "twisting" the original action of H on Y by s (i.e., the morphism $H \times Y \to Y$ defining the new action is the composition of $H \times Y \to H \times Y$, $(h, y) \mapsto (shs^{-1}, y)$ and the morphism $H \times Y \to Y$ defining the old action) then $G \times_H Y$ and $G \times_H Z$ are G-isomorphic.

6. Henceforth we will assume that G is reductive. Then X has a unique (because of Property 2) from part 2 of §1, Chapter 1) closed orbit $\mathcal{O}$. If H is any reductive subgroup of G and Y is an affine quasihomogeneous variety of it then (by Mumford's criterion, see [114]) $G \times_H Y$ is also affine. This (along with [108]) shows that the construction from part 5 reduces the classification problem to the case when $\mathcal{O}$ is just a fixed point. Indeed, by [108], there is a reductive subgroup H of G such that $X = G \times_H Y$ for some affine quasihomogeneous Y with a (unique) fixed point (in addition, $H = G_v$ for some point $v \in \mathcal{O}$). Moreover if X is smooth then Y is a (locally-transitive) finite-dimensional H-module (so that X is a homogeneous vector bundle on G/H) (see [108]). This gives us a satisfactory solution of the classification problem in the smooth case, based on the well-developed techniques for classification of reductive subgroups and locally-transitive modules of reductive groups (see [22, 52, 53, 133]). In particular, this method works when X consists of a single orbit (i.e, X is homogeneous).

7. The results of Chapter 1 allow us to accomplish the classification in the next case, i.e., when X has exactly two orbits.

THEOREM 1. *Two-orbit affine quasihomogeneous varieties X of a reductive group G are exactly the varieties $G \times_H Y$, where H is a reductive subgroup of G and Y is a two-orbit S-variety of H with a fixed point.*

PROOF. By part 6, we have $X = G\times_H Y$, where Y is an affine quasihomogeneous variety of a reductive group H with a fixed point. Because X consists of only two orbits, the same is true for Y. Therefore $\dim(Y-\mathscr{O}_Y)=0$. The assertion now follows from Corollary 2 of Theorem 2, Chapter 1. □

From part 4 one can easily obtain a description of the varieties Y mentioned in the previous theorem. Let $\rho\colon H\to \mathrm{GL}(V)$ be a rational representation, let B_H be a Borel subgroup of H, and let $v_i\in V$ be a nonzero B_H-weighted vector of weight $\Lambda_i\neq 0$, $i=1,\dots,n$. Assume also that $\Lambda_1,\dots,\Lambda_n$ are proportional to each other in $\mathrm{Hom}(B_H, k^*)\otimes_{\mathbb{Z}}\mathbb{Q}$ with positive coefficients of proportionality. Then $\overline{\rho(H)v}$, where $v=v_1+\dots+v_n$, is a two-orbit S-variety (of H) with a fixed point, and all such varieties can be constructed this way.

8. We shall now take on the subject matter of this chapter. Henceforth (until the end of the chapter) $G=\mathrm{SL}_2$. We first recall several well-known facts (we use the notation of part 5 of §2, Chapter 4).

Proper algebraic subgroups of G. Up to conjugation, their complete list is as follows:

1) (dimension 0) T_n; D_n; $\widetilde{\mathbb{T}}$ (the binary tetrahedral group), $\widetilde{\mathbb{O}}$ (the binary octahedral group), and $\widetilde{\mathbb{I}}$ (the binary icosahedral group).

2) (dimension 1) T; $N_G(T)$; $N_n=\{\left(\begin{smallmatrix}\varepsilon & 0\\ 0 & \varepsilon^{-1}\end{smallmatrix}\right) \mid \varepsilon, c\in k,\ \varepsilon^n=1\}$. In particular, $N_1=U$ is a maximal unipotent subgroup of G.

3) (dimension 2) $B=\{\left(\begin{smallmatrix}c & 0\\ d & c^{-1}\end{smallmatrix}\right) \mid c\in k^*,\ d\in k\}$ is a Borel subgroup.

Finite-dimensional G-modules. They are of the form $V=\bigoplus_{i=1}^r S_{n_i}$. A point $v\in V$ will be written as follows

$$v=v_1\oplus\dots\oplus v_r, \tag{1}$$

where $v_i\in S_{n_i}$ and if $v_i\neq 0$ then

$$v=\sum_{j=0}^{m_i}\alpha_{ij}x^{n_i-j}y^j,\qquad 0\le m_i\le n_i,\ \alpha_{im_i}\neq 0. \tag{2}$$

We also have

$$\langle x^{n-j}y^j\rangle=(S_n)_{\chi^{-n+2j}}\qquad\text{and}\qquad S_n^U=\langle x^n\rangle.$$

9. Both the algebra S and its decomposition $\bigoplus_{\Lambda\in\mathfrak{X}}S(\Lambda)$ can be easily described. In our case $\mathfrak{X}$ can be naturally identified with $\mathbb{N}$ and S is the polynomial algebra in two independent variables p and q, viewed as regular functions on the group $G=\{\left(\begin{smallmatrix}p & c\\ q & d\end{smallmatrix}\right) \mid pd-qc=1\}$. The subspace $S(n)$ of S is spanned over k by the regular function $p^n, p^{n-1}q,\dots,p^iq^{n-i},\dots, pq^{n-1}, q^n$. Thus S can be identified with $k[x,y]$ (with the action defined by formula (2.3) of Chapter 4). Notice that the grading of S defined by its isotypic components coincides with the usual grading by degrees of polynomials. (All of this follows from the facts that the generic stabilizer for G in S_1 is conjugate to U, that $S_1 - \{0\}$ is a G-orbit, and that $k[S_1 - \{0\}] = k[S_1]$). Any S-variety $X = X(\mathfrak{Z})$ has only two orbits: $\mathcal{O}_X$ and the fixed point (because the cone K_X has only two faces). If $(n_1, \dots, n_r)$ is a system of generators of the semigroup (with zero) $\mathfrak{Z}$ and $u \in \mathcal{O}_X$ is a point such that $G_u \supset U$, then $\varphi\colon X \to \bigoplus_{i=1}^r S_{n_i}$, $\varphi(gu) = gv$ for all $g \in G$, where $v = \alpha_1 x^{n_1} \oplus \cdots \oplus \alpha_r x^{n_r}$, $\alpha_i \neq 0$ for all i, is an equivariant linear embedding of X, and any embedding can be obtained this way; we call $(n_1, \dots, n_r)$ the signature of embedding φ. It is easy to see that $G_u = N_d$, where $d = \gcd(n_1, \dots, n_r)$.

10. Let now X be an affine quasihomogeneous variety of G.

If $\dim X = 0$ then X is just a point.

If $\dim X = 1$ then $\mathcal{O}_X \cong G/B \cong \mathbb{P}^1$, which is impossible because X is affine. Hence, there are no such X.

We now consider the case $\dim X = 2$.

THEOREM 2. a) *Any irreducible affine surface on which $G = SL_2$ acts nontrivially is quasihomogeneous,* b) *X is homogeneous if and only if G_v is isomorphic either to T or to $N_G(T)$ for some $v \in \mathcal{O}_X$.*

PROOF. a) This follows from the fact that any one-dimensional G-orbit is isomorphic to $\mathbb{P}^1$. b) If X is homogeneous then $\mathcal{O}_X$ is affine, and, by Matsushima's criterion, G_v, $v \in \mathcal{O}_X$, is reductive. Examining the list of subgroups of G and using the fact that N_n is not reductive, we obtain the desired implication. Conversely, if $G_v =$ either T or $N_G(T)$ then, by Matsushima's criterion, $\mathcal{O}_X$ is affine. Therefore, if $\Gamma_X \neq \varnothing$, then Γ_X is a curve (see [29]). If $u \in \Gamma_X$, then $\dim Gu \leq \dim \Gamma_X = 1$. Because the case $\dim Gu = 1$ has already been excluded (in that case $Gu \cong \mathbb{P}^1$), it follows that $\Gamma_X \subset X^G$. This contradicts the uniqueness of the closed orbit. □

COROLLARY. *If X is not homogeneous then it is an S-variety of G.*

Indeed, if $v \in \mathcal{O}_X$ then G_v must be conjugate to N_n, but $N_n \supset U$.

Thus we obtain a complete classification in the case $\dim X = 2$.

11. We shall now analyze the case $\dim X = 3$. The subcases of homogeneous and nonhomogeneous X should be treated separately. Because finite groups are reductive, the following theorem can be deduced from Matsushima's criterion and from part 8).

THEORM 3. *Three-dimensional homogeneous X are exactly G/T_n; G/D_n; $G/\widetilde{\mathbb{T}}$; $G/\widetilde{\mathbb{O}}$; $G/\widetilde{\mathbb{I}}$.*

12. Suppose now that $\dim X = 3$ and $\Gamma_X \neq \varnothing$. Because $\dim \mathcal{O}_X = \dim G = 3$, it follows that $|G_v| < \infty$ for $v \in \mathcal{O}_X$. By Matsushima's criterion, $\mathcal{O}_X$ is affine and therefore each irreducible component of Γ_X is a surface (see

[29]) that, by Theorem 2, is quasihomogeneous with respect to G (one has to take into account that X^G consists of at most one point because the closed orbit is unique).

In this section we will treat the case $X^G = \varnothing$ (the more difficult case $X^G \neq \varnothing$ will be treated in §2).

13. First we shall describe affine curves that are quasihomogeneous with respect to T and $N_G(T)$ and have a fixed point. Because T is a torus, such curves are S-varieties. We also have (in the notation of part 5 of §2, Chapter 4) that $k[T] = k[\chi, \chi^{-1}] = \bigoplus_{n\in\mathbb{Z}}\langle\chi^n\rangle$. It now follows from part 4 that all such curves for T are exactly $Y(T, \mathfrak{Z}) = \operatorname{Specm}\bigoplus_{n\in\mathfrak{Z}}\langle\chi^n\rangle$, where $\mathfrak{Z}$ is an arbitrary nonzero finitely-generated subsemigroup with zero of either $\mathbb{N}$ or $-\mathbb{N}$. Notice that $Y(T, -\mathfrak{Z})$ can be obtained from $Y(T, \mathfrak{Z})$ by "twisting" the action (see 5)) by $w = \left(\begin{smallmatrix} 0 & 1 \\ -1 & 0\end{smallmatrix}\right) \in N_G(T) - T$. For any $n \in \mathbb{Z}$ we also consider the functions φ_n and $\varphi_n \in k[N_G(T)] = k[T \cup T_w] : \varphi_n(g) = 0$ for $g \in Tw$, and $= \chi^n(g)$ for $g \in T$; $\psi_n(g) = 0$ for $g \in T$, and $= \chi^n(gw)$ for $g \in Tw$. It is easy to see that $\varphi_{n+m} = \varphi_n\varphi_m$, $\psi_{n+m} = \psi_n\psi_m$ for all $n, m \in \mathbb{Z}$ and that (with respect to the action $N_G(T)$ on $k[N_G(T)]$ by left translations) $t\varphi_n = \chi^{-n}(t)\varphi_n$, $t\psi_n = \chi^{-n}(t)\psi_n$ for all $t \in T$, and $w\varphi_n = (-1)^n\psi_{-n}$, $w\psi_n = \varphi_{-n}$. Therefore $M_n = \langle\varphi_n, \psi_{-n}\rangle$ is a submodule of the $N_G(T)$-module $k[N_G(T)]$, which is irreducible when $n \neq 0$ and isomorphic to the direct sum of $M_0^+ = \langle\varphi_0 + \psi_0\rangle$ and $M_0^- = \langle\varphi_0 - \psi_0\rangle$ for $n = 0$. Clearly, $k[N_G(T)] = \bigoplus_{n\in\mathbb{Z}} M_n$. It follows easily from this and from part 4 that for $N_G(T)$ the desired curves are exactly all possible $Y(N_G(T), \mathfrak{Z}) = \operatorname{Specm} M_0^- \oplus (\bigoplus_{n\in\mathfrak{Z}-\{0\}} M_n)$, where $\mathfrak{Z}$ is defined the same way as before. Again, $Y(N_G(T), -\mathfrak{Z})$ can be obtained from $Y(N_G(T), \mathfrak{Z})$ by "twisting" by w. It is clear that $Y(N_G(T), \mathfrak{Z})$ has exactly two irreducible components intersecting at the fixed point. With respect to the action of T, each of those components is isomorphic to $Y(T, \mathfrak{Z})$; in particular $Y(N_G(T), \mathfrak{Z})$ is always singular.

14. THEOREM 4. 1) *Three-dimensional nonhomogeneous* X *with* $X^G = \varnothing$ *are exactly the homogeneous bundles* $G \times_H Y$, *where* H = *either* T *or* $N_G(T)$, *and* Y *is the curve* $Y(H, \mathfrak{Z})$ (*see part* 13). 2) $G \times_{H_1} Y(H_1, \mathfrak{Z}_1)$ *and* $G \times_{H_2} Y(H_2, \mathfrak{Z}_2)$ *are isomorphic* $\Leftrightarrow$ $H_1 = H_2$ *and* $\mathfrak{Z}_1 = \pm\mathfrak{Z}_2$. 3) $X = G \times_H Y(H, \mathfrak{Z})$ *is normal* $\Leftrightarrow$ $H = T$ *and* $\mathfrak{Z} = \langle\langle d\rangle\rangle$ *for some* $d \in \mathbb{Z} - \{0\}$. *Under these assumptions* X *is automatically smooth.* 4) *The generic stabilizer for any* X *is a cyclic group* (*of order* n_X) *and the correspondence* $X \mapsto n_X$ *gives rise to a bijection between the set of the classes of isomorphic normal* X *and* $\mathbb{Z}^+$.

PROOF. Because the closed orbit is unique, it follows from Theorem 2 and its corollary that if $X^G = \varnothing$ then Γ_X is irreducible and homogeneous (and isomorphic to either G/T or $G/N_G(T)$). In particular, X contains exactly two orbits. Therefore 1) follows from Theorem 1 and part 13. In 2) the implication "$\Leftarrow$" follows from parts 13 and 5, and the implication "$\Rightarrow$" follows from the facts that for $X = G \times_H Y(H, \mathfrak{Z})$ the subgroup H is defined as the stabilizer of some point o in the unique two-dimensional orbit and $\pm\mathfrak{Z}$ is the semigroup corresponding (according to part 13) to the quasihomogeneous with respect to H curve $\{x \in X \mid o \in \overline{Tx}\}$. Because $Y(N_G(T), \mathfrak{Z})$ is singular, it follows from part 5 that a normal $X = G \times_H Y(H, \mathfrak{Z})$ must have $H = T$ and, furthermore (in view of the normality), $\operatorname{sing} X$ is an invariant subvariety of codimension ≥ 2, so that X is smooth. Therefore $Y(H, \mathfrak{Z})$ is also smooth. In view of Lemma 1 this yields: $\mathfrak{Z} = \langle\langle d \rangle\rangle$ for some $d \neq 0$. This proves the implication "$\Rightarrow$" in 3). The implication "$\Leftarrow$" follows from Lemma 1 and part 5. It is obvious that the generic stabilizer for T in $Y(T, \langle\langle d \rangle\rangle)$ is a cyclic group of order $|d|$. Therefore (see part 5) the generic stabilizer for G in $G \times_T Y(T, \langle\langle d \rangle\rangle)$ is also a cyclic group of order $|d|$. Now 4) follows from 3) if one takes into account that the normalization X^H of any variety X of the considered type is also of this type and that $\mathcal{O}_{X^H} \cong \mathcal{O}_X$. □

§2. The conclusion of classification

1. It remains to consider the case when X is three dimensional and has a unique fixed point o. This is the most complicated case. Let $\Gamma_1, \dots, \Gamma_d$ be all irreducible components of Γ_X. By Theorem 2 and its corollary, they are S-varieties and $\Gamma_i \cap \Gamma_j = o$ for all $i \neq j$. (Later we will show that in fact $d = 1$). We fix an equivariant linear embedding

$$X \hookrightarrow \bigoplus_{i=1}^{r} S_{n_i} = V. \tag{3}$$

We may (and will) assume that $n_i \geq 1$ for all i. Thus $o = 0$ (the zero of V).

It is clear that the normalization X^H of X is also a variety of the same type.

2. We will need the following two technical lemmas.

LEMMA 2. *For any two points $v \in \mathcal{O}_X$ and $u \in \Gamma_X^u$ there exists a sequence $b_1, b_2, \dots$ of elements of B such that $\lim_{n\to\infty} b_n v = u$ (in $\mathbb{C}$-topology).*

PROOF. It is well known (see [103]) that the closures of $Gv = \mathcal{O}_X$ in V in $\mathbb{C}$-topology and Zariski topology are the same. Therefore $\lim_{n\to\infty} g_n v = u$ for some sequence $g_1, g_2, \dots, g_i \in G$.

We now choose an SU_2-invariant Hermitian form $\|\ \|$ in V. It follows from the Iwasawa decomposition $G = SU_2 \cdot B$ that $g_n = k_n b_n$, where $k_n \in SU_2$ and $b_n \in B$ for all n. Equalities $\|u - g_n v\| = \|k_n(k_n^{-1}u - b_n v)\| = \|k_n^{-1}u - b_n v\|$ imply that the distance between the compact set $SU_2 u$ and point $b_n v$ tends to zero as $n \to \infty$. It is not difficult to deduce from this that $b_1v, b_2v, \ldots$ contains a subsequence converging to a point $u' \in SU_2u$. Therefore we may assume that $\lim_{n\to\infty} b_nv = u'$. If $u = o$ then $u' = u$ and we are done. Suppose that $u \neq o$. Consider an element $g \in G$ such that $gu' = u$. Then $\lim_{n\to\infty} gb_n = 0$, and therefore it suffices to show that $g \in B$. By construction, $u' \in \overline{Bv} - Bv$. Because $\dim G_v = 0$, $\dim G_{u'} = 1$, and $\dim B = 2$, it means that $\dim B_{u'} = 1$. Therefore $B^0_{u'} = G^0_{u'}$. Because $\overline{Gu'}$ is an S-variety of G, it follows from part 8, that $G^0_{u'}$ is a maximal unipotent subgroup of G and therefore of B. But U is the only such subgroup of B. Therefore $U = G^0_{u'}$, whence $g^{-1}Ug = G^0_u$. But, by assumption, $G^0_u \supset U$ and from the dimension count we have $G^0_u = U$. Therefore $g \in N_G(U) = B$. □

LEMMA 3. *Let $P = \sum_{i=0}^m \alpha_i x^{n-i} y^i$ and $Q = \sum_{i=0}^q \beta_i x^{p-i} y^i$ be binary forms with $0 < m < n/2$, $0 < q < p/2$ and $\alpha_m \beta_q \neq 0$ and let $\lim_{i\to\infty} b_i P = \alpha x^n$, $\alpha \in k$, for some sequence $b_i = \left(\begin{smallmatrix} a_i^{-1} & 0 \\ -d_i & a_i \end{smallmatrix}\right)$, $i = 1, 2, \ldots$. Then $\lim_{i\to\infty} a_i = 0$. If, in addition, $\alpha \neq 0$ then: 1) $\lim_{i\to\infty} b_i Q = 0$ whenever $m/n > q/p$; 2) if $m/n < q/p$ then there is no (finite) $\lim_{i\to\infty} b_i Q$; 3) if $m/n = q/p$ then there exists a subsequence $b_{i_1}, b_{i_2}, \ldots$ with $d_{i_s} \neq 0$ for all s and $\lim_{s\to\infty} d_{i_s} = \infty$, such that $\lim_{s\to\infty} b_{i_s} Q = \beta x^p$, where $\beta \in k^*$, and $(\beta/\beta_q)^m = (\alpha/\alpha_m)^q$.*

PROOF. An easy computation shows that $b_l P = \sum_{i=0}^m \gamma_{il} x^{n-i} y^i$, where

$$\gamma_{il} = \sum_{j=i}^m \alpha_i \binom{j}{j-i} a_l^{n-i-j} d_l^{j-i}. \tag{4}$$

By assumption, $\lim_{l\to\infty} \gamma_{il} = 0$ for $i = 1, \ldots, m$ and $\lim_{l\to\infty} \gamma_{0l} = \alpha$. We now have $\gamma_{ml} = \alpha_m a_l^{n-2m}$. Because $\alpha_m \neq 0$ and $n - 2m > 0$, it follows that $\lim_{l\to\infty} a_l = 0$.

Suppose now that $\alpha \neq 0$. In formula (4) we have $n - i - j > 0$ (because $n > 2m$) and $j \geq i$. Because $\alpha \neq 0$, it follows that the sequence $d_1, d_2, \ldots$ cannot be bounded. Replacing, if necessary, $b_1, b_2, \ldots$ with a suitable subsequence we may assume that $\lim_{l\to\infty} d_l = \infty$ and $d_l \neq 0$ for any l. Therefore γ_{0l} can be rewritten as $\gamma_{0l} = a_l^{n-m} d_l^m \sum_{j=0}^m \alpha_j a_l^{m-j} d_l^{j-m}$. Because $m \geq j$ the limit

$$\lim_{l\to\infty} a_l^{n-m} d_l^m = \gamma \tag{5}$$

is finite and $\alpha = \alpha_m \gamma \neq 0$ (in particular, $\gamma \neq 0$).

Consider now the form $b_l Q$; a similar argument shows that $b_l Q = \sum_{i=0}^q \lambda_{il} x^{p-i} y^i$, where $\lambda_{il} = \sum_{j=i}^q \beta_j \binom{j}{j-1} a_l^{p-i-j} d_l^{j-i}$. Because $d_l \neq 0$, we can rewrite the expression for λ_{il} as $\lambda_{il} = a_l^{p-i-q} d_l^{q-i} \sum_{j=i}^q \beta_j \binom{j}{j-i} a_l^{q-j} d_l^{j-q}$. Because $q \geq j \geq i$ and $\beta_q \neq 0$, it follows from the equalities $\lim_{l\to\infty} a_l = 0$

and $\lim_{l\to\infty} d_l = \infty$ that $\lim_{l\to\infty} \lambda_{il}$ exists if and only if $\lim_{l\to\infty} a_l^{p-i-q} d_l^{q-i}$ does; those limits (when they exist) are related by the formula $\lim_{l\to\infty} \lambda_{il} = \beta_q \binom{q}{q-i} \lim_{l\to\infty} a_l^{p-i-q} d_l^{q-i}$.

It follows at once that $\lim_{l\to\infty} \lambda_{ql} = 0$ (one should take into account that $p > 2q$). Thus the existence (and calculation) of those limits should only be considered in the remaining cases, when $i < q$. Henceforth this inequality will be automatically assumed.

Because $a_l d_l \neq 0$ for any l, we can find $\sigma_l \in \mathbb{R}$ such that $|d_l| = |a_l|^{\sigma_l}$. We now define ε_l and δ_{il} by the equalities $\sigma_l = (m - n + \varepsilon_l)/m = (q + i - p + \delta_{il})/q - i$. Then $|a_l^{n-m} d_l^m| = |a_l|^{\varepsilon_l}$ and $|a_l^{p-i-q} d_l^{q-i}| = |a_l|^{\delta_{il}}$. Because $\delta_{il} = \varepsilon_l((q-i)/m) - 2i + p + ((i-q)/m)n$, we have $|a_l^{p-i-q} d_l^{q-i}| = |a_l|^{\delta_{il}} = |a_l|^{\varepsilon_l \cdot ((q-i)/m)} \cdot |a_l|^{p-2i+((i-q)/m)\cdot n}$. Note also that $p - 2i + ((i-q)/m)n$ has the same sign as $(pm - qn) + i(n - 2m)$.

Suppose now that $m/n > q/p$. Because $n > 2m$ and $i \geq 0$, we have $p - 2i + ((i-q)/m)n > 0$. But it follows from (5) that

$$\lim_{l\to\infty} |a_l|^{\varepsilon_l \cdot (q-i)/m} = |\gamma|^{(q-i)/m} \neq 0. \tag{6}$$

Taking into account the previous equalities we have $\lim_{l\to\infty} a_l^{p-i-q} d_l^{q-i} = 0$, therefore $\lim_{l\to\infty} \lambda_{il} = 0$ for all i. But this means exactly that in the ongoing case $\lim_{i\to\infty} b_i Q = 0$.

Suppose now that $m/n < q/p$. Consider the coefficient λ_{0l}. Because $p - 2i + ((i-q)/m) \cdot n$ is negative for $i = 0$, it follows from (6) and the equality $\lim_{l\to\infty} a_l = 0$ that $\lim_{l\to\infty} a_l^{p-q} d_l^q$ cannot be finite. But then $\lim_{l\to\infty} \lambda_{0l}$ also cannot be finite. This means that the sequence of forms $b_i Q$ does not converge to any form from the space S_p.

Finally, let $m/n = q/p$. Then $p - 2i + ((i-q)/m)n > 0$ for all $i > 0$. This means that $\lim_{l\to\infty} \lambda_{il} = 0$ whenever $i > 0$, the argument being identical to the previous one. In the case $i = 0$ we have to determine whether $\lim_{l\to\infty} a_l^{p-q} d_l^q$ exists or not. Because $\lim_{l\to\infty} a_l^{n-m} d_l^m = \gamma$, there exists the limit of the mth powers of those numbers: $(\alpha/\alpha_m)^q = \lim_{l\to\infty} (a_l^{n-m} d_l^m)^q = \lim_{l\to\infty} a_l^{nq-mq} d_l^{mq} = \lim_{l\to\infty} a_l^{mp-mq} d_l^{mq} = \lim_{l\to\infty} (a_l^{p-q} d_l^q)^m \neq 0$. A simple argument then shows that the sequence $a_1^{p-q} d_1^q, a_2^{p-q} d_2^q, \ldots$ contains a subsequence $a_{i_1}^{p-q} d_{i_1}^q, a_{i_2}^{p-q} d_{i_2}^q, \ldots$ such that $\lim_{s\to\infty} a_{i_s}^{p-q} d_{i_s}^q$ exists and is equal to $\lambda \neq 0$. Then $\lim_{s\to\infty} b_{i_s} Q$ also exists and is equal to βx^p, where $\beta = \lambda \beta q$. It is clear that $(\alpha/\alpha_m)^q = (\beta/\beta_q)^m$. $\square$

3. As a first application of these lemmas we shall now calculate the signatures of the boundary varieties $\Gamma_1, \ldots, \Gamma_d$ for a given embedding (3) of X.

A point $v \in \mathscr{O}$ will be called *canonical* if $\lim_{t\to\infty} \lambda(t)v$, where $\lambda(t) = \begin{pmatrix} t^{-1} & 0 \\ 0 & t \end{pmatrix}$, $t \in k^*$, exists.

LEMMA 4. 1) *Canonical points exist.* 2) *A point* $v \in \mathscr{O}_X$ *is canonical if and only if* $m_i < n_i/2$ *in decompositions* (2) *and* (3) *for any* i. 3) *If* v *is canonical then* $\lim_{t\to 0}\lambda(t)v = o$.

PROOF. If $u = \lim_{t\to 0}\lambda(t)v$, then $G_u \supset \{\lambda(t)\,|\,t \in k^*\} = T$, so that $u \in \Gamma_X$. This yields 3) (because the stabilizers of points of $\Gamma_X - o$ do not contain maximal tori of G). Furthermore, because $x^{n-j}y^j \in (S_n)_{\chi^{-n+2j}}$ (see part 5 of §2, Chapter 4) and $\overline{Tv} - Tv = o$ for a canonical point v, we see that part 4 implies 2). Finally, by the Hilbert–Mumford theorem (see [114, 21]), there exists a point $v \in \mathscr{O}_X$ and a one-parameter subgroup $\gamma\colon k^* \to G$ such that $\lim_{t\to 0}\gamma(t)v = 0$. Because maximal tori of G are conjugate to each other, we can find $g \in G$ and $n \in \mathbb{Z}^+$ such that $g\gamma(t)g^{-1} = \lambda(t^n)$ for all t. Therefore gv is a canonical point, whence 1). □

We should remark that for a canonical point v in (2) not all m_i are equal to 0 (otherwise G_v would obviously contain U, contrary to the fact that $|G_v| < \infty$).

To facilitate the forthcoming argument we now adopt the following convention (it will remain in effect until the end of this section). Let $v \in X$ be a canonical point. We now assume that in decompositions (1) and (2)

$$h_X^V = \frac{m_1}{n_1} = \cdots = \frac{m_s}{n_s} > \frac{m_{s+1}}{n_{s+1}} \geq \cdots \geq \frac{m_r}{n_r} \tag{7}$$

(this can be achieved by reshuffling the vectors $v_1, \ldots, v_r$ because not all m_i are equal to 0).

PROPOSITION 1. *For a chosen embedding* (3) *the signature of each of the S-varieties* $\Gamma_1, \ldots, \Gamma_d$ *is equal to* $\langle\langle n_1, \ldots, n_s\rangle\rangle$.

PROOF. Let $u \in (\Gamma_1 - o)^U$. Then $u = \alpha_1 x_1^{n_1} \oplus \cdots \oplus \alpha_r x_r^{n_r}$. We shall now show that $\alpha_1 \cdots \alpha_s \neq 0$, $\alpha_i = 0$ for all $i > s$ (for the remaining Γ_j the proof is similar).

By Lemma 2, there exists a sequence $b_1, b_2, \ldots$ of elements of B such that $\lim_{i\to\infty} b_i v = u$ and therefore $\lim_{i\to\infty} b_i v_j = \alpha_j x^{n_j}$ for all j (see (1)).

Because v is a canonical point, we have $m_j < n_j/2$ for all $j = 1, \ldots, r$. Moreover it follows from (7) that $m_j > 0$ for $j = 1, \ldots, s$.

These remarks allow us to use Lemma 3. Applying it to the form $P = v_1$ we have $\lim_{i\to\infty} a_i = 0$, where $b_i = \left(\begin{smallmatrix} a_i^{-1} & 0 \\ -d_i & a_i \end{smallmatrix}\right)$. It now follows that if $m_j = 0$ then $\alpha_j = 0$ (see (2)). But $u \neq 0$ and therefore we can find j_0 for which $m_{j_0} \neq 0$ and $\alpha_{j_0} \neq 0$. We now apply Lemma 3 to the forms $P = v_{j_0}$ and $Q = v_j$, where j is any index for which $m_j = 0$, and obtain the inequality $m_{j_0}/n_{j_0} \geq m_j/n_j$. But because j can be set equal to 1, it follows from (7) that $m_{j_0}/n_{j_0} = h_X^V$. Applying Lemma 3 again we see that $\alpha_j \neq 0$ if and only if $1 \leq j \leq s$. □

4. We have two types of invariants of the embedding $X \hookrightarrow V$: the r-tuples of numbers $(n_1, \ldots, n_r)$ and $(m_1, \ldots, m_r)$ (it is easy to obtain, from assertion 2) of Lemma 4, formula (2), and the fact that G acts (by formula (2.3) of Chapter 4) by automorphisms of the algebra $k[x, y]$, an invariant description of the number $n_i - m_i$. It is the maximum of the multiplicities of the linear factors of any form in the orbit Gv_i). It turns out that with the aid of these invariants one can construct an invariant of X itself (that does not depend on the embedding): just consider the ratios of the numbers from the r-tuples. More precisely, from Theorem 5 below it will follow that the number $h_X = h_X^V = \max_{1\leq i\leq r} m_i/n_i$ (see (7)) is such an invariant. In fact, Theorem 5 not only allows us to introduce this new invariant but also essentially yields a classification of normal varieties of the considered type.

5. We need the following two lemmas.

LEMMA 5. *The generic stabilizer for G in X is a finite cyclic group.*

PROOF. It is clear that $G_v \subset G_{v_1}$. Because any finite subgroup of B is cyclic, it suffices to show that $G_{v_1} \subset B$. By Lemma 4 and formula (2), the form v_1 has a unique linear factor of the maximal multiplicity $(= n_1 - m_1)$, and this factor is x. Because $k[x, y]$ is factorial with linear forms as the homogeneous prime elements and because G acts by automorphisms of $k[x, y]$, it follows that $G_{v_1} \subset G_{\langle x\rangle} = B$. □

We denote the order of the generic stabilizer in X by n_X and call it the stabilizer of X. It is clear that n_X does not change under normalization.

LEMMA 6. *If X is normal then* $\operatorname{sing} X = o$.

PROOF. Because of normality, $\operatorname{sing} X$ is a (closed invariant) subvariety of codimension ≥ 2. Therefore $\operatorname{sing} X$ is either o or $\varnothing$. In the latter case, according to part 6 of §1, the variety X must be G-isomorphic to the G-module $\Theta_{o,X}$. But this is impossible because SL_2 has no three-dimensional locally-transitive modules. □

6. THEOREM 5. *In the notation of parts* 3 *and* 1, *let Y be another three-dimensional quasihomogeneous affine variety of G with a fixed point, w a canonical point of Y, $Y \hookrightarrow \bigoplus_{i=1}^{e} S_{p_i} = W$, $p_i \geq 1$ for all i, an equivariant linear embedding of Y, and $w = w_1 \oplus \cdots \oplus w_e$, where $w_i = \sum_{j=0}^{q_i} \beta_{ij} x^{p_i - j} y^j \in S_{p_i}$, $\beta_{iq_i} \neq 0$. Assume that (after a reshuffling, if necessary)*

$$h_Y^W = \frac{q_1}{p_1} = \cdots = \frac{q_t}{p_t} > \frac{q_{t+1}}{p_{t+1}} \geq \cdots \geq \frac{q_e}{p_e} \tag{8}$$

Then:

1) *If there exists a G-morphism $\tau\colon X \to Y$ such that $\tau(\mathscr{O}_X) = \mathscr{O}_Y$ then $n_X \mid n_Y$ and $h_X^V \geq h_Y^W$. Morphism τ has exceptional subvariety if and only if $h_X^V > h_Y^W$. This exceptional subvariety is the boundary Γ_X: its image under τ is the fixed point of Y.*

2) *Conversely, if* $n_X \mid n_Y$ *and* $h_X^V \geq h_Y^W$ *and, in addition,* X *is normal then there exists a* G*-morphism* $\tau: X \to Y$ *such that* $\tau(\mathscr{O}_X) = \mathscr{O}_Y$.

PROOF. First of all we remark that numbers n_i, m_i, p_i, and q_i (and therefore h_X^V and h_Y^W) do not depend on the choice of v and w (see part 4). Hence we can replace the canonical points v and w whenever it is convenient for our purposes.

We begin with the proof of 1). It is clear that $\tau(v)$ is a canonical point in Y. Thus we may assume that $\tau(v) = w$.

Consider another point $u \in (\Gamma_1 - o)^U$. By Lemma 2, there exists a sequence of elements $b_1, b_2, \ldots$ of B such that $\lim_{l\to\infty} b_l v = u$. Because τ is continuous in the $\mathbb{C}$-topology, $\lim_{l\to\infty} b_l w$ also exists. Passing to the coordinates, we have the existence of $\lim_{l\to\infty} b_l v_i$ and $\lim_{l\to\infty} b_l w_j$ for all i, j.

We now take into account that v and w are canonical points and that $G_u \supset U$. It follows from Proposition 1 and (7) that $\lim_{l\to\infty} b_l v_1 = \alpha x^{n_1}$, where $\alpha \neq 0$. We deduce from Lemma 4, (7) and (8) that Lemma 3 can be applied to the forms $P = v_1$ and $Q = w_1$. Because $\lim_{l\to\infty} b_l w_1$ exists, it follows from this lemma and (8) that $h_X^V \geq h_Y^W$.

If $h_X^V \geq h_Y^W$ then, applying Lemma 3 to the forms $P = v_1$ and $Q = w_j$, $j = 1, \ldots, e$ and using (8), we have $\lim_{l\to\infty} b_l w_j = 0$ for all j. This means that $\tau(u) = 0$. Because τ commutes with G, $0 \in Y^G$ and $\overline{Gu} = \Gamma_1$, we have $\tau(\Gamma_1) = 0$. A similar argument shows that $\tau(\Gamma_j) = 0$ for all j. Thus if $h_X^V > h_Y^W$ then $\tau(\Gamma_X) = 0 = Y^G$.

If $h_X^V = h_Y^W$ then, by Lemma 3 and (8), $\lim_{l\to\infty} b_l w_1 \neq 0$ and therefore $\tau(u) \neq 0$. Because τ is equivariant, it follows that $G\tau(u) = \tau(Gu)$ is a two-dimensional orbit. Hence $\tau(\Gamma_1) = \tau(\overline{Gu}) = \overline{G\tau(u)}$ is an irreducible component of Γ_Y. Taking into account that $n_X \mid n_Y$ because $\tau(\mathscr{O}_X) = \mathscr{O}_Y$, we obtain 1).

We shall now prove 2). We may (and will) assume that $G_v = T_{n_X}$ and $G_w = T_{n_Y}$. Indeed, $G_v \subset B$ (see the proof of Lemma 5). But one can see easily that any finite cyclic subgroup of B is conjugate in B to the group T_m for some m. Hence $G_{gv} = gG_v g^{-1} = T_{n_X}$ for some $g \in B$. It now suffices to note that, in view of Lemma 4, the point gv is also canonical (because $g \in G_{\langle x\rangle}$). The argument for w is similar.

Because $n_X \mid n_Y$, there exists an embedding $G_v \subset G_w$ and therefore we have a surjective G-morphism $\tau: \mathscr{O}_X \to \mathscr{O}_Y$. It induces a rational G-morphism of quasihomogeneous varieties $\tau: X \to Y$. By part 1 of §1, it suffices to show that if $h_X^V \geq h_Y^W$ and X is normal then τ is defined everywhere.

The map τ has coordinates $\tau = (f_1, \ldots, f_m)$, where $f_i \in k(X)$, $1 \leq i \leq m$. It is not defined at a point $u \in X$ if and only if at least one of the f_i is not defined at u. On a normal variety the indeterminacy set of a rational function is either empty or an unmixed subvariety of codimension 1. Therefore each f_i is either regular or undefined exactly on some irreducible components of Γ_X. We shall now prove that each f_i, $1 \le i \le m$, is regular.

Suppose, for example, that f_1 is not defined on Γ_j and let $u \in \Gamma_j$, $u \neq o$ be a point not contained in the divisor of zeros of f_1.

We can find $g \in G$ such that $G_{gu} \supset U$. By Lemma 2, we can find a sequence of elements $b_1, b_2, \dots$ of B such that $\lim_{i\to\infty} b_i v = gu$. We claim that the sequence $b_1, b_2, \dots$ can be chosen so that the sequence $b_1 w, b_2 w, \dots$ would also converge (to a point in the boundary Γ_Y).

Indeed, because v is a canonical point, we have, after applying Lemma 3 to the form $P = v_1$, that $\lim_{l\to\infty} a_l = 0$, where $b_l = \left(\begin{smallmatrix} a_l^{-1} & 0 \\ -d_l & a_l \end{smallmatrix}\right)$. It now follows that $\lim_{l\to\infty} b_l w_j = 0$ when $q_j = 0$. If $q_j \neq 0$ then we can apply Lemma 3 to the pair of forms $P = v_1$ and $Q = w_j$ and show, taking into account that $h_X^V \ge h_Y^W$, that $\lim_{l\to\infty} b_l w_j$ exists (after replacing, if necessary, the sequence $b_1, b_2, \dots$ with a suitably chosen subsequence; see Lemma 3. Henceforth we will assume that such a replacement has already been made). The group G acts by automorphisms on varieties X and Y. Therefore the sequence $g^{-1}b_1 w, g^{-1}b_2 w, \dots$ converges (its limit will be denoted by $\tilde{u}$). Moreover $\lim_{i\to\infty} g^{-1}b_i w = u$.

Because of Lemma 6 and the normality of X, we have that $u \notin \operatorname{sing} X$. Therefore there exists an affine neighborhood of u in which Γ_j is given by the local equation $f = 0$. Moreover, because of the choice of u, we may assume that in this neighborhood $f = h/f^n$, where $n \in \mathbb{Z}^+$ and h is an invertible (in this neighborhood) function. Therefore, on one hand $\lim_{i\to\infty} g^{-1}b_i w = \tilde{u} \in Y$, on the other hand the first coordinate of the point $g^{-1}b_i w$ is equal to $f_1(g^{-1}b_i w)$ (because τ commutes with G) and therefore increases without bound as $i \to \infty$. This is a contradiction. □

COROLLARY 1 (The invariance of the number h_X^V). *Let $X \hookrightarrow V$ and $X \hookrightarrow W$ be closed linear equivariant embeddings. If the G-modules V and W do not contain nonzero trivial G-submodules then $h_X^V = h_Y^W = h_X$.*

The number h_X will be called the height of the variety X; it is an invariant of X.

COROLLARY 2. *The height does not change under normalization.*

If $\pi: X^{(n)} \to X$ is the normalization then π is regular, does not have exceptional subvarieties, and $\pi(\mathscr{O}_X(n)) = \mathscr{O}_X$. Therefore the desired result follows from assertion 1) of Theorem 5. □

COROLLARY 3 (A classification). *Two three-dimensional quasihomogeneous normal affine varieties of SL_2 with fixed points are isomorphic if and only if their stabilizers and heights are the same.*

This immediately follows from part 1 of §1 and assertion 2) of Theorem 5. □

7. To complete the description of the classes of isomorphic quasihomogeneous varieties of the considered type we have to determine the possible values that the stabilizer and height can simultaneously take. We shall show that these invariants are independent and can take any value.

THEOREM 6. *For any* $h \in \mathbb{Q}$, $0 < h < 1/2$, *and* $n \in \mathbb{Z}^+$ *there exists a normal* X *such that* $n_X = n$ *and* $h_X = h$.

PROOF. Let $h = q/p$ where $p, q \in \mathbb{Z}^+$ and $\gcd(p, q) = 1$. Let l be a positive integer such that $l > p$ and $\gcd(l, p - 2q) = 1$. In the vector space $S_{np} \oplus S_{nl+2}$ we consider the vector $v = x^{n(p-q)}y^{nq} \oplus x^{nl+1}y$ and let $Y = \overline{GY}$. Clearly (in the notation of part 3) $\lim_{t\to 0}\lambda(t)v = 0$ so that $0 = Y^G \neq \varnothing$. Moreover, $\dim Y = 3$; otherwise (because $Y^G \neq \varnothing$ and in view of part 10 of §1) Y would be an S-variety of G, and then the forms x^{nl+2} and $x^{nl+1}y$ would be contained in the same G-orbit (see part 9 of §1), which is impossible (because the multiplicities of linear factors are an invariant of an orbit).

Thus Y is a variety of the type considered in this section. Because $h = nq/np > 1/(nl+2)$ it is true that $h_Y = h$. Now the standard argument of the invariance of multiplicities implies that $G_{v_1} = T_{np-2nq}$ and $G_{v_2} = T_{nl}$ so that $G_v = T_{\gcd(nl, np-2nq)} = T_n$. Therefore $n_Y = n$.

We can now choose X as the normalization of Y. □

8. For the sake of completeness we should also describe the boundary Γ_X in terms of n_X and h_X; this will be accomplished in Theorem 7. We still assume that we fixed embedding (3), and a canonical point v with decompositions (1) and (2) and relation (7); we may (and will) assume that (3) is minimal, i.e., $v_i \neq 0$ in (1) for any i.

LEMMA 7. $n_X \mid (n_i - 2m_i)$ *for all* i.

PROOF. As in the proof of part 2) of Theorem 5, we may assume that $G_v = T_{n_X}$. We shall show that $n_X \mid (n_1 - 2m_1)$, in the remaining cases the argument is similar. Let $v_1 = \alpha x^{n_1-m_1}y^l \times \prod_{i=1}^q (x + \tau_i y)^{r_i}$, where $\alpha, \tau_1, \dots, \tau_q \in k^*$, $\tau_i \neq \tau_j$ for $i \neq j$, $l + \sum_{i=1}^q r_i = m_1$ be the decomposition of v_1 into linear factors. Let $t = \left(\begin{smallmatrix} \varepsilon^{-1} & 0 \\ 0 & \varepsilon \end{smallmatrix}\right)$ be a generator of G_v. Because $tv_1 = \alpha\varepsilon^{n_1-2l}x^{n_1-m_1}y^l\prod_{i=1}^q(x + \tau_i\varepsilon^{-2}y)^{r_i}$, and $G_{v_1} \supset G_v$, it follows that $n_X \mid (n_1 - 2l)$ and the set $(\tau_1, \dots, \tau_q)$ is invariant under the multiplication by ε^{-2}. This set decomposes into orbits under the action (by multiplication) of the cyclic group generated by ε^{-2}. The stabilizer of any point of any orbit is trivial and if τ_i and τ_j are in the same orbit then $r_i = r_j$. Therefore the number of linear factors in $\prod_{i=1}^q(x + \tau_i y)^{r_i}$ (i.e., $m_1 - l$) is divisible by n_X if n_X is odd, and by $n_X/2$ if n_X is even. It now follows that $n_X \mid (n_1 - 2m_1)$. □

THEOREM 7. *The variety* Γ_X *is irreducible. If* X *is normal and* $h_X = q/p$, *where* $q, p \in \mathbb{Z}^+$ *and* $\gcd(p, q) = 1$ *then* Γ_X *is also normal and* $n_X \cdot p/\gcd(n_X, p-2q)$ *is the unique generator of the semigroup corresponding (in the sense of parts* 4 *and* 9 *of* §1) *to the* S*-variety* Γ_X.

PROOF. By Proposition 1, each one of the $\Gamma_1, \dots, \Gamma_d$ is an S-variety corresponding to the semigroup $\mathfrak{Z} = \langle\langle n_1, \dots, n_s\rangle\rangle$. In view of (7), we have $n_i = l_i p$, $m_i = l_i q$, where $l_i \in \mathbb{Z}^+$ for all $i = 1, \dots, s$. It follows from Lemma 7 that $n_X \mid l_i(p - 2q)$ and therefore $((n_X)/(\gcd(n_X, p - 2q))) \mid l_i$ for all $i = 1, \dots, s$. Consequently all of the elements of $\mathfrak{Z}$ are divisible by $pn_X/\gcd(n_X, p - 2q)$. We shall show that $pn_X/\gcd(n_X, p - 2q) \in \mathfrak{Z}$, if X is normal; this would mean that $\mathfrak{Z} = \langle\langle pn_X/\gcd(n_X, p-2q)\rangle\rangle$ and therefore (by [16]) that all of the $\Gamma_1, \dots, \Gamma_d$ are normal.

Our argument is as follows. Suppose we have constructed a variety Y of the considered type such that $n_Y = n_X$, $h_Y = h_X$, and the semigroup $\mathfrak{Z}'$ corresponding to the boundary S-varieties contains $pn_Y/\gcd(n_Y, p - 2q)$. Consider the normalization $\pi: Y^{(n)} \to Y$. We have $n_Y = n_{Y^{(n)}}$ and $h_Y = h_{Y^{(n)}}$ (see part 5 and Corollary 2 of Theorem 5). But $Y^{(n)}$ is normal. Then, by Corollary 3 and Theorem 5, the varieties $Y^{(n)}$ and X are isomorphic, and therefore the semigroup corresponding to the boundary S-varieties of $Y^{(n)}$ is $\mathfrak{Z}$. Note that the image under π of any irreducible component of $\Gamma_{Y^{(n)}}$ is an irreducible component of Γ_Y. Because π commutes with the action of G, it follows that $\mathfrak{Z}' \subset \mathfrak{Z}$. But this means that $pn_X/\gcd(n_X, p-2q) \in \mathfrak{Z}$. Thus our problem reduces to the construction of variety Y with the aforementioned properties.

This can be done as follows. Let $a = n_X/\gcd(n_X, p - 2q)$, $b = (p-2q)/\gcd(n_X, p-2q)$, and c be a positive integer such that $n_X c+2 > ap$ and $\gcd(c, b) = 1$. In $W = S_{ap} \oplus S_{n_X c+2}$ we consider the vector $u = u_1 \oplus u_2$, where $u_1 = x^{ap-aq}y^{aq}$, $u_2 = x^{n_X c+1}y$. We claim that we can set $Y = \overline{Gu}$. Indeed, using again the argument from the proof of Theorem 6, we see that $0 \in Y$, $\dim Y = 3$ and $G_{u_1} = T_{ap-2aq} = T_{n_X b}$, $G_{u_2} = T_{n_X c}$. Therefore $G_v = T_{\gcd(n_X b, n_X c)} = T_{n_X}$, i.e. $n_Y = n_X$ and, because $h_X = aq/ap > 1/(n_X c + 2)$, we see that $h_X = h_Y$. Finally it follows from Proposition 1 that the generator of the signature of the irreducible components of Γ_Y in the chosen embedding $Y \hookrightarrow W$ is equal to $ap = pn_X/\gcd(n_X, p - 2q)$, which is what we need.

It remains to prove that Γ_X is irreducible. Clearly it suffices to do this for a normal X (so we assume that X is normal). It follows from Proposition 1 that $\Gamma_i = \overline{Gw_i}$ for all $i = 1, \dots, d$, where $w_i = \beta_{1i}x^{n_1} \oplus \dots \oplus \beta_{si}x^{n_s}$, $\beta_{1i}, \dots, \beta_{si} \in k^*$. As was shown before, the signature $\mathfrak{Z}$ of the S-variety

Γ_i is $\langle\langle pn_X/\gcd(n_X, p-2q)\rangle\rangle$ for all i. Because $\mathfrak{Z} = \langle\langle n_1, \dots, n_s\rangle\rangle$, it follows that the smallest of the numbers $n_1, \dots, n_s$ (say, n_1) is equal to $pn_X/\gcd(n_X, p-2q)$. Taking into account (7) we have

$$n_i = n_1 \cdot k_i \quad \text{and} \quad m_i = m_1 \cdot k_i, \qquad k_i \in \mathbb{Z}^+ \ \forall i = 1, \dots, s. \tag{9}$$

The rest of the argument is based on these relations.

It is clear that using a translation of w_i by an element of T we can have $\beta_{1i} = \alpha_{1m_1}$ for all i; henceforth this will automatically be assumed. By Lemma 2, there exist elements $b_{li} = \begin{pmatrix} a_{li}^{-1} & 0 \\ -d_{li} & a_{li} \end{pmatrix}$; $i = 1, \dots, s$; $l = 1, 2, \dots$, such that $\lim_{l\to\infty} b_{li}v = w_i$ for all i. It follows from the proof of Lemma 3 (see (5)) that $\beta_{ji} = \alpha_{jm_j} \lim_{l\to\infty} a_{li}^{n_j - m_j} d_{li}^{m_j}$. Setting $j = 1$ and taking into account that $\beta_{1i} = \alpha_{1m_1}$ for all i we have $\lim_{l\to\infty} a_{li}^{n_1-m_1} d_{li}^{m_1} = 1$. It now follows from (9) (because k_i is a positive integer!) that $1 = 1^{k_j} = \lim_{l\to\infty}(a_{li}^{n_1-m_1} d_{li}^{m_1})^{k_j} = \lim_{l\to\infty} a_{li}^{n_j-m_j} d_{li}^{m_j}$. In view of what was proved before, this yields a key equality $\beta_{ji} = \alpha_{jm_j}$ for all $j = 1, \dots, s$. It shows that β_{ji} do not depend on i, i.e.,

$$w_1 = \cdots = w_d = \alpha_{1m_1} x^{n_1} \oplus \cdots \oplus \alpha_{sm_s} x^{n_s}.$$

Hence $d = 1$. $\square$

COROLLARY. *Under the assumptions from the beginning of part* 8, *we have* $\Gamma_X = \overline{Gw}$, *where*

$$w = \alpha_{1m_1} x^{n_1} \oplus \cdots \oplus \alpha_{sm_s} x^{n_s} \oplus \{0\} \oplus \cdots \oplus \{0\}.$$

9. In the conclusion of this section we give an example showing that a nonnormal X need not be determined uniquely by the invariants n_X, h_X, and $\mathfrak{Z}$ (the signature of Γ_X).

Let $X = \overline{Gv} \subset S_3$, where $v = x^2 y$. Arguing as in the proof of Theorem 6, we have $0 = X^G \neq \varnothing$, $\dim X = 3$, $G_v = \{e\}$, $n_X = 1$, $h_X = 1/3$, and $\mathfrak{Z} = \langle\langle 3\rangle\rangle$. Consider the normalization $\pi\colon X^{(n)} \to X$. Then $n_X = n_{X^{(n)}}$ and $h_X = h_{X^{(n)}}$. Moreover, by Theorem 7, Γ has the signature $\langle\langle 1\cdot 3/\gcd(1, 3-2\cdot 1)\rangle\rangle = \langle\langle 3\rangle\rangle$, so that Γ_X and $\Gamma_{X^{(n)}}$ are isomorphic. One the other hand, if α_0, α_1, α_2, α_3 is the basis of S_3^* dual to the basis x^3, x^2y, xy^2, y^3, then (see [48]) $k[S_3]^G = k[f]$, where

$$f = \alpha_1^2\alpha_2^2 - 4\alpha_0\alpha_2^3 + 18\alpha_0\alpha_1\alpha_2\alpha_3 - 4\alpha_1^3\alpha_3 - 27\alpha_0^2\alpha_3.$$

Because $0 \in X$ (i.e., X is contained in the null-variety of the module S_3), it follows from the irreducibility of f and the dimension count that f is a generator of the ideal $\{g \in k[S_3] \mid g \text{ vanishes on } X\}$. But $\Gamma_X = \overline{Gx^3}$ and it is easy to check that $((\partial f)/(\partial\alpha_i))(x^3) = 0$ for all i. Hence $\operatorname{sing} X = \Gamma_X$. Because $\operatorname{codim}\Gamma_X = 1$, it follows that X is not normal. Therefore $X \not\cong X^{(n)}$.

§3. Application: the structure of orbit closures in finite-dimensional rational SL_2-modules

1. Let $V = \bigoplus_{i=1}^{r} S_{n_i}$, $n_i \geq 1$ for all i. We assume that S_{n_i} are ordered so that

$$n_i = 2k_i \text{ when } i \leq s \quad \text{and} \quad n_i = 2k_i + 1 \text{ when } i > s. \tag{10}$$

We will again use decompositions (1) and (2) for a point $u \in V$. Each element of S_n is a product of n elements of S_1 that are defined uniquely up to proportionality (because $k[x, y]$ is factorial). Therefore we can introduce the following notation for a point $v = v_1 \oplus \cdots \oplus v_r \in V$:

Let M_i be the maximum of the multiplicities of the linear factors of the form $v_i \neq 0$ (if $v_i = 0$ we set $M_i = n_i$); $h = \max_{1 \leq i \leq r} n_i/M_i$.

Let γ be the number of distinct pairwise nonporportional linear forms $w \in k[x, y]$ with the following property: if v_i is any nonzero form in decomposition (1) then w is a factor of v_i of multiplicity M_i. If $v = 0$ then we set $\gamma = 0$.

Because G acts on $k[x, y]$ by homogeneous automorphisms, the numbers M_i, h, and γ are invariants of the orbit Gv, i.e., they are the same for any two points in Gv.

It turns out that the closure of Gv can be described in terms of the invariants h and γ.

2. THEOREM 8. *The orbit of any nonzero vector $v \in V$ is either two dimensional or three dimensional. We also have:*

1) *Gv is two dimensional and closed if and only if $\gamma = 2$ and $h = 2$. In this case Gv contains a vector of the form $\alpha_1 x^{k_1} y^{k_1} \oplus \cdots \oplus \alpha_s x^{k_s} y^{k_s} \oplus \{0\} \oplus \cdots \oplus \{0\}$, where $\alpha_i \neq 0 \Leftrightarrow v_i \neq 0$.*

2) *Gv is two dimensional and nonclosed if and only if $\gamma = 1$ and $h = 1$. In this case $\overline{Gv}$ contains only Gv and the zero of V. In Gv there exists a vector of the form $\alpha_1 x^{n_1} \oplus \cdots \oplus \alpha_r x^{n_r}$, where $\alpha_i \neq 0 \Leftrightarrow v_i \neq 0$.*

3) *Gv is three dimensional and nonclosed if and only if $\gamma = 1$ and $1 < h \leq 2$. In this case we can have the following possibilities:*

a) *if $\gamma = 1$ and $h = 2$ then $\overline{Gv}$ contains, besides Gv, another closed two-dimensional orbit Gw, where $w = \alpha_1 x^{k_1} y^{k_1} \oplus \cdots \oplus \alpha_s x^{k_s} y^{k_s} \oplus \{0\} \oplus \cdots \oplus \{0\}$, and $\alpha_i = 0 \Leftrightarrow k_i = M_i$;*

b) *if $\gamma = 1$ and $1 < h < 2$ then $\overline{Gv}$ contains, besides Gv, the zero of V and a nonclosed two-dimensional orbit Gw, where $w = \alpha_1 x^{n_1} \oplus \cdots \oplus \alpha_r x^{n_r}$, and $\alpha_i = 0 \Leftrightarrow n_i/M_i < h$.*

4) *In all other cases Gv is closed and of dimension* 3.

PROOF. The first assertion was proved in §1 (because $\overline{Gv}$ is a quasihomogeneous affine variety of G and $\dim \overline{Gv} = \dim Gv$).

1) Let $\overline{Gv} = Gv$ and $\dim Gv = 2$. By Theorem 2, there exists a vector of the T-weight 0 in Gv, i.e., a vector of the form $\alpha_1 x^{k_1} y^{k_1} \oplus \cdots \oplus \alpha_s x^{k_s} y^{k_s} \oplus \cdots \oplus \{0\} \oplus \cdots \oplus \{0\}$, where $\alpha_i = 0 \Leftrightarrow v_i = 0$ (remember that S_{n_i} contains no vector of weight 0 when $i > s$). Therefore in this case $\gamma = h = 2$.

Conversely, let $\gamma = h = 2$. It is clear that $v \neq 0$. Because $\gamma = 2$, there exist two nonproportional linear forms w_1 and $w_2 \in k[x, y]$ with the following property: if v_i is any nonzero form in decomposition (1) then w_1 and w_2 are factors of the form v_i of multiplicity M_i. Replacing v with gv for a suitable $g \in G$ we may assume that $w_1 = x$ and $w_2 = y$ so that $v_i = x^{M_i} y^{M_i} u_i$ for all i, where u_i is a form from $k[x, y]$. But $h = 2$ and therefore $2M_i \geq n_i$ for all i. This means that $u_i = \alpha_i \in k$ for all i. Moreover $\alpha_i = 0$ if and only if either $i > s$ or $v_i = 0$. Therefore either $G_v = T$ or $G_v = N_G(T)$. Applying Theorem 2 to $\overline{Gv}$ we see that $\overline{Gv} = Gv$.

2) Suppose now that Gv is nonclosed and of dimension 2. Then, by the corollary of Theorem 2, $\overline{Gv}$ is an S-variety of G and therefore (see part 9) Gv contains a vector of the form $\alpha_1 x^{n_1} \oplus \cdots \oplus \alpha_r x^{n_r}$, where $\alpha_i \neq 0 \Leftrightarrow v_i \neq 0$. It follows that $\gamma = h = 1$.

Conversely let $\gamma = h = 1$. It is clear that $v \neq 0$. Because $\gamma = 1$ there exists a linear form $w \in k[x, y]$ with the following property: if v_i is any nonzero form in decomposition (1) then w is a factor of the form v_i of multiplicity M_i. Replacing v with gv for a suitable g we may assume that $w = x$ and $v_i = x^{M_i} u_i$, where $u_i \in k[x, y]$. But $h = 1$ and therefore $M_i \geq n_i$, whence $u_i = \alpha_i \in k$ for all i, where $\alpha_i \neq 0$ if and only if $v_i \neq 0$. Hence $(G_v)^0 = U$ and therefore, by Theorem 2, Gv is nonclosed (and of dimension 2).

3) Suppose that Gv is nonclosed and of dimension 3. It follows from Theorem 4 and Lemma 4 that we may assume, after replacing, if necessary, v with gv for a suitable $g \in G$, that $(\overline{Gv} - Gv) \cap \overline{Tv} \neq \varnothing$. Hence $\overline{Tv} \neq Tv$ and therefore (see part 4 of §1) in the decomposition of v into the sum of T-weighted vectors all of the weights are of the same sign. We may assume, by replacing v with gv for a suitable $g \in G$, that they are all nonpositive. Therefore in decomposition (2) $m_i \leq n_i/2$ for all i.

Because $n_i - m_i$ is the multiplicity of the factor x of the form v_i, it follows that $M_i = n_i - m_i$ and $1 < h \leq 2$. Because $n_i - m_i \geq n_i/2$, it is true that $1 \leq \gamma \leq 2$. However the case $\gamma = 2$ should be excluded. Otherwise the inequality $M_i \geq n_i/2$ would imply that $h = 2$, which, as was shown before, contradicts the fact that $\dim Gv = 3$. Thus $\gamma = 1$ and $1 < h \leq 2$.

If $\lambda(t) = \left(\begin{smallmatrix} t^{-1} & 0 \\ 0 & t \end{smallmatrix}\right)$ then it follows from (1), (2), and the inequality $m_i \leq n_i/2$ for all i that $\lim_{t \to 0} \lambda(t) v = u$, where $u = \alpha_1 x^{k_1} y^{k_1} \oplus \cdots \oplus \alpha_s x^{k_s} y^{k_s} \oplus \{0\} \oplus \cdots \oplus \{0\}$ and $\alpha_i = \alpha_{im_i} \neq 0$ if $m_i = k_i$, and $\alpha_i = 0$ if $m_i < k_i$. Therefore if $h = 2$ then $u \neq \{0\}$ and if $1 < h < 2$ then $u = \{0\}$. Because $T \subset G_u$, it follows from Theorem 4 and the corollary of Theorem 7 that if

$h = 2$ then $\overline{Gv} - Gv = Gu$ and if $1 < h < 2$ then $\overline{Gv} - Gv = \overline{Gw}$, where $w = \alpha_1 x^{n_1} \oplus \cdots \oplus \alpha_r x^{n_r}$, $\alpha_i = \alpha_{im_i}$ for $h = n_i/M_i$ and $\alpha_i = 0$ for $h > n_i/M_i$ (remember that $\max_{1 \le i \le r} m_i/n_i = 1 - 1/h$, and $M_i = n_i - m_i$).

Conversely, suppose that $1 < h \le 2$ and $\gamma = 1$. Then, as was shown before, $\dim Gv = 3$. Because $\gamma = 1$, there exists a linear form $w \in k[x, y]$ with the following property: if v_i is any nonzero form in (1) then w is a factor of the form v_i of multiplicity M_i. Replacing v with gv for a suitable $g \in G$ we may assume that $w = x$. But then $h \le 2$ implies $n_i - m_i \ge n_i/2$ in (2). Therefore $\lim_{t \to 0} \lambda(t)v = u$ exists. We have $u \in \overline{Gv}$ and it is clear that $T \subset G_u$. Because Gv is of dimension 3, it means that $u \in \overline{Gv} - Gv$. □

CHAPTER 7

Quasihomogeneous Curves, Surfaces, and Solids

In §1, except in parts 7 and 11, $\operatorname{char} k$ is arbitrary, and in those two parts $\operatorname{char} k = 0$. In §§2 and 3 the assumptions on the field k are the same as in Chapter 3.

§1. A classification of irreducible quasihomogeneous curves

1. In this section we will in fact solve an a priori more general problem: we will obtain a classification of all irreducible curves with infinite automorphism groups and describe those automorphism groups. Each such curve turns out to be quasihomogeneous with respect to the corresponding algebraic group. We should remark that because in this case the dimension is one, each irreducible curve with a nontrivial regular action of a nontrivial connected algebraic group is automatically quasihomogeneous with respect to this action. Thus our results yield a classification of quasihomogeneous irreducible curves.

Henceforth X will denote an irreducible curve with infinite automorphism group $\operatorname{Aut} X$, and $\pi\colon X^{(n)} \to X$ the normalization of X. We will identify $\operatorname{Aut} X$ with a subgroup of $\operatorname{Aut} X^{(n)}$, so that $\pi \circ a = a \circ \pi$ for all $a \in \operatorname{Aut} X$.

2. Constructing a singular curve by its normal model. Irreducible curves with given normal model Y are described by the following construction of Rosenlicht–Serre (see [45, 46]).

Let $S \subset Y$ be a finite set and let R be an equivalence relation on points of Y such that $x = \{y \in Y \mid y \overset{R}{\sim} x\}$ for all $x \in Y - S$. Let $p\colon Y \to Y/R$ be the canonical projection. We endow Y/R with a topology in which proper closed subsets are exactly finite subsets. For each point $z \in Y/R$ consider the semilocal ring $\mathscr{O}_z = \bigcap_{p(y)=z} \mathscr{O}_{y,Y}$ and its radical $\mathscr{W}_z = \bigcap_{p(y)=z} \mathfrak{m}_{y,Y}$. On Y/R we define a sheaf $\mathscr{O}'$ of rings $\mathscr{O}'_z$, $z \in Y/R$, by setting $\mathscr{O}'_z = \mathscr{O}_z$ for any $z \notin p(S)$ and for $z \in p(S)$ the ring $\mathscr{O}'_z$ is an arbitrary subring of $\mathscr{O}_z$ satisfying

$$k + \mathscr{W}_z \supset \mathscr{O}'_z \supset k + \mathscr{W}_z^n \tag{1}$$

for some $n > 0$ depending on z.

Then the ringed space $(Y/R, \mathscr{O}')$ is an irreducible algebraic curve, $p: Y \to Y/R$ is its normalization, and, if $\Sigma = \{s \in S \mid \mathscr{O}'_{p(s)} \neq \mathscr{O}_{p(s)}\}$, then $p(\Sigma) = \operatorname{sing}(Y/R, \mathscr{O}')$ (we do not exclude the possibility that $\Sigma = \varnothing$).

PROPOSITION 1.1. *Let* $h \in \operatorname{Aut} Y$. *Then* $h \in \operatorname{Aut}(Y/R, \mathscr{O}') \Leftrightarrow$ 1) h *preserves* R, *i.e.*, $x \overset{R}{\sim} y \Rightarrow hx \overset{R}{\sim} hy$; 2) h *induces an isomorphism between* $\mathscr{O}'_z$ *and* $\mathscr{O}'_{hz}$ *for any* $z \in Y/R$.

The proof is obvious. □

Σ is invariant under $\operatorname{Aut}(Y/R, \mathscr{O}')$. Suppose that $\Sigma \neq \varnothing$ and let $\operatorname{Aut}_\Sigma Y = \{h \in \operatorname{Aut} Y \mid hs = s \text{ for all } s \in \Sigma\}$, $\operatorname{Aut}_\Sigma(Y/R, \mathscr{O}') = \operatorname{Aut}(Y/R, \mathscr{O}') \cap \operatorname{Aut}_\Sigma Y$. Then

$$[\operatorname{Aut}(Y/R, \mathscr{O}') \colon \operatorname{Aut}_\Sigma(Y/R, \mathscr{O}')] < \infty. \tag{2}$$

3. The normalization of the curve X. Let $g(Z)$ denote the geometric genus of the variety Z.

PROPOSITION 1.2. *Either* $g(X) = 0$ (*and then* $X \cong \mathbb{P}^1$, $\mathbb{A}^1$, *or* $\mathbb{A}^1_*$) *or* $g(X) = 1$ (*and then* X *is complete and smooth*).

PROOF. It is known that if $g(Y) \geq 2$ for a curve Y then $|\operatorname{Aut} Y| < \infty$. Let $\overline{X}$ be the complete nonsingular model of the field $k(X)$. We identify $X^{(n)}$ with an open subset of $\overline{X}$ and $\operatorname{Aut} X^{(n)}$ with the subgroup of $\operatorname{Aut} \overline{X}$ leaving $\overline{X} - X^{(n)}$ invariant. If $g(X) = 0$ then $\overline{X} \cong \mathbb{P}^1$, and because $\operatorname{Aut} \mathbb{P}^1$ acts one-transitively on triples of distinct points in $\mathbb{P}^1$, it follows that $|\overline{X} - X^{(n)}| \leq 2$, i.e., $X^{(n)} = \mathbb{P}^1$, $\mathbb{A}^1$, or $\mathbb{A}^1_*$. If $g(x) = 1$ then choosing a point of $\overline{X}$ as the zero of the group structure we obtain an elliptic curve $E(\overline{X})$. $\operatorname{Aut} \overline{X}$ is a semidirect product of the finite automorphism group of the algebraic group $E(\overline{X})$ and $E(\overline{X})$. Because $E(\overline{X})$ acts (by translations) on $\overline{X}$ transitively and with finite stabilizer, it is true that $X^{(n)} = \overline{X}$. But X is nonsingular, otherwise $X = (\overline{X}/R, \mathscr{O}')$ and (see part 2 and (2)) $\operatorname{Aut}_\Sigma(\overline{X}/R, \mathscr{O}')$ is an infinite subgroup of $\operatorname{Aut} \overline{X}$ contained in the stabilizer of some point of $\overline{X}$, which is impossible. Therefore $X = X^{(n)} = \overline{X}$. □

Thus it is the rational curves that are of the most interest for us and henceforth we will assume that X is a rational curve.

4. Smooth curves. They were already described in Proposition 2; here we list them again, along with some formulas and notation that we will need later.

The curve $\mathbb{P}^1$. Let $(\xi : \eta)$ be homogeneous coordinates on $\mathbb{P}^1$. Let $\left[\begin{smallmatrix} a & b \\ c & d \end{smallmatrix}\right]$ denote the image of $\left(\begin{smallmatrix} a & b \\ c & d \end{smallmatrix}\right) \in \mathrm{GL}_2$ under the natural projection $\mathrm{GL}_2 \to \mathrm{PGL}_1$. The action of PGL_1 on $\mathbb{P}^1$ defined by the formula

$$(\xi : \eta) \overset{\left[\begin{smallmatrix} a & b \\ c & d \end{smallmatrix}\right]}{\longmapsto} (a\xi + b\eta \colon c\xi + d\eta) \tag{3}$$

identifies PGL_1 with $\mathrm{Aut}\,\mathbb{P}^1$. $T = \left\{\begin{bmatrix}1&0\\0&a\end{bmatrix} \mid a \in k^*\right\}$; $U = \left\{\begin{bmatrix}1&0\\u&1\end{bmatrix} \mid u \in k_+\right\}$; $B = TU = \left\{\begin{bmatrix}1&0\\u&a\end{bmatrix} \mid a \in k^*, u \in k_+\right\}$ and $N(T) = \left\{\begin{bmatrix}1&0\\0&a\end{bmatrix}, \begin{bmatrix}0&1\\a&0\end{bmatrix} \mid a \in k^*\right\}$ are, respectively, a maximal torus, a maximal unipotent subgroup, a Borel subgroup, and the normalizer of T in PGL_1. Henceforth U will be identified with k_+ via the isomorphism $\begin{bmatrix}1&0\\u&1\end{bmatrix} \mapsto u$, and T with k^* via the isomorphism $\begin{bmatrix}1&0\\0&a\end{bmatrix} \mapsto a$. The morphism $\chi: T \to k^*$, $\chi(t) = t$ is a basis character in $\mathrm{Hom}(T, k^*)$. T has two fixed points in $\mathbb{P}^1$: $0 = (1:0)$ and $\infty = (0:1)$ whereas U and B have only one (the point ∞). T, B, and U act transitively on the complement of the set of their fixed points. The functions $t = \eta/\xi$ and $s = \xi/\eta$ are local parameters at 0 and, respectively, ∞.

The curve $\mathbb{A}^1$ can be identified with $\mathbb{P}^1 - \infty$ and $\mathrm{Aut}\,\mathbb{A}^1$ with B. The function t is a coordinate on $\mathbb{A}^1$.

The curve $\mathbb{A}^1_*$ can be identified with $\mathbb{P}^1 - \{0 \cup \infty\}$ and $\mathrm{Aut}\,\mathbb{A}^1_*$ with $N(T)$. Henceforth all these identifications will be automatically assumed.

Thus $\mathrm{Aut}\,X$ for a smooth X is a zero-dimensional linear algebraic group. Henceforth we will assume that $\mathrm{sing}\,X \neq \emptyset$.

5. A reduction of the problem. In the notation of part 2, $X = (X^{(n)}/R, \mathscr{O}')$ for some S, R, and $\mathscr{O}'$. For $z \in \mathrm{sing}\,X$ we consider the finite-dimensional vector space $(k + \mathscr{W}_z)/(k + \mathscr{W}_z^n)$ and its subspace $\mathscr{O}'_z/(k + \mathscr{W}_z^n)$. $\mathrm{Aut}_\Sigma X^{(n)}$ is an algebraic subgroup of $\mathrm{Aut}\,X^{(n)}$ acting by k-automorphisms of the ring $k + \mathscr{W}_z$; this action gives rise to a rational representation $\mathrm{Aut}_\Sigma X^{(n)} \to \mathrm{GL}(\mathscr{O}'_z(k + \mathscr{W}_z^n))$. It follows from (1) that $\mathscr{O}'_z$ is invariant under $h \in \mathrm{Aut}_\Sigma X^{(n)} \Leftrightarrow \mathscr{O}'_z/(k + \mathscr{W}_z^n)$ is invariant under h. Therefore it follows from Proposition 1 and (2) that $\mathrm{Aut}_\Sigma X$ is a linear algebraic group of dimension ≥ 1 and of finite index in $\mathrm{Aut}\,X$. Notice that T acts on $\mathbb{A}^1_*$ with trivial stabilizer and any regular action of k_+ on $\mathbb{A}^1$ is transitive. Thus we have a proof of

PROPOSITION 1.3. $\mathrm{Aut}\,X$ *is a linear algebraic group of dimension* ≥ 1 *and* $X^{(n)}$ *is isomorphic to either* $\mathbb{P}^1$ *or* $\mathbb{A}^1$.

COROLLARY 1. *If* $X^{(n)} \cong \mathbb{P}^1$ *then either* k^* *or* k_+ *acts regularly and nontrivially on* X, *and if* $X^{(n)} \cong \mathbb{A}^1$ *then* k^* *acts on* X *(and* k_+ *does not). In particular,* X *is quasihomogeneous with respect to* $\mathrm{Aut}\,X$.

Using the conjugacy theorems for algebraic groups we fix an isomorphism between $X^{(n)}$ and either $\mathbb{P}^1$ or $\mathbb{A}^1$ (by which these curves will be identified), so that $\mathrm{Aut}\,X$ contains either T or U.

COROLLARY 2. $|\mathrm{sing}\,X| \leq 2$ *and if* $|\mathrm{sing}\,X| = 2$ *then* X *is a complete curve.*

Thus the problem reduces to describing all curves X such that either $\operatorname{Aut} X \supset T$ or $\operatorname{Aut} X \supset U$. It follows from parts 2, 4, and Proposition 1.1 that:

1) $\operatorname{Aut} X \supseteq U \Leftrightarrow X^{(n)} = \mathbb{P}^1$, $S = \Sigma = \{\infty\}$ and for $p(\infty) = z$ the ring $\mathscr{O}_z'$ is a U-invariant subring of $\mathscr{O}_z = \mathscr{O}_{\infty, \mathbb{P}^1}$ satisfying Condition (1).

2) $\operatorname{Aut} X \supseteq T$ if and only if one of the following conditions is satisfied:

a) $X^{(n)} = \mathbb{A}^1$, $S = \Sigma = \{0\}$ and for $p(0) = z$ the ring $\mathscr{O}_z'$ is a T-invariant subring of $\mathscr{O}_z = \mathscr{O}_{0, \mathbb{A}^1}$ satisfying Condition (1);

b) $X^{(n)} = \mathbb{P}^1$, $S = \Sigma = \{\infty\}$ and for $p(\infty) = z$ the ring $\mathscr{O}_z'$ is a T-invariant subring of $\mathscr{O}_z = \mathscr{O}_{\infty, \mathbb{P}^1}$ satisfying Condition (1);

c) $X^{(n)} = \mathbb{P}^1$, $S = \Sigma = \{0\}$ and for $p(0) = z$ the ring $\mathscr{O}_z'$ is a T-invariant subring of $\mathscr{O} = \mathscr{O}_{0, \mathbb{P}^1}$ satisfying Condition (1);

d) $X^{(n)} = \mathbb{P}^1$, $S = \Sigma = \{0 \cup \infty\}$, $p(0) \neq p(\infty)$ and for $p(0) = z$ (resp., $p(\infty) = z$) the ring $\mathscr{O}_z'$ is a T-invariant subring of $\mathscr{O}_z = \mathscr{O}_{0, \mathbb{P}^1}$ (resp., $\mathscr{O}_z = \mathscr{O}_{\infty, \mathbb{P}^1}$) satisfying Condition (1);

e) $X^{(n)} = \mathbb{P}^1$, $S = \Sigma = \{0 \cup \infty\}$, $p(0) = p(\infty) = z$ and $\mathscr{O}_z'$ is a T-invariant subring of $\mathscr{O}_z = \mathscr{O}_{0, \mathbb{P}^1} \cap \mathscr{O}_{\infty, \mathbb{P}^1}$ satisfying Condition (1).

In all of these cases, except e), the curve X has only one branch (i.e., p is bijective), and in the case e) the curve has two branches (i.e., $|p^{-1}(z)| = 2$).

Thus the problem reduces to describing T- and U-invariant subrings of the corresponding rings $\mathscr{O}_z$ satisfying Condition (1). Later we will need a different formulation of the problem. Namely, $(k + \mathscr{W}_z)/(k + \mathscr{W}_z^n)$ is a finite-dimensional commutative k-algebra and $\mathscr{O}_z'/(k + \mathscr{W}_z^n)$ is a subalgebra of it. Conversely, if A is a subalgebra of $(k + \mathscr{W}_z)/(k + \mathscr{W}_z^n)$ (different from $k + \mathscr{W}_z$ when $\mathscr{O}_z = k + \mathscr{W}_z$), then the preimage of A under the quotient morphism $k + \mathscr{W}_z \to (k + \mathscr{W}_z)/(k + \mathscr{W}_z^n)$ will be a subring $\mathscr{O}_z'$ of $\mathscr{O}_z$ satisfying Condition (1). Our problem can also be viewed as the classification problem of T- and U-invariant subalgebras of the algebra $(k + \mathscr{W}_z)/(k + \mathscr{W}_z^n)$.

Henceforth we will use the following notation: $\mathscr{O}_0 = \mathscr{O}_{0, \mathbb{A}^1} = \mathscr{O}_{0, \mathbb{P}^1}$, $\mathfrak{m}_0 = \mathfrak{m}_{0, \mathbb{A}^1} = \mathfrak{m}_{0, \mathbb{P}^1}$, $\mathscr{O}_\infty = \mathscr{O}_{\infty, \mathbb{P}^1}$, $\mathfrak{m}_\infty = \mathfrak{m}_{\infty, \mathbb{P}^1}$, $\mathscr{W}_{0,\infty} = \mathfrak{m}_0 \cap \mathfrak{m}_\infty$.

6. The curves $\mathbb{P}^1_{n,m}(c_0, \dots, c_p; d_0, \dots, d_q)$. Let $\mathfrak{z}$ be the subsemigroup with zero of the additive semigroup $\mathbb{N}$ of natural numbers containing all integers that are bigger than some constant. Let n be the smallest integer for which $\mathfrak{z}$ contains $\{m \in \mathbb{N} \mid m \geq n\}$ and let $0 = d_0 < \dots < d_r$ be all numbers in $\mathfrak{z}$ that are smaller than n. It is clear that if $\mathfrak{z} \neq \mathbb{N}$ then $d_1 \geq 2$ and $d_r \leq n - 2$. The set of integers $n, d_0, \dots, d_r$ completely determines the semigroup $\mathfrak{z}$ and we denote this semigroup $\mathfrak{z}_n(d_0, \dots, d_r)$.

DEFINITION. The curve $\mathbb{P}^1_{n,m}(c_0, \dots, c_p; d_0, \dots, d_q)$ is a complete rational irreducible curve obtained from $\mathbb{P}^1$ by the Rosenlicht–Serre construction, where $S = \{0 \cup \infty\}$, $p(0) \neq p(\infty)$ and the k-vector space $\mathscr{O}_{p(0)}'$ (resp., $\mathscr{O}_{p(\infty)}'$) is spanned by the functions t^i (resp., s^i), where i runs through the semigroup $\mathfrak{z}_n(c_0, \dots, c_p)$ (resp., $\mathfrak{z}_m(d_0, \dots, d_q)$).

Clearly $\mathbb{P}^1_{n,m}(c_0, \dots, c_p; d_0, \dots, d_q) \cong \mathbb{P}^1_{m,n}(d_0, \dots, d_q; c_0, \dots, c_p)$. If $n = 1$ and $m \geq 2$ (resp., $n \geq 2$ and $m = 1$) $p(\infty)$ (resp., $p(0)$) is the only singular point of this curve. If $n \geq 2$ and $m \geq 2$ then this curve has two singular points: $p(0)$ and $p(\infty)$.

DEFINITION. The curve $\mathbb{A}^1_n(c_0, \dots, c_p)$ is the curve obtained from $\mathbb{P}^1_{n,1}(c_0, \dots, c_p; 0)$ by deleting the point $p(\infty)$.

7. Unibranched curves with a k_+-action.

THEOREM 1.1. *The curve $\mathbb{P}_{1,m}(0; 0)$ is a unibranched irreducible rational curve that admits a nontrivial action of k_+. If* $\operatorname{char} k = 0$ *then the converse is also true: any irreducible singular rational curve admitting a nontrivial k_+-action is isomorphic to $\mathbb{P}^1_{1,m}(0; 0)$ for some $m \geq 2$.*

PROOF. Let $\bar{a}$ denote the image of $a \in \mathcal{O}_\infty$ under the quotient homomorphism $\mathcal{O}_\infty \to \mathcal{O}_\infty/(k+m^n_\infty)$. Then $\bar{s}^1, \dots, \bar{s}^{n-1}$ is a basis of $\mathcal{O}_\infty/(k+m^n_\infty)$. It follows from (3) that $u \in U$ acts on $s \in \mathcal{O}_\infty$ according to the formula $s \mapsto s(us+1)^{-1}$. Therefore u acts on $s \in \widehat{\mathcal{O}}_\infty = k[[s]]$ according to the formula $s \xrightarrow{u} s - us^2 + u^2s^3 - \cdots$. Hence $s^i \xrightarrow{u} s^i - ius^{i+1} + \cdots$ for all $i = 1, 2, \dots$, and therefore u has the following matrix in the given basis:

$$\begin{pmatrix} 1 & & & & \\ -2u & 1 & & 0 & \\ & -3u & 1 & & \\ & & & \ddots & \\ & * & & -(n-1)u & 1 \end{pmatrix}.$$

If $\operatorname{char} k = 0$ then it follows that $(\mathcal{O}_\infty/(k+m^n_\infty))^u = \langle \bar{s}^{n-1} \rangle$. Hence by the Lie–Kolchin theorem, any nonzero U-invariant subspace of $\mathcal{O}/(k+m^n_\infty)$ contains $\bar{s}^{n-1}$. The assertion of the theorem now follows from part 5. □

REMARK. If $\operatorname{char} k \neq 0$ then the assertion of the theorem is no longer true. For example, suppose that $\operatorname{char} k = 2$ and $m = 4$. Then the matrix of u in the basis $\bar{s}, \bar{s}^2, \bar{s}^3$ is

$$\begin{pmatrix} 1 & 0 & 0 \\ u & 1 & 0 \\ u^2 & 0 & 1 \end{pmatrix}.$$

Thus $U \subset \operatorname{Aut} \mathbb{P}^1_{1,4}(0; 0, 2)$. But $\mathbb{P}^1_{1,4}(0; 0, 2) \not\cong \mathbb{P}^1_{1,m}(0; 0)$.

8. Unibranched curves with k^*-action.

THEOREM 1.2. *Unibranched complete (resp., noncomplete) singular rational curves admitting a nontrivial k^*-action are exactly the curves*

$$\mathbb{P}^1_{n,m}(c_0, \dots, c_p; d_0, \dots, d_q), \qquad n+m \geq 3$$

(respectively, $\mathbb{A}^1_n(c_0, \dots, c_p)$, $n \geq 2$).

PROOF. It follows from (3) that t^i and s^i are eigenvectors for the actions of T in $\mathscr{O}_0$ and $\mathscr{O}_\infty$ with weights χ^i and χ^{-i}, respectively. Because those weights are different for different values of i, it follows that any T-invariant subspace of $\mathscr{O}_0$ (resp., $\mathscr{O}_\infty$) is spanned over k by t^i (resp., s^i), where i runs through a subset $\mathfrak{z}$ of $\mathbb{N}$. It is clear that this subspace is a subalgebra satisfying (1) if and only if $\mathfrak{z} = \mathfrak{z}_n(c_0, \dots, c_p)$. □

We shall now consider the case 2e) of part 5 when X has two branches and $T \subset \operatorname{Aut} X$. First we are going to investigate the T-weighted decomposition of $(k + \mathscr{W}_{0,\infty})/(k + \mathscr{W}_{0,\infty}^n)$. We will show that all of the weights in this decomposition are of multiplicity one. This remarkable fact allows us to describe easily all T-invariant subalgebras of the algebra $(k + \mathscr{W}_{0,\infty})/(k + \mathscr{W}_{0,\infty}^n)$ and therefore (see part 5) the desired curves X.

9. The T-weighted decomposition of $(k+\mathscr{W}_{0,\infty})/(k+\mathscr{W}_{0,\infty}^n)$ with respect to T. Let $\omega = \xi/(\xi - \eta) \in k(\mathbb{P}^1)$ and $\Omega = \{(\xi : \eta) \in \mathbb{P}^1 \mid \xi \neq \eta\}$. Then Ω is isomorphic to $\mathbb{A}^1$ and ω is a coordinate on Ω (i.e., $k[\Omega] = k[\omega]$). Because $\omega(0) = 1$ and $\omega(\infty) = 0$, it follows that

$$\{f \in k[\Omega] \mid f(0) = f(\infty) = 0\}$$

is a principal ideal generated by the function $\nu = \omega(\omega - 1)$. Notice that $\omega = (1-t)^{-1}$ and $\nu = t(1-t)^{-2}$. One can easily prove the following

LEMMA 1.1. $\mathscr{W}_{0,\infty} = \nu \cdot (\mathscr{O}_0 \cap \mathscr{O}_\infty)$. □

PROPOSITION 1.4. *The dimension of the space $(k + \mathscr{W}_{0,\infty}^l)/(k + \mathscr{W}_{0,\infty}^{l+1})$ is 2 and the classes of the functions $\omega\nu^l$ and $(\omega - 1)\nu^l$ form a T-weighted basis of it. The first vector is of weight χ^l and the second is of weight χ^{-l}.*

PROOF. First we shall show that the classes of the functions ν^l and $\omega\nu^l$ form a basis of $\mathscr{W}_{0,\infty}^l/\mathscr{W}_{0,\infty}^{l+1} = (k + \mathscr{W}_{0,\infty}^l)/(k + \mathscr{W}_{0,\infty}^{l+1})$.

By the lemma, any $f \in \mathscr{W}_{0,\infty}^l$ is of the form $\nu^l p q^{-1}$, where p and q are relatively prime elements of $k[\omega]$, $q(0) \neq 0$ and $q(\infty) \neq 0$. The classes f and $\nu^l(a\omega + b)q^{-1}$ modulo $\mathscr{W}_{0,\infty}^{l+1}$ coincide for some values of $a, b \in k$ (just divide with remainder p into ν). In their turn, the classes $\nu^l(a\omega+b)q^{-1}$ and $\nu^l(c\omega+d)$ modulo $\mathscr{W}_{0,\infty}^{l+1}$ also coincide for some values of c and $d \in k$; to see this we assign values to c and d so that the polynomial $a\omega + b - q(cu+d)$ be divisible by ν or, which is equivalent, so that this polynomial vanishes at 0 and ∞. The last condition gives rise to a system of linear equations in c and d, which, as is easily checked, always has a solution. Thus the classes of ν^l and $\omega\nu^l$ generate $\mathscr{W}_{0,\infty}^l/\mathscr{W}_{0,\infty}^{l+1}$. But they are also linearly independent over k: if $a\nu^l + b\omega\nu^l \in \mathscr{W}_{0,\infty}^{l+1}$ for some $a, b \in k$ then, by the lemma, $a + b\omega \in \mathscr{W}_{0,\infty}$ and (because $\omega(0) = 1$ and $\omega(\infty) = 0$) we have $a = b = 0$. Therefore the classes of $\omega\nu$ and $(\omega - 1)\nu^l$ form a basis of $(k + \mathscr{W}_{0,\infty}^l)/(k + \mathscr{W}_{0,\infty}^{l+1})$.

If $a \in k^* = T$, then $\omega\nu^l = t^l(1-t)^{-2l-1} \xrightarrow{a} (at)^l(1-at)^{-2l-1}$. But $(at)^l(1-at)^{-2l-1} - a^l t^l(1-t)^{-2l-1} = a^l\nu^l f$ where, as is easily seen, $f(0) = f(\infty) = 0$. Hence $f \in \mathscr{W}_{0,\infty}$ and, by the lemma, this difference is contained in $\mathscr{W}_{0,\infty}^{l+1}$. Therefore the class of $\omega\nu^l$ modulo $k + \mathscr{W}_{0,\infty}^{l+1}$ is an eigenvector of weight χ^l with respect to T. The argument for $(\omega-1)\nu^l$ is similar. □

COROLLARY. *The spectrum of T in $(k + \mathscr{W}_{0,\infty})/(k + \mathscr{W}_{0,\infty}^n)$ consists of the weights $\chi, \chi^{-1}, \chi^2, \chi^{-2}, \dots, \chi^{n-1}, \chi^{-n+1}$, and therefore*

$$\dim(k + \mathscr{W}_{0,\infty})/(k + \mathscr{W}_{0,\infty}^n) = 2n-2.$$

REMARKS. 1) The classes $\omega\nu^l$ and $(\omega-1)\nu^l$, $1 \le l \le n-1$, modulo $k + \mathscr{W}_{0,\infty}^n$ form a basis of $(k + \mathscr{W}_{0,\infty})/(k + \mathscr{W}_{0,\infty}^n)$. However if $l < n-1$ then these vectors are not weighted with respect to T.. For example, if $n = 3$ then the class of $\omega\nu - 3\omega\nu^2 - (\omega-1)\nu^2$ and not the class of $\omega\nu$, will be an eigenvector of weight χ. The problem of the explicit determination of the eigenvectors for each particular n reduces to solving the corresponding system of linear equations. 2) The ring $\mathscr{O}_0 \cap \mathscr{O}_\infty$ has no nonconstant eigenvectors of T.

10. The algebra $(k + \mathscr{W}_{0,\infty})/(k + \mathscr{W}_{0,\infty}^n)$. It follows from Proposition 4 and its corollary that for each positive integer n there exists a set of elements $f_{n,l} \in \mathscr{W}_{0,\infty}$, $l = \pm 1, \dots, \pm(n-1)$ satisfying the following conditions:

a) $f_{n,l} \in \mathscr{W}_{0,\infty}^{|l|}$, $f_{n,l} \notin \mathscr{W}_{0,\infty}^{|l|+1}$,

b) the class modulo $k + \mathscr{W}_{0,\infty}^n$ of $f_{n,l}$ is an eigenvector of weight χ^l with respect to the action of T on $(k + \mathscr{W}_{0,\infty})/(k + \mathscr{W}_{0,\infty}^n)$,

c) the class modulo $\mathscr{W}_{0,\infty}^{|l|+1}$ of $f_{n,l}$ coincides with the class of $\omega\nu^l$ when $l > 0$ and with the class of $(\omega-1)\nu^l$ when $l < 0$.

The class of $f_{n,l}$ modulo $k + \mathscr{W}_{0,\infty}^n$ will be denoted $\overline{f}_{n,l}$. We also set $f_{n,0} = 0$ and $\overline{f}_{n,0} = 0$.

PROPOSITION 1.5. *The elements $\overline{f}_{n,l}$, $l = \pm 1, \dots, \pm(n-1)$ form a T-weighted basis of the algebra $(k + \mathscr{W}_{0,\infty})/(k + \mathscr{W}_{0,\infty}^n)$. The multiplication table of this algebra in this basis is as follows:*

$$\overline{f}_{n,l} \cdot \overline{f}_{n,m} = \begin{cases} 0, & \text{if } lm < 0, \\ 0, & \text{if } lm > 0,\ |l+m| \ge n, \\ \overline{f}_{n,l+m}, & \text{if } l > 0,\ m > 0,\ l+m < n, \\ -\overline{f}_{n,l+m}, & \text{if } l < 0,\ m < 0,\ l+m > -n. \end{cases}$$

PROOF. The first assertion follows from Condition b) and the corollary in part 9. We shall now prove the second assertion.

By Condition b), for any $a \in T$ we have $f_{n,l} \xrightarrow{a} a^l f_{n,l} + f_l$, where $f_l \in \mathscr{W}_{0,\infty}^n$. Therefore $f_{n,l} \cdot f_{n,m} \xrightarrow{a} a^{l+m} f_{n,l} \cdot f_{n,m} + f$, where $f \in \mathscr{W}_{0,\infty}^n$. Therefore the class of $f_{n,l} \cdot f_{n,m}$ modulo $k + \mathscr{W}_{0,\infty}^n$ is an eigenvector of weight

χ^{l+m} with respect to T. By Condition a), this vector is nonzero if and only if $f_{n,l} \cdot f_{n,m} \notin \mathscr{W}_{0,\infty}^{|l+m|+1}$. It follows from the corollary in part 9 that this vector is zero if $lm > 0$ and $|l+m| \geq n$. If $lm < 0$ then $|l+m| < |l| + |m|$ and, because, by Condition a), $f_{n,l} \cdot f_{n,m} \in \mathscr{W}_{0,\infty}^{|l|+|m|}$, this vector is again equal to zero. Suppose now that $lm > 0$ and $|l+m| < n$. Consider the case $l > 0$, $m > 0$. Then, by Condition c), $f_{n,l} = \omega\nu^l + f$, $f_{n,m} = \omega\nu^m + h$, where $f \in \mathscr{W}_{0,\infty}^{l+1}$, $h \in \mathscr{W}_{0,\infty}^{m+1}$. Therefore $f_{n,l} \cdot f_{n,m} = \omega^2\nu^{l+m} + r$, where $r \in \mathscr{W}_{0,\infty}^{l+m+1}$. But $\omega^2 = \nu + \omega$ so that $f_{n,l} \cdot f_{n,m} = \omega\nu^{l+m} + g$, where $g \in \mathscr{W}_{0,\infty}^{l+m+1}$ and therefore $f_{n,l} \cdot f_{n,m} \notin \mathscr{W}_{0,\infty}^{l+m+1}$ so that in this case $\overline{f}_{n,l} \cdot \overline{f}_{n,m} = \overline{f}_{n,l+m}$. If $l < 0, m < 0$ and $l + m > -n$, then $f_{n,l} = (\omega - 1)\nu^{|l|} + f$, $f_{n,m} = (\omega - 1)\nu^{|m|} + h$, where $f \in \mathscr{W}_{0,\infty}^{|l|+1}$, $h \in \mathscr{W}_{0,\infty}^{|m|+1}$, and therefore $f_{n,l} \cdot f_{n,m} = (\omega - 1)^2\nu^{|l|+|m|} + r$, where $r \in \mathscr{W}_{0,\infty}^{|l+m|+1}$. But $(\omega - 1)^2 = \nu - (\omega - 1)$, hence $f_{n,l} \cdot f_{n,m} = -(w - 1)\nu^{|l+m|} + g$, where $g \in \mathscr{W}_{0,\infty}^{|l+m|+1}$ and therefore, $f_{n,l} \cdot f_{n,m} \notin \mathscr{W}_{0,\infty}^{|l+m|+1}$ so that in this case $\overline{f}_{n,l} \cdot \overline{f}_{n,m} = -\overline{f}_{n,l+m}$. □

11. Two-branch curves with k^*-actions. Consider the following sets of integers: $\{p_i\}_{i=0}^m$ and $\{q_j\}_{j=0}^l$: $0 = p_0 < \cdots < p_m \leq n-1$ and $0 = q_0 > \cdots > q_l \geq -(n-1)$, and let $\mathfrak{z}$ be their union. The linear hull of the vectors $\overline{f}_{n,i}$, where i runs through $\mathfrak{z}$, is a T-invariant subspace of $(k + \mathscr{W}_{0,\infty})/(k + \mathscr{W}_{0,\infty}^n)$ and, because the weights of the vectors $\overline{f}_{n,i}$ and $\overline{f}_{n,j}$ are different when $i \neq j$, $i, j \neq 0$, it follows that any T-invariant subspace can be obtained this way (for a suitable choice of $\mathfrak{z}$). It follows from Proposition 1.5 that such a subspace is a subalgebra if and only if

$$\forall i, j \in \mathfrak{z}, \text{ such that } ij > 0, |i+j| < n, \text{ it is true that } i + j \in \mathfrak{z}. \tag{4}$$

The preimage of this subalgebra under the quotient homomorphism $(k + \mathscr{W}_{0,\infty}) \to (k + \mathscr{W}_{0,\infty})/(k + \mathscr{W}_{0,\infty}^n)$ is a T-invariant subring $\mathscr{O}_z'$ of $\mathscr{O}_z = \mathscr{O}_0 \cap \mathscr{O}_\infty$ satisfying Condition (1). The k-vector space $\mathscr{O}_z'$ is spanned by $k + \mathscr{W}_{0,\infty}^n$ and the functions $f_{n,i}$, $i \in \mathfrak{z}$. We now require that

$$p_m - q_l \neq 2n - 2. \tag{5}$$

Then different subalgebras will give rise to different rings.

DEFINITION. Suppose that $\mathfrak{z}$ satisfies Conditions (4) and (5). The curve $\mathbb{P}_n^1(p_0, \ldots, p_m; q_0, \ldots, q_l)$ is the complete irreducible rational curve obtained from $\mathbb{P}^1$ by the Rosenlicht–Serre construction where $S = \{0 \cup \infty\}$, $p(0) = p(\infty) = z$ and the k-vector space $\mathscr{O}_z'$ is spanned by $k + \omega_{0,\infty}^n$ and the functions $f_{n,i}$, where i runs through $\mathfrak{z}$ (see part 10).

Taking into account the earlier remarks and the results of part 5, we obtain a proof of

THEOREM 1.3. *Two-branch curves with a nontrivial action of* k^* *are exactly the curves* $\mathbb{P}^1_n(p_0, \ldots, p_m; q_0, \ldots, q_l)$.

12. The automorphism groups.

THEOREM 1.4. *Type* I. *Let* $X = \mathbb{P}^1_{n,m}(c_0, \ldots, c_p; d_0, \ldots, d_q)$. *Then*:

a) *when* $n \geq 2$, $m \geq 2$ *then* $\operatorname{Aut} X = N(T)$ *if (see part* 6) $\mathfrak{z}_n(c_0, \ldots, c_p) = \mathfrak{z}_m(d_0, \ldots, d_q)$ *and* $\operatorname{Aut} X = T$ *otherwise.*

b) *when* $n = 1$ *and* $m \geq 2$ (*resp.,* $n \geq 2$ *and* $m = 1$) *and* $\operatorname{char} k = 0$ *then* $\operatorname{Aut} X = T$ *if* $q > 0$ (*resp.,* $p > 0$) *and* $\operatorname{Aut} X = B$ *if* $q = 0$ (*resp.,* $p = 0$).

Type II. *Let* $X = \mathbb{A}^1_n(c_0, \ldots, c_p)$. *Then* $\operatorname{Aut} X = T$.

Type III. *Let* $X = \mathbb{P}^1_n(p_0, \ldots, p_m; q_0, \ldots, q_l)$. *Then* $\operatorname{Aut} X = N(T)$ *if* $l = m$ *and* $p_i + q_i = 0$ *for all* $i = 0, 1, \ldots, m$ *and* $\operatorname{Aut} X = T$ *otherwise.*

PROOF. Type Ia). $S = \Sigma = \{0 \cup \infty\}$ so that $\operatorname{Aut} X$ is contained in the subgroup of PGL_1, which leaves S invariant. But this subgroup is $N(T)$. The assertion now follows from the relations $t \overset{w_0}{\to} s$ and $s \overset{w_0}{\to} t$, where $w_0 = \begin{bmatrix} 0 & 1 \\ 1 & 0 \end{bmatrix}$.

Type Ib). $\Sigma = \{\infty\}$ so that $\operatorname{Aut} X \subseteq B$. By Proposition 3 and Theorem 2, $\operatorname{Aut} X$ is either T or B, and in the latter case $\operatorname{Aut} X \supseteq U$. The assertion now follows from Theorem 1.

Type II. $\Sigma = \{0\}$ so that $\operatorname{Aut} X \subseteq T$ and the assertion now follows from Theorem 2.

Type III. As in Ia), $S = \Sigma = \{0 \cup \infty\}$, and the assertion follows from the equality $\omega_0 \chi = \chi^{-1}$. □

13. A classification.

THEOREM 1.5. *Curves of different types* (*in the notation of Theorem* 4) *are not isomorphic.*

$\mathbb{P}^1_{n,m}(c_0, \ldots, c_p; d_0, \ldots, d_q) \cong \mathbb{P}^1_{n',m'}(c'_0, \ldots, c'_{p'}; d'_0, \ldots, d'_{q'}) \Leftrightarrow$ *either* $n = n'$, $m = m'$, $p = p'$, $q = q'$ *and* $c_i = c'_i$ *for all* i, $d_j = d'_j$ *for all* j, *or* $n = m'$, $m = n'$, $p = q'$, $q = p'$ *and* $c_i = d'_i$ *for all* i, $d_j = c'_j$ *for all* j.

$\mathbb{A}^1_n(c_0, \ldots, c_p) \cong \mathbb{A}^1_{n'}(c'_0, \ldots, c'_{p'}) \Leftrightarrow n = n'$, $p = p'$, $c_i = c'_i$ *for all* i.

$\mathbb{P}^1_n(p_0, \ldots, p_m; q_0, \ldots, q_l) \cong \mathbb{P}^1_{n'}(p'_0, \ldots, p'_{m'}; q'_0, \ldots, q'_{l'}) \Leftrightarrow n' = n'$ *and either* $m = m'$, $l = l'$, $p_i = p'_i$ *for all* i, $q_j = q'_j$ *for all* j, *or* $m = l'$, $l = m'$, $p_i + q'_i = 0$ *for all* i, $q_j + p_j = 0$ *for all* j.

PROOF. Isomorphisms preserve the property of a curve to be complete, the number of branches, the number of singular points, as well as the property of $(\operatorname{Aut} X)^0$ be a (not necessarily one-dimensional) torus. If $x \in X$ and $\mathscr{W}_x$ is the radical of the integral closure of $\mathscr{O}_{x,X}$ in $k(X)$, then (see part 2) $k + \mathscr{W}_x \supset \mathscr{O}_{x,X} \supset k + \mathscr{W}_x^n$ and the smallest n with this property is preserved (as well as the equality $\mathscr{O}_{x,X} = k + \mathscr{W}_x^n$ if it holds) under isomorphisms.

If $(\mathrm{Aut}\, X)^0$ is a torus, χ_0 is a basis character of it, x is a point fixed by $(\mathrm{Aut}\, X)^0$, and $\{\chi_0^i \mid i \in \mathfrak{z} \subset \mathbb{N}\}$ is the system of weights of $(\mathrm{Aut}\, X)^0$ in $\mathscr{O}_{x,X}/(k + \mathscr{W}_x^n)$ then, up to a simultaneous change of the signs of all of the numbers, $\mathfrak{z}$ does not change under isomorphisms and a change of χ_0. All the assertions now follow. □

14. Examples (char $k \neq 2$).

1. $\mathbb{P}_1^1(0; 0)$ is the simplest example of a singular curve with infinite automorphism group. It is easy to see that $\mathbb{P}_1^1(0; 0)$ is the *folium cartesii* $x_0 x_1^2 = x_0 x_2^2 + x_2^3$ (here $(x_0 : x_1 : x_2)$ are coordinates on $\mathbb{P}^2$).

2. Suppose that X is obtained from $\mathbb{P}^1$ by the Rosenlicht–Serre construction, where $S = \{\infty\}$ and $\mathscr{O}'_{p(\infty)}$ is spanned over k by 1, $s^2 + s^3$, and s^i, $i \geq 4$. It is easy to check that in this case $\mathrm{Aut}\, X \cong \mathbb{Z}_2$. In general, if X is unibranched, $|\mathrm{sing}\, X| = 1$ and $|\mathrm{Aut}\, X| < \infty$, then $\mathrm{Aut}\, X$ is a cyclic group.

15. Applications. The following theorems on actions of a linear algebraic group on an irreducible quasiprojective variety Y are proved in [152].

THEOREM A. *Suppose that Y is normal. Then there exist an embedding $i: Y \hookrightarrow \mathbb{P}^n$ and a homomorphism $\rho: G \to \mathrm{PGL}_n$ such that $\rho(g)(i(x)) = i(g(x))$ for all $x \in Y$, $g \in G$.*

THEOREM B. *Suppose that Y is normal and G is a torus. Then any point $x \in Y$ has a nonempty invariant affine neighborhood.*

The example $Y = \mathbb{P}_n^1(p_0, \ldots, p_m; q_0, \ldots, q_l)$ and $G = (\mathrm{Aut}\, Y)^0$ shows that the assertion of Theorem B may be false for a nonnormal Y.

THEOREM 1.6. *Suppose that G is reductive, Y is projective, and there exists an equivariant embedding of Y in $\mathbb{P}^n$. Then any fixed point $x \in Y$ has a nonempty invariant affine neighborhood.*

PROOF. Let Γ be an invariant hypersurface in $\mathbb{P}^n$ such that $x \notin \Gamma$ (it always exists, see [78]). Then $Y - \Gamma$ is the desired neighborhood. □

The example $Y = \mathbb{P}_n^1(p_0, \ldots, p_m; q_0, \ldots, q_l)$ and $G = (\mathrm{Aut}\, G)^0$ shows that, in view of Theorem 1.6, the assertion of Theorem A may be false for a nonnormal Y.

§2. A classification of irreducible affine surfaces with algebraic groups of automorphisms acting transitively on the complement of a finite number of points

1. Let X be an irreducible affine variety. In the remainder of this chapter we shall investigate the following problem: for which X does there exist an algebraic subgroup of $\mathrm{Aut}\, X$ acting transitively on the complement of a finite set of points? In other words, which X admit the structure of a quasihomogeneous (with respect to a suitable algebraic group G) variety such that $\dim \Gamma_X \leq 0$?

Because X is irreducible and affine, we may (and will) assume that G is a connected linear group (see [127]); for the sake of convenience we will also assume that G is simply connected and the noneffectivity kernel of the action of G on X is finite.

In this section we shall obtain a classification of such surfaces.

2. We begin with general results, assuming nothing about $\dim X$. Let P be the semisimple Levi component of G, so that $G = P \cdot \operatorname{Rad} G$.

If $\Gamma_X \neq 0$ then, by Corollary 2 of Proposition 2, Chapter 1, G is reductive, Γ_X consists of only one point, and X is an S-variety of G. Such X were constructed in part 7 of §1, Chapter 6, and classified (up to G-isomorphism, but not isomorphism!) in part 4 of §1, Chapter 6. According to part 6 of §1, Chapter 6, if, in addition, X is smooth then $X \cong \mathbb{A}^n$, and, in particular, X is homogeneous with respect to an algebraic group (different from G).

Therefore, the smooth varieties of the type in question are automatically homogeneous with respect to some algebraic group, and the singular ones are S-varieties.

Proposition 2.1. *If* $\dim X \geq 2$ *and* $P = \{e\}$ *then* $X \cong \mathbb{A}^p \times (\mathbb{A}^1_*)^q$ *and* G *acts transitively on* X.

Proof. Because $G = \operatorname{Rad} G$, G is reductive. Therefore $\mathscr{O}_X \cong \mathbb{A}^p \times (\mathbb{A}^1_*)^q$ (see [138]). In particular, $\mathscr{O}_X$ is affine and therefore (see [29, 30]) either $\Gamma_X = \varnothing$ or $\dim \Gamma_X = \dim X - 1 \geq 1$, the latter being impossible because $\dim \Gamma_X \leq 0$. □

Remark. Evidently, for any $p, q \in \mathbb{N}$ the variety $\mathbb{A}^p \times (\mathbb{A}^1_*)^q$ is homogeneous with respect to an algebraic group (for example, with respect to the natural action of $(k_+)^p \times (k^*)^q$).

Theorem 2.1. *Let* $R \neq \{e\}$ *be a connected semisimple group and* Q *be an observable subgroup of* R *(see part* 1 *of* §2, *Chapter* 1) *that contains no nonzero normal subgroups of* R. *Then* $\operatorname{codim}_R Q \geq \operatorname{rk} R + 1$.

Proof. R acts naturally on R/Q with finite noneffectivity kernel. By [70, 12], $\operatorname{codim}_R Q > \operatorname{rk} R$. Suppose that $\operatorname{codim}_R Q = \operatorname{rk} R$. Then R is a semisimple subgroup of $\operatorname{Aut} R/Q$ of maximal rank. But, according to [70, 12], this is possible only when R/Q is a product of projective spaces (and R is the product of the corresponding projective groups). This contradicts the observability of Q. □

Theorem 2.2. *Suppose that a semisimple group* $R \neq \{e\}$ *acts regularly on an irreducible quasiaffine variety* Y *with finite noneffectivity kernel. Then* $\operatorname{rk} R \leq \dim Y - 1$.

PROOF. Because the noneffectivity kernel is finite and the number of nonzero-dimensional normal subgroups of R is also finite, we can find a point $y \in Y$ such that R_y would contain no nonzero normal subgroups of R. The assertion now follows from Theorem 2.1, because R_y is observable (this follows from the fact that Y is quasiaffine). □

COROLLARY. $\operatorname{rk} P \leq \dim X - 1$.

3. We shall now turn to the classification problem for surfaces and assume that $\dim X = 2$.

By the corollary of Theorem 2.2, either $P = \{e\}$ or $P = \mathrm{SL}_2$. By Proposition 2.1 and the remark after it, the case $P = \{e\}$ gives rise only to the surfaces $\mathbb{A}^2$, $\mathbb{A}^1 \times \mathbb{A}^1_*$ and $\mathbb{A}^1_* \times \mathbb{A}^1_*$. If $P = \mathrm{SL}_2$ then, because the noneffectivity kernel is finite, it follows from Theorem 2, Chapter 6, the corollary of this theorem, and part 9 of §1, Chapter 6 that P itself acts transitively on the complement of a finite set of points of X. Thus it remains to consider the case $G = \mathrm{SL}_2$ (henceforth this equality will be automatically assumed; we will also use the notation from part 5 of §2, Chapter 4 and part 8 of §1 of Chapter 6).

4. By Theorem 2 of Chapter 6, $\Gamma_X = \varnothing \Leftrightarrow X \cong G/T$ or $G/N_G(T)$.

LEMMA 2.1. a) *The surface G/T is isomorphic to $\mathbb{P}^1 \times \mathbb{P}^1 - \Delta$, where Δ is the diagonal.* b) *The surface $G/N_G(T)$ is isomorphic to $\mathbb{P}^2 - C$, where C is a nondegenerate quadric.*

PROOF. a) Let p be the point corresponding to the coset B in $G/B = \mathbb{P}^1$, $w = \begin{pmatrix} 0 & -1 \\ 1 & 0 \end{pmatrix} \in G$, and $x = p \times wp \in \mathbb{P}^1 \times \mathbb{P}^1 - \Delta$.

Then $G_x = B \cap wBw^{-1} = T$. On the other hand, G acts transitively on $\mathbb{P}^1 \times \mathbb{P}^1 - \Delta$ (because, in view of $N_G(B) = B$, we have $\dim Gy = 1$ for all $y \in \mathbb{P}^1 \times \mathbb{P}^1 - \Delta$.

b) It is clear that $\mathbb{P}^2 - C$ is one of the orbits of the natural action of G on the space $\mathbb{P}^2$ of binary quadratic forms (considered up to proportionality) in x and y; here C is the nondegenerate quadric defined by the vanishing of the discriminant of a quadratic form. It remains to remark that $G_{xy} = N_G(T)$. □

5. Suppose now that $\Gamma_X \neq \varnothing$. According to the corollary of Theorem 2, Chapter 6 as well as parts 9 and 4 of §1, Chapter 6, this happens exactly when $X = \operatorname{Specm} A$, where A is a subalgebra of $k[x, y] = \bigoplus_{n \in \mathbb{N}} S_n$ of the form $A = \bigoplus_{n \in \mathfrak{z}} S_n$ and $\mathfrak{z}$ is a nonzero finitely-generated subsemigroup with zero of $\mathbb{N}$ (the action of G on $\operatorname{Specm} A$ is defined by formula (2.3) of Chapter 4). The geometric meaning of such $\operatorname{Specm} A$ is clear. Namely, consider an r-tuple $n_1, \dots, n_r \in \mathbb{Z}^+$ and the map $v_{n_1 \cdots n_r} \colon \mathbb{P}^1 \to \prod_{i=1}^r \mathbb{P}^{n_i}$, $v_{n_1 \cdots n_r} = v_{n_1}(x) \times \cdots \times v_{n_r}(x)$ where $v_m \colon \mathbb{P}^1 \to \mathbb{P}^m$ is the Veronese embedding. Let

$\mathbb{V}(n_1, \ldots, n_r)$ denote the affine surface, which is the cone over the curve $v_{n_1 \cdots n_r}(\mathbb{P}^1)$ in $\prod_{i=1}^r \mathbb{P}^{n_i}$ (more precisely: we identify $\mathbb{P}^n$ with the variety of lines in $\mathbb{A}^{n+1}$ passing through 0; if the homogeneous coordinates in $\mathbb{P}^n$ are viewed as coordinates in $\mathbb{A}^{n+1}$ then $v_{n_1 \cdots n_r}$ gives rise to a morphism $\mathbb{A}^2 \to \prod_{i=1}^r \mathbb{A}^{n_i+1}$ and $\mathbb{V}(n_1, \ldots, n_r)$ is its image). It is now clear that the surfaces $\operatorname{Specm} A$ are exactly all possible $\mathbb{V}(n, \ldots, n_r)$ (with $\mathfrak{z} = \langle\langle n_1, \ldots, n_r \rangle\rangle$).

6. We can now prove the main theorem of this section.

THEOREM 2.3. (A classification). *The following is the complete list* (*up to isomorphism*) *of irreducible affine algebraic surfaces admitting an algebraic group of automorphisms that acts transitively on the complement of a finite set of points*:

1) *Smooth surfaces*: a) $\mathbb{A}^2$; b) $\mathbb{A}^1 \times \mathbb{A}^1_*$; c) $\mathbb{A}^1_* \times \mathbb{A}^1_*$; d) $\mathbb{P}^2 - C$, *where* C *is a nondegenerate quadric*; e) $\mathbb{P}^1 \times \mathbb{P}^1 \times \mathbb{P}^1 - \Delta$, *where* Δ *is a diagonal. The surfaces* a)–e) *are pairwise nonisomorphic.*

2) *Singular surfaces*: *all possible* $\mathbb{V}(n_1, \ldots, n_r)$ *where* $n_i \geq 2$ *for any* i. *Here* $\mathbb{V}(n_1, \ldots, n_r) \cong \mathbb{V}(m_1, \ldots, m_s)$ *if and only if* $\langle\langle n_1, \ldots, n_r \rangle\rangle = \langle\langle m_1, \ldots, m_s \rangle\rangle$.

PROOF. By Lemma 1, Chapter 6 (and because $\dim S_n = n+1$), $\mathbb{V}(n_1, \ldots, n_r)$ is nonsingular (and isomorphic to $\mathbb{A}^2$) if and only if $n_i = 1$ for some i. It now follows from parts 3, 4, and 5 that our list is complete. Thus it remains to check its irredundancy.

It is known that for any irreducible variety X the group $k[X]^*/k^*$, where

$$k[X]^* = \{f \in k[X] \mid f^{-1} \in k[X]\}$$

is a free abelian group of finite rank (see [128]); we denote this rank by $\operatorname{rk}(X)$. By [129], $\operatorname{rk}(X \times Y) = \operatorname{rk}(X) + \operatorname{rk}(Y)$ and if X is an algebraic group then $\operatorname{rk}(X) = \operatorname{rk}\operatorname{Hom}(X, k^*)$. We should also mention that, by [33], for any connected simply connected algebraic group H and any algebraic subgroup $L \subset H$ we have

$$\operatorname{Pic} H/L \cong \operatorname{Hom}(L, k^*)/\operatorname{Hom}_H(L, k^*),$$

where $\operatorname{Hom}_H(L, k^*)$ is the subgroup of those characters of L that can be lifted to H. In view of this and Lemma 2.1 we now obtain Table 1:

TABLE 1

X	$\mathbb{A}^2$	$\mathbb{A}^1 \times \mathbb{A}^1_*$	$\mathbb{A}^1_* \times \mathbb{A}^1_*$	$\mathbb{P}^2 - C$	$\mathbb{P}^1 \times \mathbb{P}^1 - \Delta$
$\operatorname{rk}(X)$	0	*I*	2	0	0
$\operatorname{Pic} X$	0	0	0	$\mathbb{Z}_2$	$\mathbb{Z}$

Therefore surfaces a)–e) are pairwise nonisomorphic. It remains to prove a similar result for singular surfaces. If $\langle\langle n_1, \dots, n_r\rangle\rangle = \langle\langle m_1, \dots, m_s\rangle\rangle$ then, in view of part 4 of §1, Chapter 6, $\mathbb{V}(n_1, \dots, n_r)$ and $\mathbb{V}(m_1, \dots, m_s)$ are G-isomorphic, and therefore isomorphic. Let us now prove the converse: if $X = \mathbb{V}(n_1, \dots, n_r)$ and $Y = \mathbb{V}(m_1, \dots, m_s)$ are isomorphic, then $\langle\langle n_1, \dots, n_r\rangle\rangle = \langle\langle m_1, \dots, m_s\rangle\rangle$.

Assume the opposite. Then we can find $\lambda \in \mathbb{Z}^+$ contained in one semigroup (say $\langle\langle m_1, \dots, m_s\rangle\rangle$) but not in the other. Let $\varphi\colon X \to Y$ be an isomorphism and $d_1 = \gcd(n_1, \dots, n_r)$, $d_2 = \gcd(m_1, \dots, m_s)$. By [16], $\mathbb{V}(d_1)$ and $\mathbb{V}(d_2)$ are, respectively, the normalizations of X and Y and we have the commutative diagram

$$\begin{array}{ccc} \mathbb{V}(d_1) & \xrightarrow{\psi} & \mathbb{V}(d_2) \\ {\scriptstyle\pi_1}\downarrow & & \downarrow{\scriptstyle\pi_2} \\ X & \xrightarrow{\varphi} & Y \end{array} \tag{1}$$

where π_i are the normalization maps and ψ is an isomorphism. Let $a \in X$ and $b \in Y$ be singular points. Because $a = \operatorname{sing} X$ and $b = \operatorname{sing} Y$ (see part 9 of §1, Chapter 6) we have $\varphi(a) = b$. Because π_i commute with the action of G, it follows again from part 9 of §1, Chapter 6 that $|\pi_1^{-1}(a)| = |\pi_2^{-1}(b)| = 1$. Let $p = \pi_1^{-1}(a)$ and $q = \pi_2^{-1}(b)$. Because diagram (1) commutes, it is true that $\psi(p) = q$. We also set $\mathfrak{m}_p = \mathfrak{m}_{p, \mathbb{V}(d_1)} \cap k[\mathbb{V}(d_1)]$, $\mathfrak{m}_q = \mathfrak{m}_{q, \mathbb{V}(d_2)} \cap k[\mathbb{V}(d_2)]$ and identify $k[X]$ and $k[Y]$ with subalgebras of $\mathbb{V}(d_1)$ and, respectively, $\mathbb{V}(d_2)$.

We have $\lambda = d_2 l$ for some $l \in \mathbb{Z}^+$. We claim that there exists $f \in k[\mathbb{V}(d_2)]$ with the following properties: (f_1) $f \in \mathfrak{m}_q$; (f_2) $f \notin \mathfrak{m}_q^2$; (f_3) $f^l \in k[X]$. Indeed $k[\mathbb{V}(d_2)] = \bigoplus_{n \in \langle\langle d_2\rangle\rangle} S_n \supset k[Y] = \bigoplus_{n \in \langle\langle m_1, \dots, m_s\rangle\rangle} S_n$ and, by Lemma 1, Chapter 6, $\mathfrak{m}_q = \bigoplus_{n \in \langle\langle d_2\rangle\rangle_0} S_n$, $\mathfrak{m}_q^2 = \bigoplus_{n \in 2\langle\langle d_2\rangle\rangle_0} S_n$. Therefore for f we can take any nonzero element of S_{d_2}.

Because ψ is an isomorphism, the element $h = \psi^*(f) \in k[\mathbb{V}(d_1)]$ has the following properties: (h_1) $h \in \mathfrak{m}_p$; (h_2) $h \notin \mathfrak{m}_p^2$; (h_3) $h^l \in k[X]$. As before, we have $k[\mathbb{V}(d_1)] = \bigoplus_{n \in \langle\langle d_1\rangle\rangle} S_n \supset k[X] = \bigoplus_{n \in \langle\langle n_1, \dots, n_r\rangle\rangle} S_n$, $\mathfrak{m}_p = \bigoplus_{n \in \langle\langle d_1\rangle\rangle_0} S_n$, $\mathfrak{m}_q^2 = \bigoplus_{n \in 2\langle\langle d_2\rangle\rangle_0} S_n$. Therefore $h = h_1 + h_2$, where $h_1 \in S_{d_1}$, $h_1 \neq 0$, and $h_2 \in \bigoplus_{n \in 2\langle\langle d_2\rangle\rangle_0} S_n$. Hence $h^l = h_1^l + \widetilde{h}$, where $\widetilde{h} = \sum_{i=1}^{l} \binom{l}{i} h_1^{l-i} h_2$. It is clear that $h_1^l \in S_{d,l}$ and $\widetilde{h} \in \bigoplus_{n \in (l+1)\langle\langle d_1\rangle\rangle_0} S_n$. In view of (h_3), this yields $S_{d,l} \subset k[X]$ and therefore $d_1 l \in \langle\langle n_1, \dots, n_r\rangle\rangle$.

Because $\mathbb{V}(d_1)$ and $\mathbb{V}(d_2)$ are isomorphic, their Weil divisor class groups are isomorphic. By [16], those groups are isomorphic, respectively, to $\mathbb{Z}_{d_1}$ and $\mathbb{Z}_{d_2}$. Hence $d_1 = d_2$. Therefore $\lambda = d_2\lambda = d_1\lambda \in \langle\langle n_1, \dots, n_r\rangle\rangle$, a contradiction. □

§3. A classification of irreducible affine solids with algebraic groups of automorphisms acting transitively on the complement of a finite number of points

1. In this section we use the notation and results from parts 1 and 2 of §2. From now on we assume that $\dim X = 3$. By the corollary of Theorem 2.2, P can only be one of the following groups: $\{e\}$, SL_2, $SL_2 \times SL_2$, SL_3, G_2, and Sp_4. The case $P = \{e\}$ yields, in view of Proposition 2.1 and the following remark, only the solids $\mathbb{A}^p \times (\mathbb{A}^1_*)^q$, $p+q=3$. Henceforth we will assume that $P \neq \{e\}$.

PROPOSITION 3.1. *$P \neq G_2$ and $P \neq Sp_4$.*

PROOF. Suppose that $P = G_2$. As in the proof of Theorem 2.2, we can find a proper observable subgroup H of P with $\operatorname{codim}_P H \leq 3$. Let $\widetilde{H}^0$ be the maximal connected proper subgroup of P containing H^0. Then $\widetilde{H}^0$ is either semisimple (of rank ≤ 2) or parabolic of the form $P[\alpha]$, where α is a simple root (see [22, 50]). The dimension count gives $\operatorname{codim}_P \widetilde{H}^0 > 3$, a contradiction.

Let now $P = Sp_4$ and H and $\widetilde{H}^0$ be the same as above. If $\widetilde{H}^0$ is semisimple, then, because $\operatorname{codim}_P \widetilde{H}^0 \leq 3$, it follows that $\widetilde{H}^0$ is locally isomorphic to SL_3. But $\operatorname{rk} P = \operatorname{rk} \widetilde{H}^0$ so that $\widetilde{H}^0$ is regular. This contradicts Table 9 from [22]. If $\widetilde{H}^0 = P[\alpha]$ then (because, as is easily verified, in our case $\operatorname{codim}_P P[\beta] = 3$ for all β) $H^0 = \widetilde{H}^0$ contrary to the fact that H^0 is observable. □

2. The remaining possibilities for P, namely SL_2, $SL_2 \times SL_2$, and SL_3 can indeed be realized, as will be shown later. The case $P = SL_2$ is the main source of smooth X, whereas the cases $P = SL_2 \times SL_2$, SL_3 yield singular X.

LEMMA 3.1. *Let $D = SL_2 \times SL_2$ and C be a proper observable subgroup of D such that $\operatorname{codim}_D C \leq 3$. Then $\dim C =$ either 3 or 4 and: a) if $\dim C = 3$ then either D/C is affine and C is locally isomorphic to SL_2, or D/C is not affine, D/C is the open orbit in a quasihomogeneous affine solid of D that has zero-dimensional boundary and C^0 is a solvable group with two-dimensional unipotent radical; b) if $\dim C = 4$ then either D/C is affine, one of the factors of D is a semisimple Levi subgroup of C^0, and $\operatorname{Rad} C^0 \cong k^*$, or D/C is not affine, D/C is the open orbit in a quasihomogeneous affine surface of D with zero-dimensional boundary, one of the factors of D is a semisimple Levi subgroup of C^0 and $\operatorname{Rad} C^0 \cong k_+$.*

PROOF. Let $C^0 = L \cdot \operatorname{Rad} C^0$. If C is semisimple then $L = C^0$ is locally isomorphic to SL_2 and therefore D/C is affine by Matsushima's criterion.

If $\operatorname{Rad} C^0 \neq \{e\}$ then either $L = \{e\}$ or L is locally isomorphic to SL_2.

Let $L = \{e\}$. Because $\operatorname{rk} D = 2$ and the dimension of a maximal unipotent subgroup of D is equal to 2, it follows from $\dim C^0 \geq 3$ that $1 \leq \operatorname{rk} C^0 \leq 2$. If $\operatorname{rk} C^0 = 2$ then D/C, being quasiaffine, is affine (see [110]) and therefore, by Matsushima's criterion, C^0 is a torus, i.e., $\dim C = 2$, a contradiction. If $\operatorname{rk} C^0 = 1$ then C^0 contains a maximal unipotent subgroup of D and therefore D/C is the open orbit in an S-variety Y of D. Because in this case $\operatorname{rk} C = \operatorname{rk} D - 1$, it follows (see [16]) that the complement of the open orbit in Y is a single point. Because C is not reductive, D/C is not affine. It is easy to see that such S-varieties do exist (see below).

Suppose now that L is locally isomorphic to SL_2. Then $1 \leq \dim \operatorname{Rad} C^0 \leq 2$, $0 \leq \operatorname{rk} \operatorname{Rad} C^0 \leq 1$ and $0 \leq \dim \operatorname{Rad}_u C^0 \leq 1$. If $\operatorname{rk} \operatorname{Rad} C^0 = 1$ then $\operatorname{rk} C^0 = 3$ and therefore D/C, being quasiaffine, is also affine (see [110]), i.e., $\operatorname{Rad} C^0 \cong k^*$ (by Matsushima's criterion) and $\dim D/C = 2$. In this case C^0 contains a maximal torus of D. Hence (see [22]) L is a regular subgroup of D and therefore (see [22]) L is one of the factors of D. It is clear that the subgroups of this type do exist.

If $\operatorname{rk} \operatorname{Rad} C^0 = 0$ then $\operatorname{Rad} C^0 \cong k_+$ and C^0 contains a maximal unipotent subgroup of D. Thus D/C is the open orbit in a two-dimensional S-variety of D, and because $\operatorname{rk} C^0 = 1$, it follows that the complement of this orbit is a single point. By [16], L is a regular subgroup of D; therefore (see [22]) L is one of the factors of D. It is clear that such subgroups do exist. □

COROLLARY. *Any three-dimensional S-variety of $\mathrm{SL}_2 \times \mathrm{SL}_2$ is, with respect to this group, a quasihomogeneous, but not homogeneous, solid with zero-dimensional boundary.*

PROOF. This follows from [16] because, in view of part a) of Lemma 3.1, the rank of the generic stabilizer in such an S-variety is equal to 1. □

LEMMA 3.2. *Let D be a connected simple linear algebraic group and C be a subgroup of $D \times D$ locally isomorphic to D. Then one of the factors of $D \times D$ acts transitively on $(D \times D)/C$.*

PROOF. Because D is simple, one of the homomorphisms $p_1: C \to D \times \{e\}$ and $p_2: C \to \{e\} \times D$, where p_i is the projection onto the ith factor, is surjective. Now the assertion follows. □

LEMMA 3.3. *Let D be a connected linear algebraic group whose semisimple Levi subgroup L is isomorphic to $\mathrm{SL}_2 \times \mathrm{SL}_2$ to Y a solid on which D acts regularly, transitively and with finite noneffectivity kernel. Suppose that L_y is locally isomorphic to SL_2 for some $y \in Y$. Then L acts on Y transitively.*

PROOF. Let $C = L \cdot \operatorname{Rad}_u D$. Because $Cy \supset Ly$ and Ly is open in Y, it follows that Cy is also open in Y. We shall show that $Cy = Ly$.

Because $L_y \subset C_y \subset C$, it is true that $1 \le \operatorname{rk} C_y \le 2$ and the semisimple Levi subgroup H of C_y contains a subgroup locally isomorphic to SL_2. Therefore if $\operatorname{rk} C_y = 1$ then H is locally isomorphic to SL_2 and $\operatorname{Rad} C_y = \operatorname{Rad}_u C_y$. Therefore in this case $|\operatorname{Hom}(C_y, k^*)| < \infty$.

If $\operatorname{rk} C_y = 2$ then a priori we can have the two possibilities: either $\operatorname{rk} H = 2$ or $\operatorname{rk} H = 1$. In the former case $\operatorname{Rad} C_y^0 = \operatorname{Rad}_u C_y^0$ and, as before, $|\operatorname{Hom}(C_y, k^*)| < \infty$. In the latter case the reductive Levi subgroup of C_y^0 is locally isomorphic to $\mathrm{SL}_2 \times k^*$ and in C_y^0 we can find a one-dimensional torus S centralizing L_y^0 (so that $S \cdot L_y^0$ is the reductive Levi subgroup of C_y^0). If $\pi \colon C \to L$ is the epimorphism defined by the formula $\pi(ln) = l$ for all $l \in L$, $n \in \operatorname{Rad}_u D$, then $\pi|_{L_y^0} = \mathrm{Id}$, $\operatorname{Ker} \pi|_S = \{e\}$. Therefore $\operatorname{Ker} \pi|_{S \cdot L_y^0} = \{e\}$ and $\pi(S) \cdot L_y^0$ is a reductive subgroup of L of rank 2. Because $\operatorname{rk} L = 2$, it follows that $\pi(S) \cdot L_y^0$ is a regular subgroup of L and therefore L_y^0 coincides with one of the factors of L (see [22]). This contradicts the finiteness of the noneffectivity kernel, because Ly is open in Y. Thus we have shown that $|\operatorname{Hom}(C_y, k^*)| < \infty$.

By Matsushima's criterion, Ly is affine. Therefore (see [30]), if $Cy \neq Ly$ then $Cy - Ly$ is an unmixed subvariety of codimension 1 in Cy. Suppose that $Cy \neq Ly$. It follows from [33] that $\operatorname{Pic} C = 0$ and $\operatorname{Pic} Cy \cong \operatorname{Hom}(C_y, k^*)$ is a finite group. Therefore there exists a function $f \in k[Cy]$ such that $Cy - Ly = \{x \in Cy | f(x) = 0\}$. If $\alpha \colon k(Cy) \to k(Ly)$ is the isomorphism induced by the embedding $Ly \hookrightarrow Cy$ then $\alpha(f) \in k[Ly]^*$ and $\alpha(f) \notin k^* \subset k[Ly]^*$. But the natural surjection $L \to Ly$ induces an embedding $k[Ly]^* \hookrightarrow k[L]^*$, and we see that $k[L]^* \neq k^*$. But $k[L]^*/k^* \cong \operatorname{Hom}(L, k^*)$ is a trivial group (see [128]), a contradiction. Therefore $Cy = Ly$.

We shall now show that $Ly = Y$. Let T be a torus in $\operatorname{Rad} D$ such that $T \cdot L$ is the reductive Levi subgroup of D. Then $Y = Dy = T \cdot Cy = T \cdot Ly$. Then $\operatorname{Pic} Y = \operatorname{Pic}(T \cdot L/(T \cdot L)_y) \cong \operatorname{Hom}((T \cdot L)' \cap (T \cdot L)_y, k^*)$ as follows from [33]. Because $(T \cdot L)' = L$, we have $(T \cdot L)' \cap (T \cdot L)_y = L_y$ and therefore $\operatorname{Pic} Y = 0$.

Taking into account that Ly is affine and $\operatorname{Hom}(L, k^*)$ is trivial and arguing as before we deduce that $Ly = Y$. □

PROPOSITION 3.2. *If $P = \mathrm{SL}_2 \times \mathrm{SL}_2$ then either X is an S-variety of P or one of the factors of P acts on X transitively.*

PROOF. We remark first that $m_p(X) = \max_{x \in X} \dim Px = 3$. Indeed if this is not the case then it follows from part b) of Lemma 3.1 that for any $x \in X$ the stabilizer P_x contains one of the factors of P. It follows from the irreducibility of X that one of the factors of P is contained in $\{g \in G \mid gx = x \text{ for all } x \in X\}$, which is a contradiction. Thus $m_p(X) = 3$.

It follows from part a) of Lemma 3.1 and the corollary of this lemma that either X is an S-variety of P or P_x is locally isomorphic to SL_2 for some $x \in X$. Assume the latter. Then $Gx = Px$ by Lemma 3.3. If $Gx = X$ the assertion follows from Lemma 3.2. But the case $Gx \neq X$ is impossible, otherwise $X - Px$ would be a single point (see part 2 of §2) and, by Theorem 1 of §1, Chapter 6, X would be an S-variety of P. But then P_x could not be locally isomorphic to SL_2. Thus $X = Gx = Px$ and, by Lemma 3.2, one of the factors of P acts transitively on X. □

COROLLARY. *If $P = \mathrm{SL}_2 \times \mathrm{SL}_2$ and X is smooth then X is isomorphic to the solid SL_2/F, where F is a finite subgroup of SL_2. If X is singular then it is an S-variety of P.*

PROOF. It follows from [16] and part 6 of §1, Chapter 6 that any smooth S-variety of a semisimple group is isomorphic to $\mathbb{A}^n$ with a linear action of this group. But, as is easily seen, $\mathrm{SL}_2 \times \mathrm{SL}_2$ has no three-dimensional linear representations such that $\mathrm{SL}_2 \times \mathrm{SL}_2$ acts transitively on the complement to 0 in the corresponding space. Now the assertion follows from Proposition 3.2. □

Thus the case of a smooth X and $P = \mathrm{SL}_2 \times \mathrm{SL}_2$ reduces to the case $P = \mathrm{SL}_2$, which will be studied later.

LEMMA 3.4. *Let $D = \mathrm{SL}_3$ and C be a proper observable subgroup of D. Then $\operatorname{codim}_D C \geq 3$, and if $\operatorname{codim}_D C = 3$ then D/C is the open orbit in some S-variety of D and is not affine. Any three-dimensional S-variety of D is quasihomogeneous with respect to D and has zero-dimensional boundary.*

PROOF. Let L be the semisimple Levi subgroup of C^0. Then it follows from the dimension count that L is locally isomorphic to either $\{e\}$, or SL_2, or $\mathrm{SL}_2 \times \mathrm{SL}_2$. The last case is impossible because SL_3 contains no regular subgroups locally isomorphic to $\mathrm{SL}_2 \times \mathrm{SL}_2$ (see [22]). If $L = \{e\}$ then C is solvable and, being observable, must have codimension in D greater than the codimension of a Borel subgroup of D (which is equal to 3). Finally, if L is locally isomorphic to SL_2 then $\operatorname{rk}\operatorname{Rad} C \leq 1$. If $\operatorname{rk}\operatorname{Rad} C = 1$ then C contains a maximal torus of D and therefore C, being observable, must be reductive (see [110]), so that $\operatorname{codim}_D C = 4$. If $\operatorname{rk}\operatorname{Rad} C = 0$ then, because the dimension of a maximal unipotent subgroup of D is equal to 3 and the dimension of a maximal unipotent subgroup of SL_2 is equal to 1, we have $\dim \operatorname{Rad} C \leq 2$, i.e., $\operatorname{codim} C_D \geq 3$. Therefore $\operatorname{codim}_D C = 3$ only if C contains a maximal unipotent subgroup of D and $L = \mathrm{SL}_2$. In this case D/C is the open orbit in some S-variety of D (see [16]). The assertion about the boundary follows from the fact that D contains no observable subgroups of codimension 1 and 2. □

Because a smooth S-variety of a semisimple group is isomorphic to $\mathbb{A}^n$, we have the following

COROLLARY. *If $P = \mathrm{SL}_3$ and X is smooth then $X = \mathbb{A}^3$.*

3. We shall now summarize the results of part 2. So far we only have the following four trivial examples of smooth affine solids X of the desired type: $\mathbb{A}^3$, $\mathbb{A}^2 \times \mathbb{A}^1_*$, $\mathbb{A}^1 \times \mathbb{A}^1_* \times \mathbb{A}^1_*$, $\mathbb{A}^1_* \times \mathbb{A}^1_* \times \mathbb{A}^1_*$, and the classification problem is reduced to the case $P = \mathrm{SL}_2$. This problem will be solved in parts 6–12.

Singular solids X, as varieties with group action, can be described as follows.

PROPOSITION 3.3. *Singular X are exactly three-dimensional singular S-varieties of the groups $\mathrm{SL}_2 \times \mathrm{SL}_2$ and SL_3.*

PROOF. It follows from the corollaries of Theorem 2.2, Proposition 3.2, Lemma 3.1, and Lemma 3.4 that we only need to show that if $\operatorname{sing} X \neq \varnothing$ then $P \not\cong \mathrm{SL}_2$. Suppose that $P \cong \mathrm{SL}_2$. Because $\Gamma_x \supset \operatorname{sing} X \neq \varnothing$, it follows (see part 2 of §2) that X is a two-orbit S-variety of a reductive group G. It follows from their general construction (see the end of part 7 of §1, Chapter 6) and the finiteness of the set $\{g \in G \mid gx = x \text{ for all } x \in X\}$ that $\dim Z_G \leq 1$. Because $\dim X = \dim P = 3$, it means that $G = \mathrm{SL}_2 \cdot k^*$ and $G_x^0 \cong k_+$ where $x \in \mathscr{O}_X$. It follows, in view of [16], that (in the notation of part 4 of §1, Chapter 6) the cone K_X is of dimension 2 and has three proper faces (two in dimension one and a vertex). This contradicts (see part 4 of §1, Chapter 6) the fact that X has only two G-orbits. □

This proposition yields a geometric description of singular solids X of the desired type (similar to what was done for surfaces in part 5 of §2) and a classification theorem for them.

First we shall introduce a special type of affine varieties.

4. Varieties $\mathbb{V}_{n_1\cdots n_s}(A)$. Let $n_1, \ldots, n_s$ and $m_1, \ldots, m_s$ be two sets of positive integers. If $x_1, \ldots, x_{n_1}$ are coordinates in $\mathbb{A}^{n_1}$, $y_1, \ldots, y_{n_2}$ are coordinates in $\mathbb{A}^{n_2}$, ..., and $z_1, \ldots, z_{n_s}$ are coordinates in $\mathbb{A}^{n_s}$ then let $v_{n_1\cdots n_s}^{m_1\cdots m_s}\colon \mathbb{A}^{n_1} \times \cdots \times \mathbb{A}^{n_s} \to \mathbb{A}^N$, where $N = \prod_{i=1}^s \binom{n_i+m_i-1}{m_i}$ denote the map defined by the lexicographically-ordered set of all monomials of the form $x_1^{i_1}\cdots x_{n_1}^{i_{n_1}} y_1^{j_1}\cdots y_{n_2}^{j_{n_2}} \cdot\cdots\cdot z_1^{l_1}\cdots z_{n_s}^{l_{n_s}}$ with $i_1+\cdots+i_{n_1} = m_1$, $j_1+\cdots+j_{n_2} = m_2, \ldots, l_1+\cdots+l_{n_s} = m_s$. Thus $v_{n_1\cdots n_s}^{m_1\cdots m_s}$ is the composition of the affine Veronese and Segre maps.

DEFINITION. Let $A = (a_{ij})$, $1 \leq i \leq r$, $1 \leq j \leq s$, be a matrix with elements $a_{ij} \in \mathbb{Z}^+$ (where i is the row number and j is the column number) and $n_1, \ldots, n_s$ is a set of positive integers. We define the variety $\mathbb{V}_{n_1\cdots n_s}(A)$ as the closure of the image of the map

$$v_{n_1\cdots n_s}^{a_{11}\cdots a_{1s}} \times \cdots \times v_{n_1\cdots n_s}^{a_{r1}\cdots a_{rs}}\colon \mathbb{A}^{n_1} \times \cdots \times \mathbb{A}^{n_s} \to \mathbb{A}^{N_1} \times \cdots \times \mathbb{A}^{N_r},$$

where $N_j = \prod_{i=1}^s \binom{n_i+a_{ji}-1}{a_{ji}}$, $j = 1, \ldots, r$.

(One can show that this image is actually closed.)

For example, $\mathbb{V}_n(m)$ is the affine cone over the Veronese variety $v_m(\mathbb{P}^{n-1})$ and the surface $\mathbb{V}(n_1, \dots, n_r)$ from §2 is $\mathbb{V}_2\left(\begin{pmatrix} n_1 \\ \vdots \\ n_r \end{pmatrix}\right)$; the varieties just defined are their generalizations.

Henceforth the rows of matrix A will be viewed as elements of the corresponding coordinate space. When referring to the semigroup generated by the rows of A we will mean the semigroup in the additive group of vectors of the coordinate space.

5. We can now prove a classification theorem for singular solids X of the considered type.

THEOREM 3.1. *Singular affine solids admitting an algebraic automorphism group acting transitively on the complement of a finite set of points are exactly the varieties*

$$\mathbb{V}_{2,2}\left(\begin{pmatrix} n_1 & m_1 \\ & \vdots & \\ n_s & m_s \end{pmatrix}\right), \quad \textit{where } 0 < n_1/m_1 = \cdots = n_s/m_s \leq 1,$$

and

$$\mathbb{V}_3\left(\begin{pmatrix} n_1 \\ \vdots \\ n_s \end{pmatrix}\right), \quad \textit{where } n_i \geq 2 \textit{ for all } i.$$

The solids $\mathbb{V}_{2,2}(A)$ and $\mathbb{V}_3(B)$ are not isomorphic. Two singular solids $\mathbb{V}_{2,2}(A)$ and $\mathbb{V}_{2,2}(C)$, where

$$A = \begin{pmatrix} n_1 & m_1 \\ & \vdots & \\ n_s & m_s \end{pmatrix}, \qquad C = \begin{pmatrix} p_1 & q_1 \\ & \vdots & \\ p_t & q_t \end{pmatrix}$$

and $n_1/m_1 = \cdots = n_s/m_s \leq 1$, $p_1/q_1 = \cdots = p_t/q_t \leq 1$, (resp., $\mathbb{V}_3(B)$ and $\mathbb{V}_3(D)$) are isomorphic if and only if the additive semigroup generated by the rows of A (resp., B) coincides with the additive semigroup generated by the rows of C (resp., D).

PROOF. By Proposition 3.3, we have to classify up to isomorphism (but not G-isomorphism!) singular three-dimensional S-varieties X of groups $G = \mathrm{SL}_3$ and $\mathrm{SL}_2 \times \mathrm{SL}_2$.

We begin with the case $G = \mathrm{SL}_3$. Because $\dim X = 3$, it follows from [16] that the cone K_X coincides with one of the two faces of the Weyl chamber of a fixed maximal torus of G, so that $X = X(\mathfrak{z})$, where $\mathfrak{z} = \langle\langle n_1\omega_i, \dots, n_s\omega_i\rangle\rangle$, $i = 1$ or 2 and $n_j \in \mathbb{Z}^+$ for all j (henceforth we use the notation from part 4 of §1, Chapter 6). Conversely, any S-variety of this type is an affine solid on which $G = \mathrm{SL}_3$ acts so that this action is transitive on the complement of some point o (see [16] and part 4 of §1, Chapter 6). We remark that it suffices to consider only the solids of the form $X(\langle\langle\langle n_1\omega_1, \dots, n_s\omega_1\rangle\rangle\rangle)$. Indeed, if σ is an outer automorphism of G then, after defining a new action of G on $X(\langle\langle\langle n_1\omega_1, \dots, n_s\omega_1\rangle\rangle\rangle)$ by the formula $x \overset{g}{\mapsto} \sigma(g)x$ we obtain the G-variety $X(\langle\langle\langle n_1\omega_2, \dots, n_s\omega_2\rangle\rangle\rangle)$ (so that the varieties $X(\langle\langle\langle n_1\omega_1, \dots, n_s\omega_1\rangle\rangle\rangle)$ and $X(\langle\langle\langle n_1\omega_2, \dots, n_s\omega_2\rangle\rangle\rangle)$ are isomorphic but not G-isomorphic).

Because G acts transitively on the complement in $R(\omega_2)$ of zero, it follows that $X(\langle\langle\langle \omega_1\rangle\rangle\rangle) = R(\omega_2)$ (see part 4 of §1, Chapter 6). Let x_1, x_2, x_3 be coordinates in $R(\omega_2)$. Then the G-module $k[x_1, x_2, x_3]_d$ of degree d forms in x_1, x_2, x_3 is isomorphic to $R(d\omega_1)$. In view of part 4 of §1, Chapter 6, we have

$$\begin{aligned} X(\langle\langle\langle n_1\omega_1, \dots, n_s\omega_1\rangle\rangle\rangle) &= \operatorname{Specm} \bigoplus_{d\in\langle\langle n_1, \dots, n_s\rangle\rangle} k[x_1, x_2, x_3]_d \\ &= \mathbb{V}_3\left(\begin{pmatrix} n_1 \\ \vdots \\ n_s \end{pmatrix}\right). \end{aligned} \tag{2}$$

Because $\dim R(d\omega_1) = (d+1)(d+2)/2$, it follows from (2) and part 3 of Lemma 1 of §1, Chapter 6, that $\operatorname{sing} \mathbb{V}_3(B) = \varnothing \Leftrightarrow \dim \Theta_{o, \mathbb{V}_3(B)} = 3$ if and only if one of the matrix elements of B is equal to 1 (if and only if $\mathbb{V}_3(B) = \mathbb{A}^3$, see part 6 of §1, Chapter 6). Let

$$B = \begin{pmatrix} n_1 \\ \vdots \\ n_s \end{pmatrix}.$$

Then, in view of (2) and [16], the normalization of $\mathbb{V}_3(B)$ is $\mathbb{V}_3((d))$, where $d = \gcd(n_1, \dots, n_s)$, and the Weil divisor class group of $\mathbb{V}_3((d))$ is $\mathbb{Z}_d$. These properties are completely analogous to the corresponding properties of the surfaces $\mathbb{V}(n_1, \dots, n_r)$ from §2. It is easy to check that the proof of the implication "$\Rightarrow$" in part 2 of Theorem 2.3 uses only those properties and not the dimension of the varieties in question. It now follows from this observation (and (2)) that the proof also works for the solids $\mathbb{V}_3(B)$, i.e., for those solids the last assertion of our theorem is true.

Suppose now that $G = \mathrm{SL}_2 \times \mathrm{SL}_2$. Because $\dim X = 3$, it follows from [16] that the cone K_X is of dimension 1 and is not orthogonal to any of the simple roots. Therefore $X = X(\mathfrak{z})$, where $\mathfrak{z} = \langle\langle n_1\omega_1 + m_1\omega_2, \dots, n_s\omega_1 + m_s\omega_2\rangle\rangle$, $n_i, m_i \in \mathbb{Z}^+$ for all i and $n_1/m_1 = \dots = n_s/m_s$. Conversely, any S-variety of this form is an affine solid on which $G = \mathrm{SL}_2 \times \mathrm{SL}_2$ acts so that this action is transitive on the complement of some point o. It suffices to consider only the solids $X(\langle\langle\langle n_1\omega_1 + m_1\omega_2, \dots, n_s\omega_1 + m_s\omega_2\rangle\rangle\rangle)$ with $n_1/m_1 = \dots = n_s/m_s \le 1$ as before, $X(\langle\langle\langle m_1\omega_1 + n_1\omega_2, \dots, m_s\omega_1 + n_s\omega_2\rangle\rangle\rangle)$

can be obtained from $X(\langle\langle n_1\omega_1 + m_1\omega_2, \dots, n_s\omega_1 + m_s\omega_2\rangle\rangle)$ by an outer automorphism of G.

It is easy to see that G_v, where v is the sum of the highest weight vectors of the irreducible components of $R(\omega_1) \oplus R(\omega_2)$, is a maximal unipotent subgroup of G. It follows now from the dimension count that $\overline{Gv} = R(\omega_1) \oplus R(\omega_2)$. Therefore $X(\langle\langle \omega_1, \omega_2\rangle\rangle) = R(\omega_1) \oplus R(\omega_2)$ (see part 4 of §1, Chapter 6). Let x_1, x_2 and y_1, y_2 be coordinates on $R(\omega_1)$ and $R(\omega_2)$, respectively. Then x_1, x_2, y_1, y_2 can be viewed as coordinates on $R(\omega_1) \oplus R(\omega_2)$ in an obvious way, and the G-module $k[x_1, x_2, y_1, y_2]_{(n,m)}$ of forms of degree n in x_1 and x_2 and degree m in y_1 and y_2 is isomorphic to $R(m\omega_1 + n\omega_2)$. Therefore, in view of part 4 of §1, Chapter 6, we have

$$\begin{aligned} X(\langle\langle n_1\omega_1 + m_1\omega_2, \dots, n_s\omega_1 + m_s\omega_2\rangle\rangle) &= \operatorname{Specm} \bigoplus_{(n,m)\in\langle\langle (n_1,m_1),\dots,(n_s,m_s)\rangle\rangle} k[x_1, x_2, y_1, y_2]_{(n,m)} \\ &= \mathbb{V}_{2,2}\left(\left(\begin{matrix} n_1 & m_1 \\ & \vdots & \\ n_s & m_s \end{matrix}\right)\right). \end{aligned} \tag{3}$$

Let

$$A = \begin{pmatrix} n_1 & m_1 \\ & \vdots & \\ n_s & m_s \end{pmatrix},$$

where $n_i, m_i \in \mathbb{Z}^+$ for all i and $n_1/m_1 = \dots = n_s/m_s \leq 1$. Because $\dim R(n\omega_1 + m\omega_2) = (n+1)(m+1)$, and $\dim \mathbb{V}_{2,2}(A) = 3$, it follows from (3) and part 3 of Lemma 1 of §1, Chapter 6 that $o = \operatorname{sing} \mathbb{V}_{2,2}(A)$. Let $p = \gcd(n_1, \dots, n_s)$ and $q = \gcd(m_1, \dots, m_s)$. Then it follows from (3) and [16] that $\mathbb{V}_{2,2}((pq))$ is the normalization of $\mathbb{V}_{2,2}(A)$, and the Weil divisor class group of $\mathbb{V}_{2,2}((pq))$ is $\mathbb{Z}_{\gcd(p,q)}$. In particular, we see that $\mathbb{V}_{2,2}(A)$ and its normalization have the same cone K_X. It means that the argument from Theorem 2.3 can be used to prove the last assertion of the theorem (for solids X of the form $\mathbb{V}_{2,2}(A)$) if we only establish its validity for normal solids X. However this cannot be done as in Theorem 2.3: the divisor class group does not determine the corresponding semigroup (in this case the embedding of this semigroup in $\mathfrak{X}$ is determined not only by its abstract structure but also by the direction of the ray K_X). The desired result will be proved by a different argument using Poincaré series of local rings of singular points. More precisely, by a simple calculation we deduce from (3) and part 6) of Lemma 1 of §1, Chapter 6 that

$$P(\mathscr{O}_{o,\mathbb{V}_{2,2}((pq))}, t) = \frac{1 + (pq - p - q - 2)t + (pq + p + q + 1)t^2}{(1-t)^3}.$$

Suppose now that $\mathbb{V}_{2,2}((ab)) \cong \mathbb{V}_{2,2}((cd))$, where $a \le b$, $c \le d$. Then, because such varieties have only one singular point o, it follows that $P(\mathscr{O}_{o,\mathbb{V}_{2,2}((ab))}, t) = P(\mathscr{O}_{o,\mathbb{V}_{2,2}((cd))}, t)$. Combining this with the previous formula we have $ab-a-b-2 = cd-c-d-2$ and $ab+a+b+1 = cd+c+d+1$, whence $a = c$ and $b = d$.

It remains to prove that the solids $\mathbb{V}_{2,2}(A)$ and $\mathbb{V}_3(B)$ are not isomorphic. Let

$$A = \begin{pmatrix} n_1 & m_1 \\ & \vdots & \\ n_s & m_s \end{pmatrix}, \qquad 0 < n_1/m_1 = \cdots = n_s/m_s \le 1$$

and

$$B = \begin{pmatrix} b_1 \\ \vdots \\ b_r \end{pmatrix}, \qquad b_i \ge 2 \text{ for all } i.$$

Suppose that $\mathbb{V}_{2,2}(A)$ and $\mathbb{V}_3(B)$ are isomorphic. Then their normalizations are also isomorphic, and therefore the local rings of the (unique) singular points of those normalizations are isomorphic too. Hence these local rings have the same Poincaré series. It easily follows from (2) and part 6) of Lemma 1 of §1, Chapter 6 that

$$P(\mathscr{O}_{o,\mathbb{V}_3((d))}, t) = \frac{2 + (d^2+3d-4)t + (d^2-3d+2)t^2}{2(1-t)^3}.$$

Because the normalizations of $\mathbb{V}_{2,2}(A)$ and $\mathbb{V}_3(B)$ are $\mathbb{V}_{2,2}((pq))$ and, respectively, $\mathbb{V}_3((d))$, where $p = \gcd(n_1, \ldots, n_s)$, $q = \gcd(m_1, \ldots, m_s)$, $d = \gcd(b_1, \ldots, b_r)$ it follows, after equating the corresponding rational functions that $d^2+3d-4 = 2(pq-p-q-2)$ and $d^2-3d+2 = 2(pq+p+q+1)$. An easy calculation shows now that $3d = -2p-2q$, contrary to the fact that p, q, and d are positive integers. □

6. From now on we may (and will) assume that X is smooth and G acts transitively on X (see part 2 of §2). Before classifying those X, we want to introduce several special types of solids.

We fix an isomorphism $\varphi\colon \operatorname{Pic}(\mathbb{P}^1 \times \mathbb{P}^1 - \Delta) \to \mathbb{Z}$ (see the proof of Theorem 2.3) and identify the groups in question by this isomorphism.

DEFINITION. The variety $\mathbb{X}_n$, $n \in \mathbb{Z}$, is the total space of the one-dimensional locally trivial (in the Zariski topology) vector bundle on $\mathbb{P}^1 \times \mathbb{P}^1 - \Delta$ corresponding to the element $n \in \operatorname{Pic}(\mathbb{P}^1 \times \mathbb{P}^1 - \Delta)$.

DEFINITION. The variety $\mathbb{X}_n^*$, $n \in \mathbb{Z}$, is the complement of the zero section of $\mathbb{X}_n$.

These definitions may seem dependent on the isomorphism φ. We will show however that $\mathbb{X}_n \cong \mathbb{X}_{-n}$ and $\mathbb{X}_n^* \cong \mathbb{X}_{-n}^*$ (as varieties but not as bundles!).

Let 0 and 1 be, respectively, the identity and the nonidentity element of $\mathrm{Pic}(\mathbb{P}^2 - C) \cong \mathbb{Z}_2$ (see the proof of Theorem 2.3).

DEFINITION. The variety $\mathbb{Y}_n$, $n = 0, 1$, is the total space of the one-dimensional locally-trivial (in the Zariski topology) vector bundle on the surface $\mathbb{P}^2 - C$ corresponding to the element $n \in \mathrm{Pic}(\mathbb{P}^2 - C)$.

DEFINITION. The variety $\mathbb{Y}_n^*$, $n = 0, 1$, is the complement of the zero section of Y_n.

Later we will show that $\mathbb{X}_n$, $\mathbb{X}_n^*$, $\mathbb{Y}_n$, and $\mathbb{Y}_n^*$ are affine solids. Finally we shall now introduce several other types of solids whose affinity would follow directly from their definition and Matsushima's criterion.

DEFINITION. The varieties $\mathbb{W}_n$, $n \in \mathbb{Z}^+$; $\mathbb{S}_3$; $\mathbb{S}_4$; and $\mathbb{S}_5$ are, respectively, the homogeneous spaces SL_2/D_n; $\mathrm{SL}_2/\widetilde{\mathbb{T}}$; $\mathrm{SL}_2/\widetilde{\mathbb{Q}}$; and $\mathrm{SL}_2/\widetilde{\mathbb{I}}$ (see parts 8 and 11 of §1, Chapter 6).

It follows from the definition that $\mathbb{W}_n$, $\mathbb{S}_3$, $\mathbb{S}_4$, and $\mathbb{S}_5$ should be added to our list of smooth affine solids (see part 3) that is being described in this section. Later we will show that the same is true for $\mathbb{X}_n$, $\mathbb{X}_n^*$, $\mathbb{Y}_0$, $\mathbb{Y}_0^*$, and $\mathbb{Y}_1^*$ (but not $\mathbb{Y}_1$).

7. As the results of parts 1 and 2 show, we may (and will) assume that X is homogeneous with respect to G and that $P = \mathrm{SL}_2$. Because the noneffectivity kernel of the action of G on X is finite and SL_2 contains no two-dimensional observable subgroups, it follows that $m_P(X) = \max_{x\in X} \dim Px = 2$ or 3. We will see later that both possibilities can be realized. First we shall find out what X looks like if

$$m_P(X) = 3. \tag{4}$$

In view of (4), the variety X is a smooth affine solid quasihomogeneous with respect to SL_2. By Theorems 3 and 4, Chapter 6, such solids are exactly $\mathbb{W}_n$ for all $n \in \mathbb{Z}^+$, $\mathbb{S}_3$, $\mathbb{S}_4$, $\mathbb{S}_5$, as well as SL_2/T_n and $F_n = \mathrm{SL}_2 \times_T \mathbb{A}^1$ where the action of T on $\mathbb{A}^1$ is linear and is given (in the notation of part 5 of §2, Chapter 4) by the character χ^n, $n \in \mathbb{Z}^+$.

PROPOSITION 3.4. *For any $n \in \mathbb{Z}$, $n \neq 0$, we have the isomorphisms $\mathbb{X}_n \cong F_{|n|}$ and $\mathbb{X}_n^* \cong \mathrm{SL}_2/T_n$.*

PROOF. By Lemma 2.1, $\mathrm{SL}_2/T \cong \mathbb{P}^1 \times \mathbb{P}^1 - \Delta$ so that F_n is a one-dimensional vector bundle on $\mathbb{P}^1 \times \mathbb{P}^1 - \Delta$ (corresponding to the character χ^n). Because $\mathrm{Pic}\,\mathrm{SL}_2 = 0$ (see [33]) and $\mathrm{Hom}(\mathrm{SL}_2, k^*) = \{e\}$, it follows that any one-dimensional vector bundle on SL_2/T has a unique structure of a homogeneous bundle. Moreover, assigning to this bundle the character of the torus T by which this torus acts on the fiber over the coset eT we obtain an isomorphism $\mathrm{Pic}\,\mathrm{SL}_2/T = \mathrm{Hom}(T, k^*)$ (see [33]). We now remark that, according to Theorem 4, Chapter 6, the total spaces of the bundles corresponding to the characters χ^n and χ^{-n} are isomorphic (and even

SL_2-isomorphic) for any $n \in \mathbb{Z}^+$ and the complement of the zero section in such a space is an SL_2-orbit isomorphic to SL_2/T_n. □

COROLLARY. *For any n, the solids $\mathbb{X}_n$ and $\mathbb{X}_n^*$ are affine, and $\mathbb{X}_n \cong \mathbb{X}_{-n}$, $\mathbb{X}_n^* \cong \mathbb{X}_{-n}^*$ for all $n \in \mathbb{Z}$.*

Thus $\mathbb{X}_n^*$, $n \in \mathbb{Z}$, should also be added to the list of smooth solids in question: if $n \neq 0$ then $\mathbb{X}_n^*$ is homogeneous with respect to SL_2 (in view of Proposition 3.4) and $\mathbb{X}_0^* = SL_2/T \times \mathbb{A}_*^1$ is obviously homogeneous with respect to $SL_2 \times k^*$.

It turns out that the same is true for the solids $\mathbb{X}_n$: we will show that they are homogeneous with respect to certain extensions of SL_2 (we should remark here that the action of SL_2 on $\mathbb{X}_n$ given by Proposition 3.4 is not transitive on the complement of any finite set of points).

8. The groups $SL_2(n)$, $n \in \mathbb{N}$. We keep the notation of part 5 of §2, Chapter 4.

DEFINITION. The group $SL_2(n)$ is the semidirect product of the group SL_2 and the additive group S_n, given by the SL_2-module structure of S_n.

REMARKS. a) In other words, the elements of $SL_2(n)$ are the pairs (g, u), $g \in SL_2$ and $u \in S_n$, with multiplication given by the formula $(g_1, u_1) \cdot (g_2, u_2) = (g_1 g_2, g_1 u_2 + u_1)$. Henceforth we will view the groups SL_2 and S_n as subgroups of $SL_2(n)$ (with respect to the their natural embeddings in $SL_2(n)$).

b) It is clear that $SL_2(n)$ is a connected linear algebraic group, SL_2 is its semisimple Levi subgroup, and $\operatorname{Rad} SL_2(n) = \operatorname{Rad}_u SL_2(n) = S_n$.

9. THEOREM 3.2. *For any $n \in \mathbb{Z}$ the group $SL_2(|n|)$ acts regularly and transitively on $\mathbb{X}_n$.*

PROOF. By the corollary of Proposition 3.4, we may (and will) assume that $n \geq 0$. Assume first that $n > 0$.

Let $V_0 = \bigoplus_{i=1}^n (S_n)_{\chi^{n-2i}}$ and $H_n = \{(t, u) \in SL_2(n) \mid t \in T, u \in V_0\}$. It is clear that H_n is a connected solvable group of rank 1 and $\operatorname{Rad}_u H_n = V_0$. We have $\dim SL_2(n) = n + 4$, $\dim H_n = n + 1$ and $\operatorname{Rad}_u H_n \subset \operatorname{Rad}_u SL_2(n)$. The variety $SL_2(n)/\operatorname{Rad}_u H_n$ is obviously affine (it is isomorphic to $SL_2 \times \mathbb{A}^1$). Because $H_n/\operatorname{Rad}_u H_n \cong k^*$ is reductive, it follows that $SL_2(n)/H_n \cong (SL_2(n)/\operatorname{Rad}_u H_n)/(H_n/\operatorname{Rad}_u H_n)$ is also affine (and is a solid). The group $SL_2(n)$ acts transitively on this variety. Consider the restriction of this action to SL_2. By Theorem 4, Chapter 6 and Proposition 3.4, it suffices to show that $SL_2(n)/H_n$ is not homogeneous with respect to SL_2, that $(SL_2(n)/H_n)^{SL_2} = \varnothing$, and that the SL_2-stabilizer of some point of $SL_2(n)/H_n$ is T_n.

Let $x_0 \in SL_2(n)/H_n$ be the point corresponding to the coset H_n. Then $(SL_2)_{gx_0} = SL_2 \cap g H_n g^{-1}$ for all $g \in SL_2(n)$. If $g = (e, 0)$ then $(SL_2)_{gx_0} = T$ so that the action of SL_2 on $SL_2(n)/H_n$ is not transitive. Let now $u \neq 0$ be an element of $(S_n)_{\chi^n}$ and $g = (e, u)$. A simple calculation shows that

$(\mathrm{SL}_2)_{gx_0} = T_n$. Finally $\mathrm{SL}_2 \neq \mathrm{SL}_2 \cap gH_ng^{-1}$ for all $g \in \mathrm{SL}_2(n)$ because gH_ng^{-1} is isomorphic to H_n, and therefore its Levi subgroup is one dimensional.

If $n = 0$ then $\mathrm{SL}_2(0) = \mathrm{SL}_2 \times k_+$ and $H_0 = T$, so that $\mathrm{SL}_2(0)/H_0 \cong (\mathrm{SL}_2/T) \times \mathbb{A}^1 = X_0$ (by Lemma 2.1). $\square$

Thus the solids X satisfying Condition (4) are exactly $\mathbb{W}_n$ for all $n \in \mathbb{Z}^+$; $\mathbb{X}_n$ and $\mathbb{X}_n^*$ for all $n \in \mathbb{Z}$; $\mathbb{S}_3$; $\mathbb{S}_4$ and $\mathbb{S}_5$.

10. Now we shall find out what X looks like if

$$m_p(X) = 2. \tag{5}$$

It is convenient to start out with the case

$$\mathrm{Rad}_u G = \{e\}. \tag{6}$$

Let $Z = \mathrm{Rad}\, G$. Then $G = P \times Z$, $P = \mathrm{SL}_2$ and Z is a torus. But $\dim X = 3$, the action of G on X is locally effective, and $\mathrm{rk}\, P = 1$. Therefore it follows from (5) that $1 \leq \dim Z \leq 2$.

PROPOSITION 3.5. *We have* $\dim Z = 1$.

PROOF. Let $Y = \{x \in X \mid \dim Zx < \max_{y \in X} \dim Zy\}$. Then Y is a closed, possibly empty, subvariety of X. Because $X \lhd G$, this subvariety is G-invariant. Therefore (because the action of G on X is transitive) $Y = \varnothing$, so that all of the orbits of Z on X are of the same dimension, so they are closed in X. Because Z is reductive, it means that the fibers of the canonical morphism $\pi_{X,Z}\colon X \to X/Z$ are orbits. But $\dim X = 3$, $1 \leq \dim Z \leq 2$, and $\pi_{X,Z}$ is a surjection, therefore X/Z is an irreducible affine variety with $1 \leq \dim X/Z \leq 2$. Because G acts transitively on X, it follows that $G/Z = \mathrm{SL}_2$ acts transitively on X/Z. But SL_2 contains no two-dimensional observable subgroups. Therefore $\dim X/Z = 1$ and the desired assertion follows from the finiteness of the noneffectivity kernel of the action of G on X. $\square$

Thus $G = \mathrm{SL}_2 \times k^*$ and $X \cong G/\Phi$, where Φ is a one-dimensional subgroup of G. Because X is affine and G is reductive, it follows from Matsushima's criterion that Φ is also reductive. Moreover, by (5), we must have

$$\dim(P \cap g\Phi g^{-1}) = 1 \quad \forall g \in G. \tag{7}$$

Conversely, it follows from Matsushima's criterion that for any one-dimensional subgroup Φ of G, satisfying condition (7), G/Φ is an affine G-homogeneous solid satisfying conditions (5) and (6).

Thus we have reduced our problem to the classification of such subgroups Φ. We set $\mu_n = \{\alpha \in k^* \mid \alpha^n = 1\}$ for all $n \in \mathbb{Z}^+$. Let $\mathrm{SL}_2 \overset{p_1}{\leftarrow} \mathrm{SL}_2 \times k^* \overset{p_2}{\rightarrow} k^*$ be the natural projections. We identify SL_2 and k^* with the subgroups $\mathrm{SL}_2 \times \{1\}$ and $\{e\} \times k^*$ of $\mathrm{SL}_2 \times k^*$.

PROPOSITION 3.6. *One-dimensional reductive subgroups* Φ *of* $G = \mathrm{SL}_2 \times k^*$ *satisfying the condition*

$$\dim(\mathrm{SL}_2 \cap g\Phi g^{-1}) = 1 \quad \forall g \in G, \tag{8}$$

are exactly the subgroups conjugate to one of the following subgroups of $\mathrm{SL}_2 \times k^*$: 1) $\Phi_1(n) = T\mu_n$; 2) $\Phi_2(n) = N_G(T)\mu_n$; 3) $\Phi_3(n) = \{gs \in G|$ *either* $s \in \mu_n$ *and* $g \in T$, *or* $s \in \mu_{2n} - \mu_n$ *and* $g \in N_G(T) - T\}$ (*here* $n \in \mathbb{Z}^+$ *in all cases*).

PROOF. Because Φ is reductive, we have $\Phi^0 \cong k^*$, and by (8) we have $\Phi^0 \subset \mathrm{SL}_2$. Replacing Φ with a conjugate subgroup we may (and will) assume that $\Phi^0 = T$. Because $p_2(\Phi^0) = \{e\}$, it follows that $p_2(\Phi) = \mu_n$ for some $n \in \mathbb{Z}, n \geq 1$. Because $\Phi^0 \triangleleft \Phi$ and $[N_G(T) : T] = 2$, we see that either $p_1(\Phi) = T$ or $p_1(\Phi) = N_G(T)$. If $p_1(\Phi) = T = \Phi^0$ then $p_2(\Phi) \subset \Phi$ and therefore $\Phi = \Phi_1(n)$. Suppose now that $p_1(\Phi) = N_G(T)$. Because $[N_G(T) : T] = 2$ it follows that $\mu_n^2 = \{g^2 \mid g \in \mu_n\} \subset \Phi$. Hence if n is odd then $\mu_n^2 = \mu_n \subset \Phi$ and therefore $\Phi = \Phi_2(n)$. If n is even then $\mu_n^2 = \mu_{n/2} \subset \mu_n$ and $[\mu_n : \mu_n^2] = 2$. We now consider those $gs \in \Phi$ for which $s \in \mu_n$ but $s \notin \mu_{n/2}$. If we can find among them an element for which $g \in T$ then $\mu_n \subset \Phi$ and therefore $\Phi = \Phi_2(n)$. If no such element exists then $\Phi = \Phi_3(n/2)$. The rest is obvious. □

PROPOSITION 3.7. *We have the following isomorphisms* $G/\Phi_1(n) \cong \mathbb{X}_0^*$, $G/\Phi_2(n) \cong \mathbb{Y}_0^*$, $G/\Phi_3(n) \cong \mathbb{Y}_1^*$.

PROOF. The first two isomorphisms can be obtained from the definition of the subgroups $\Phi_1(n)$ and $\Phi_2(n)$ and Lemma 2.1.

Let now ψ be the (only) nontrivial character of $N_G(T)$ and $\varphi: k^* \to k^*$, $\varphi(t) = t$ for all $t \in k^*$. Consider the character θ_n of the group $H = N_G(T) \times k^* \subset G$ defined by the formula $\theta_n(us) = \psi(u)\varphi^n(s)$ for all $u \in N_G(T), s \in k^*$.

Because $\operatorname{Pic} G = 0$ (see [33]) and $\theta_n \notin \operatorname{Hom}_G(H, k^*)$ it follows from [33] that θ_n gives rise to a nontrivial one-dimensional bundle on G/H, which can be assumed to be homogeneous. But $G/H = \mathbb{P}^2 - C$ (see Lemma 2.1) and it follows from the nontriviality of θ_n that the action of G on this one-dimensional bundle is transitive on the complement of the zero section. By definition this complement is the solid $\mathbb{Y}_1^*$. If $o \in G/H$ is the point corresponding to the coset H then for any point $x \in \mathbb{Y}_1^*$ contained in the fiber over o we have $G_x = H_x = \operatorname{Ker}\theta_n = \Phi_3(n)$. □

Thus the solids X satisfying Conditions (5) and (6) are exactly $\mathbb{X}_0^*$, $\mathbb{Y}_0^*$, and $\mathbb{Y}_1^*$.

11. Finally we investigate the structure of the solids X in the remaining case when Condition (5) is satisfied and $U = \operatorname{Rad}_u G \neq \{e\}$.

By Theorem 6, either $X^P = \varnothing$ and all of the orbits of P in X are of dimension 2 or $X^P \neq \varnothing$. Because X is homogeneous with respect to G, it follows that in the latter case X is also homogeneous with respect to the solvable group $\operatorname{Rad} G$ and, in view of Proposition 2.1, $X \cong \mathbb{A}^p \times (\mathbb{A}^1_*)^q$, $p+q = 3$. Thus we are reduced to the former case. Then all of the orbits of P in X are closed and therefore coincide with the fibers of the canonical morphism $\pi_{X,P}\colon X \to X/P$. Because those fibers are of dimension 2, it follows that X/P is an irreducible affine curve. Because X is a normal variety, it follows from [29] that X/P is a normal (and therefore smooth) curve. Finally, because X is quasihomogeneous with respect to G, we have a dominant morphism $G \to X$. Composing this morphism with $\pi_{X,P}$ we obtain a dominant morphism $G \to X/P$. It is known that the group variety G is rational, therefore X/P is a unirational and, by the Luroth theorem, rational curve. Therefore $X = \mathbb{A}^1 - Y$, where Y is either finite or empty.

Let x be a point of X. The orbit Ux is closed in X because U is unipotent and X is affine (see [151]). The canonical morphism $U \to Ux$, $u \mapsto ux$, induces an embedding $k[Ux]^* \hookrightarrow k[U]^*$. But $k[U]^* = k^*$ so that $\operatorname{Hom}(U, k^*) = \{e\}$ (see the proof of Theorem 2.3). Therefore $k[Ux]^* = k^*$. It now follows easily that $\pi_{X,P}(Ux)$ is either a single point or $\pi_{X,P}(Ux) = X/P = \mathbb{A}^1$.

Suppose that $\pi_{X,P}(Ux)$ is a single point for any $x \in X$. Then $Ux \subset \pi^{-1}_{X,P}(\pi_{X,P}(x)) = Px$ for all $x \in X$. Because X is homogeneous with respect to G, it means that X is already homogeneous with respect to the reductive Levi subgroup of G, and therefore we are in the situation studied in part 10.

Thus from now on we may assume that for some point $x_0 \in X$ the following condition is satisfied:

$$\pi_{X,P}(Ux_0) = X/P \cong \mathbb{A}^1. \tag{9}$$

It follows from (9) that $P \cdot U$ acts transitively on X. Therefore we may (and will) assume that $\operatorname{Rad} G = U$. It also follows from (9) that $1 \leq \dim Ux_0 \leq 3$. We shall investigate these three possibilities respectively.

The case $\dim Ux_0 = 3$. Because Ux_0 is closed in X, we have $Ux_0 = X$. It now follows from [138] that $X \cong \mathbb{A}^3$.

The case $\dim Ux_0 = 2$. The smooth solid X contains two irreducible closed surfaces: Ux_0 and $Px_0 = \pi^{-1}_{X,P}(\pi_{X,P}(x_0))$. By (9), they do not coincide, and the point x_0 is contained in both of them. Therefore they intersect along some (possibly reducible) curve Γ.

LEMMA 3.5. *There exists a one-dimensional unipotent subgroup N of P such that $Nx_0 \not\subset \Gamma$.*

PROOF. Let $\mathfrak{N}$ denote the set of unipotent elements of SL_2; it is a hypersurface in SL_2. As is easily checked, if H is one of the subgroups T

and $N_G(T)$ and $\tau\colon \mathrm{SL}_2 \to \mathrm{SL}_2/H$ is the natural morphism then $\tau(\mathfrak{N})$ is a nonempty open subset of SL_2/H (one should take into account that τ is an open map). But $P = \mathrm{SL}_2$ and Px_0 is a two-dimensional orbit closed in X, therefore, by Theorem 2, Chapter 6, P_{x_0} is conjugate to either T or $N_G(T)$. Hence for some point $x \in Px_0$ the set $\mathfrak{N}x = \{ux \mid u \in \mathfrak{N}\}$ is open in P_{x_0}. But $x = gx_0$ for some $g \in P$, and we have, taking into account that $\mathfrak{N}$ is invariant under the action of P on itself by conjugations, that $\mathfrak{N}x = \mathfrak{N}gx_0 = g\mathfrak{N}x_0$ and thus $\mathfrak{N}x_0 = g^{-1}\mathfrak{N}x$ is open in P_{x_0}. Because Γ is a curve and P_{x_0} is a surface, it is true that $ux_0 \notin \Gamma$ for some $u \in \mathfrak{N}$. Therefore for N we can take a one-dimensional unipotent subgroup of P containing u (such a subgroup exists (see [50]). □

We now consider the group $V = N \cdot U$, where N is the group from Lemma 3.5. The group V is unipotent and, by construction, $Vx_0 \supset Ux_0$ and $Vx_0 \neq Ux_0$. Therefore $\dim Vx_0 = 3$ and, as before, $X \cong \mathbb{A}^3$.

The case $\dim Ux_0 = 1$.

LEMMA 3.6. *For some n the group $\mathrm{SL}_2(n)$ acts on X transitively and regularly.*

PROOF. Consider the action of Z_U on X. Then $Z_U \not\subset G_{x_0}$ because $Z_U \triangleleft G$, $\dim Z_U \geq 1$ and the noneffectivity kernel of the action of G on X is finite. Because Z_U and U are unipotent groups, it follows that both Ux_0 and Z_Ux_0 are closed in X (see [151]). Taking into account that $\dim Ux_0 = 1$ and $Z_U \subset U$ we have $Ux_0 = Z_Ux_0$. Therefore the group $P \cdot Z_U$ acts transitively on X (remember that $\pi_{X,P}(Z_Ux_0) = X/P$ and P is transitive on the fibers of $\pi_{X,P}$).

The group Z_U is isomorphic to $(k_+)^m$ for some m and the adjoint action of P on Z_U defines on Z_U a structure of an SL_2-module. This module is semisimple and decomposes into a direct sum of simple SL_2-modules. The same arguments as above show that for one of those simple SL_2-modules, say S_n, we have $S_nx_0 = Z_Ux_0$. Therefore $P \cdot S_n$ acts transitively on X. □

Thus we may assume that X is homogeneous with respect to $G = P \cdot S_n = \mathrm{SL}_2(n)$. Let $H = G_{x_0}$. Then H has the following properties:

$$G/H \text{ is an affine solid;} \tag{10}$$

$$\text{any orbit of } P = \mathrm{SL}_2 \text{ in } G/H \text{ is of dimension 2;} \tag{11}$$

$$\dim(H \cap S_n) = n. \tag{12}$$

Let also $V_j = \bigoplus_{i=0,\, i\neq j}^{n} (S_n)_{\chi^{n-2i}}$.

PROPOSITION 3.8. *Let $G = \mathrm{SL}_2(n)$ and H be a subgroup of G satisfying Conditions* (10), (11), *and* (12). *Then n is even and H is conjugate to either the group $T \cdot V_{n/2}$ or the group $N_G(T) \cdot V_{n/2}$. Each of these groups also satisfies Properties* (10), (11), *and* (12).

PROOF. First of all we may (and will) assume (in addition to Conditions (10), (11), and (12)) that

$$\mathrm{SL}_2 \cap H = \text{either } T \text{ or } N_G(T). \tag{13}$$

Let $V = H \cap S_n$. The group V is unipotent, $\dim V = n$, because of (12), and $V \triangleleft H$. Consider the subgroup $(\mathrm{SL}_2 \cap H) \cdot V$ of H; by (11), $\dim(\mathrm{SL}_2 \cap H) \cdot V = n+1$. It now follows from (10) that $[H : (\mathrm{SL}_2 \cap H) \cdot V] < \infty$. Combining this with (13) we have $H^0 = T \cdot V$ and therefore V is T-invariant. Hence $V = V_{i_0}$ for some i_0.

Consider the homomorphism $\varphi: G \to \mathrm{SL}_2$, $(g, u) \overset{\varphi}{\mapsto} g$ for all $g \in \mathrm{SL}_2$, $u \in S_n$. Because $\varphi(H)^0 = \varphi(H^0) = T$, we have $T \subset \varphi(H) \subset N_G(T)$ and therefore $\varphi(H) =$ either T or $N_G(T)$.

Suppose that $\varphi(H) = T$. Then $H = H^0 = T \cdot V = T \cdot V_{i_0}$. We shall show that in this case n is even and $i_0 = n/2$. Indeed it follows from (11) that $\dim \mathrm{SL}_2 \cap gHg^{-1} = 1$ for each $g \in G$. Suppose that $u \in (S_n)_{\chi^{n-2i_0}}$, $u \neq 0$. Then, as an easy calculation shows that for $t \in T$, $v \in V$ the element $(e, u)(t, v)(e, -u)$ is contained in SL_2 if and only if $v = 0$ and $\chi^{n-2i_0}(t) = 1$. Because the last relation, viewed as an equation in t, has only finitely many solutions when $n - 2i_0 \neq 0$, it follows that n is even and $i_0 = n/2$.

Suppose now that $\varphi(H) = N_G(T)$. Notice that $w(S_n)_{\chi^{n-2j}} w^{-1} = (S_n)_{\chi^{-n+2j}}$ $\left(\text{here } w = \begin{pmatrix} 0 & 1 \\ -1 & 0 \end{pmatrix}\right)$. But H contains an element of the form (w, u), where $u \in S_n$. Moreover, because $H^0 = TV$, we may assume that $u \in (S_n)_{\chi^{n-2i_0}}$. Taking into account that $(w, u)^2 \in H$ we have $wu + u \in V_{i_0}$. It now follows from the definition of V_{i_0} that $wu = -u$, that n is even, and that $i_0 = n/2$.

If $u = 0$ then $(w, 0) \in H$ and therefore $H = N_G(T) \cdot V_{i_0}$. If $u \neq 0$ then, because w acts on $(S_n)_{\chi^0}$ by the multiplication by $(-1)^{n/2}$, it follows from the relation $wu = -u$ that in this case $n \not\equiv 0 \bmod 4$. Setting $g = (e, 2^{-1}u)$ and using the fact that $H^0 = TV$, we easily check that g centralizers H^0, so that if $\widetilde{H} = g^{-1}Hg$ then $\widetilde{H}^0 = H^0$. It is also easy to check that $(e, -2^{-1}u)(w, u)(e, 2^{-1}u) = (w, 0) \in \widetilde{H}$. It follows from the previous results that $\widetilde{H} = N_G(T) \cdot V_{n/2}$ and therefore H is conjugate to $N_G(T) \cdot V_{n/2}$.

Finally we shall show that the groups $T \cdot V_{n/2}$ and $N_G(T) \cdot V_{n/2}$ satisfy Conditions (10), (11), and (12). Condition (12) is obviously satisfied. Moreover, because the unipotent radicals of the groups in question are contained in $\mathrm{Rad}_u G$, it follows that the quotient of G by each of those subgroups is affine (see [30]). Because this quotient is obviously a solid, it follows that (10) is also satisfied. Finally, (11) should be verified only for $T \cdot V_{n/2}$. Thus we have to show $\dim(\mathrm{SL}_2 \cap p(T \cdot V_{n/2})p^{-1}) = 1$ for all $p \in G$.

Let $p = (g, u)$, where $g \in \mathrm{SL}_2$ and $u \in S_n$. For $t \in T$ denote by a_t the element $(t, tg^{-1}u - g^{-1}u)$. It is not difficult to see that $a_t \in T \cdot V_{n/2}$;

moreover, a direct computation shows that $pa_tp^{-1} = (gtg^{-1}, 0) \in \mathrm{SL}_2$ for all $t \in T$. Therefore $\dim(\mathrm{SL}_2 \cap p(T \cdot V_{n/2})p^{-1}) \geq 1$. Because the reductive Levi subgroup of $p(T \cdot V_{n/2})p^{-1}$ is a one-dimensional torus, it follows that $\dim(\mathrm{SL}_2 \cap p(T \cdot V_{n/2})p^{-1}) \leq 2$. However this intersection cannot be of dimension 2, otherwise $G/T \cdot V_{n/2}$ would contain a point whose SL_2-orbit would be isomorphic to $\mathbb{P}^1$, contrary to the fact that $G/T \cdot V_{n/2}$ is affine. □

Now we can describe the solids X in the case in question (see the beginning of part 11).

PROPOSITION 3.9. *Let $m = 2n$, where $n \in \mathbb{Z}^+$. Then* $\mathrm{SL}_2(m)/T \cdot V_n \cong \mathbb{X}_0$, $\mathrm{SL}_2(m)/N_G(T) \cdot V_n \cong \mathbb{Y}_0$.

PROOF. We have $V_n \triangleleft T \cdot V_n$. Therefore

$$\mathrm{SL}_2(m)/T \cdot V_n \cong (\mathrm{SL}_2(m)/V_n)/(T \cdot V_n/V_n).$$

Consider the morphism $\mathrm{SL}_2(m) \to \mathrm{SL}_2 \times (S_m)_{\chi^0}$ defined as follows: if $g \in \mathrm{SL}_2$, $u \in S_m$ and $u = u_1 + u_2$, where $u_1 \in (S_m)_{\chi^0}$, $u_2 \in V_n$ (such decomposition exists and is unique), then $\tau((g, u)) = g \times u_1$. It is easy to see that τ is the quotient morphism for the action of V_n on $\mathrm{SL}_2(m)$ by right translations. Thus we can identify $\mathrm{SL}_2(m)/V_n$ with $\mathrm{SL}_2 \times (S_m)_{\chi^0}$. The groups $T \cdot V_n/V_n$ and T can also be identified via the isomorphism $(t, v) \mapsto t$ for all $t \in T$, $v \in V_n$. Then the natural action of $T \cdot V_n/V_n$ on $\mathrm{SL}_2(m)/V_n$ induces an action of T on $\mathrm{SL}_2 \times (S_m)_{\chi^0}$. Because T acts trivially on $(S_m)_{\chi^0}$, it follows easily that the aforementioned action of T on $\mathrm{SL}_2 \times (S_m)_{\chi^0}$ is defined as follows: $(g, u) \overset{t}{\mapsto} (gt, u)$ for all $g \in \mathrm{SL}_2$, $u \in (S_m)_{\chi^0}$. Taking into account that $\dim(S_m)_{\chi^0} = 1$ and $\mathrm{SL}_2/T \cong \mathbb{P}^1 \times \mathbb{P}^1 - \Delta$, we have $(\mathrm{SL}_2(m)/V_n)/(T \cdot V_n/V_n) \cong (\mathrm{SL}_2/T) \times (S_m)_{\chi^0} \cong (\mathbb{P}^1 \times \mathbb{P}^1 - \Delta) \times \mathbb{A}^1 = \mathbb{X}_0$.

The argument in the case of the group $N_G(T) \cdot V_n$ is similar. One only has to remember that $\mathrm{SL}_2/N_G(T) \cong \mathbb{P}^2 - C$. □

12. Now we can prove a classification theorem for solids X of the type considered in this section.

THEOREM 3.3. *Smooth affine solids admitting an algebraic group of automorphisms acting transitively on the complement of a finite set of points are exactly* $\mathbb{X}_n$, $n \in \mathbb{Z}$; $\mathbb{X}_n^*$, $n \in \mathbb{Z}$; $\mathbb{W}_n$, $n \in \mathbb{Z}^+$; $\mathbb{Y}_0$; $\mathbb{Y}_0^*$; $\mathbb{Y}_1^*$; $\mathbb{S}_3$; $\mathbb{S}_4$; $\mathbb{S}_5$; $\mathbb{A}^3$; $\mathbb{A}^2 \times \mathbb{A}^1_*$; $\mathbb{A}^1 \times (\mathbb{A}^1_*)^2$; $(\mathbb{A}^1_*)^3$. *These solids are in fact homogeneous with respect to the corresponding algebraic automorphism group. The only isomorphic solids in this list are $\mathbb{X}_n$ and $\mathbb{X}_{-n}$ for any $n \in \mathbb{Z}$ and $\mathbb{X}_n^*$ and $\mathbb{X}_{-n}^*$, also for any $n \in \mathbb{Z}$.*

PROOF. The completeness of this list is a direct consequence of the results of parts 6–11. The same is true for the homogeneity. In view of Proposition 3.4, it remains to prove that no two solids from the list, different from $\mathbb{X}_n$

and $\mathbb{X}_{-n}$, and $\mathbb{X}_n^*$ and $\mathbb{X}_{-n}^*$, are isomorphic. We shall establish this by computing, as in the proof of Theorem 2.3, certain numerical invariants of those solids. More precisely, we will have Table 2 (here $\pi_i(X)$ denote the ith homotopy group of X):

TABLE 2

No.	X	$\operatorname{Pic} X$	$\pi_1(X)$	$\pi_2(X)$	$\operatorname{rk}(X)$
1	$\mathbb{X}_n$	$\mathbb{Z}$	0	$\mathbb{Z}$	0
2	$\mathbb{X}_n^*$, $n \neq 0$	$\mathbb{Z}_{\lvert n\rvert}$	$\mathbb{Z}_{\lvert n\rvert}$	0	0
3	$\mathbb{X}_0^*$	$\mathbb{Z}$	$\mathbb{Z}$	$\mathbb{Z}$	1
4	$\mathbb{W}_n$, $2 \mid n$	$\mathbb{Z}_2 \oplus \mathbb{Z}_2$	$D_{\lvert n\rvert}$	0	0
5	$\mathbb{W}_n$, $2 \nmid n$	$\mathbb{Z}_4$	$D_{\lvert n\rvert}$	0	0
6	$\mathbb{Y}_0$	$\mathbb{Z}_2$	$\mathbb{Z}_2$	$\mathbb{Z}$	0
7	$\mathbb{Y}_0^*$	$\mathbb{Z}_2$	$\mathbb{Z} \oplus \mathbb{Z}_2$	$\mathbb{Z}$	1
8	$\mathbb{Y}_1^*$	0	$\mathbb{Z}$	$\mathbb{Z}$	1
9	$\mathbb{S}_3$	$\mathbb{Z}_3$	$\widetilde{\mathbb{T}}$	0	0
10	$\mathbb{S}_4$	$\mathbb{Z}_2$	$\widetilde{\mathbb{O}}$	0	0
11	$\mathbb{S}_5$	0	$\widetilde{\mathbb{I}}$	0	0
12	$\mathbb{A}^3$	0	0	0	0
13	$\mathbb{A}^2 \times \mathbb{A}_*^1$	0	$\mathbb{Z}$	0	1
14	$\mathbb{A}^1 \times \mathbb{A}_*^1 \times \mathbb{A}_*^1$	0	$\mathbb{Z} \oplus \mathbb{Z}$	0	2
15	$\mathbb{A}_*^1 \times \mathbb{A}_*^1 \times \mathbb{A}_*^1$	0	$\mathbb{Z} \oplus \mathbb{Z} \oplus \mathbb{Z}$	0	3

1) The value of $\operatorname{rk}(X)$ can be obtained by the same argument as in the proof of Theorem 2.3 (one should take into account Propositions 3.4, 3.7, 3.9, and Theorem 3.2, as well as that for any algebraic group G and any subgroup $\operatorname{Hom}(\mathrm{SL}_2, k^*) \cong \operatorname{Hom}(\mathrm{SL}_2(n), k^*) \cong \{0\}$, $\operatorname{Hom}(\mathrm{SL}_2 \times k^*, k^*) \cong \mathbb{Z}$, and $\operatorname{rk}(G/H) \leq \operatorname{rk}(G)$. This way we can compute $\operatorname{rk}(X)$ for all X except $\mathbb{Y}_1^*$. But $\mathbb{Y}_1^* \cong (\mathrm{SL}_2 \times k^*)/\Phi_3(n)$ (see Proposition 3.7). The function $f \in k[\mathrm{SL}_2 \times k^*]^*$, $f((g, a)) = a^{2n}$ for all $(g, a) \in \mathrm{SL}_2 \times k^*$ is invariant under right translations by elements of $\Phi_3(n)$, and therefore it is a nonconstant element of $k[\mathbb{Y}^*]^*$. Because $\operatorname{rk}(\mathrm{SL}_2 \times k^*) = 1$, it follows that $\operatorname{rk}(\mathbb{Y}_1^*) = 1$.

2) The group $\operatorname{Pic} X$ can also be computed with the aid of Propositions 3.4, 3.7, 3.9, Theorem 3.2, and the argument from the proof of Theorem 2.3: one

should only take into account that $\operatorname{Pic}(Y \times \mathbb{A}^1) \cong \operatorname{Pic} Y$ for any Y, as well as that $\operatorname{Hom}(H_n, k^*) \cong \mathbb{Z}$ (see the proof of Theorem 3.2); $\operatorname{Hom}(D_n, k^*) \cong \mathbb{Z}_2 \oplus \mathbb{Z}_2$ if $2 \mid n$ and $\cong \mathbb{Z}_4$ if $2 \nmid n$; $\operatorname{Hom}(\widetilde{\mathbb{T}}, k^*) \cong \mathbb{Z}_3$; $\operatorname{Hom}(\widetilde{\mathbb{O}}, k^*) \cong \mathbb{Z}_2$; $\operatorname{Hom}(\widetilde{\mathbb{I}}, k^*) = \{0\}$; $\operatorname{Hom}(\Phi_1(n), k^*) \cong \mathbb{Z} \oplus \mathbb{Z}_n$; $\operatorname{Hom}_{\mathrm{SL}_2 \times k^*}(\Phi_1(n), k^*) \cong \mathbb{Z}_n$; $\operatorname{Hom}(\Phi_2(n), k^*) \cong \mathbb{Z}_2 \oplus \mathbb{Z}_n$; $\operatorname{Hom}_{\mathrm{SL}_2 \times k^*}(\Phi_2(n), k^*) \cong \mathbb{Z}_n$; $\operatorname{Hom}(\Phi_3(n), k^*) = \operatorname{Hom}_{\mathrm{SL}_2 \times k^*}(\Phi_3(n), k^*) \cong \mathbb{Z}_{2n}$. For binary polyhedral groups these isomorphisms are well known (and can be easily proved (see [48, 35])), and then for the remaining groups, except the case $\Phi_3(n)$, they are obvious. To treat the group $\Phi_3(n)$, we check, by direct computation, that $\Phi_3(n)' = T = \Phi_3^0(n)$, so that $\operatorname{Hom}(\Phi_3(n), k^*) \cong \operatorname{Hom}(\Phi_3(n)/T, k^*)$. But $\Phi_3(n)/T = \mathbb{Z}_{2n}$. On the other hand, for any $l \in \mathbb{Z}$ the morphism $\chi_l \colon \Phi_3(2n) \to k^*$, $gs \overset{\chi_l}{\mapsto} s^l$ for all $g \in N_{\mathrm{SL}_2 \times k^*}(T)$, $s \in \mu_{2n}$ is a character, and all χ_l, $l = 1, \dots, 2n$, are pairwise distinct. Thus $\{\chi_l \mid 1 \le l \le 2n\} = \operatorname{Hom}(\Phi_3(n), k^*)$. But it is clear that all χ_l can be extended to characters of $\mathrm{SL}_2 \times k^*$.

3) Because $\mathbb{A}^1_*$ is homotopy equivalent to the circle and $\mathbb{A}^1$ is contractible, we have $\pi_1(\mathbb{A}^1_*) \cong \mathbb{Z}$ and $\pi_1(\mathbb{A}^1) = 0$. Because $\pi_n(Y \times Z) \cong \pi_n(Y) \oplus \pi_n(Z)$, we can compute $\pi_1(X)$ for the solids X nos. 12–15. Because $\pi_1(\mathrm{SL}_2) = 0$, it follows that the group $\pi_1(X)$ for the solids X nos. 2, 4, 5, 9, 10, 11 can be computed with the aid of Proposition 3.4 and the definition of the varieties $\mathbb{W}_n$, $\mathbb{S}_3$, $\mathbb{S}_4$, $\mathbb{S}_5$. Because $\mathbb{P}^1 \times \mathbb{P}^1 - \Delta \cong \mathrm{SL}_2/T$, it follows from the homotopy sequence of the fibration $\mathrm{SL}_2 \overset{T}{\to} \mathrm{SL}_2/T$ that $\pi_1(\mathbb{P}^1 \times \mathbb{P}^1 - \Delta) = 0$. Similarly, from the isomorphism $\mathbb{P}^2 - C \cong \mathrm{SL}_2/N_G(T)$, we have $\pi_1(\mathbb{P}^2 - C) = \mathbb{Z}_2$. This yields $\pi_1(X)$ for solids X nos. 3, 6, 7. Furthermore, it follows from Theorem 3.2 and the homotopy sequence of the fibration $\mathrm{SL}_2(n) \overset{H_n}{\to} \mathrm{SL}_2(n)/H_n$ that $\pi_1(X_n) = 0$. It remains to determine $\pi_1(\mathbb{Y}_1^*)$. Because $\Phi_1(1) \triangleleft \Phi_3(1)$, it follows from Proposition 3.7 that the solid $\mathbb{Y}_1^*$ is the quotient of $\mathbb{X}_0^*$ by the action of $\mathbb{Z}_2$ defined by the formula $(p, q) \times a \overset{\zeta}{\mapsto} (q, p) \times (-a)$, where $p, q \in \mathbb{P}^1$, $p \neq q$, $a \in \mathbb{A}^1_*$, and ζ is the generator of $\mathbb{Z}_2$. Consider the two-fold covering $\tau \colon (\mathbb{P}^1 \times \mathbb{P}^1 - \Delta) \times \mathbb{A}^1_* \to \mathbb{Y}_1^*$. We fix some point $y_0 \in \mathbb{Y}_1^*$, and let $x_1 = (p, q) \times a$ and $x_2 = (q, p) \times (-a)$ be points from $\tau^{-1}(y_0)$. The group $\pi_1((\mathbb{P}^1 \times \mathbb{P}^1 - \Delta) \times \mathbb{A}^1_*) \cong \mathbb{Z}$ is a subgroup of $\pi_1(\mathbb{Y}_1^*)$ of index 2, where the embedding is induced by τ. Because $\pi_1(\mathbb{P}^1 \times \mathbb{P}^1 - \Delta) = 0$, it follows that any loop in $(\mathbb{P}^1 \times \mathbb{P}^1 - \Delta) \times \mathbb{A}^1_*$ starting and ending at x_1 can be continuously deformed into the loop $\lambda_n(t)$, $0 \le t \le 1$, $n \in \mathbb{Z}$, defined by the conditions $p_1(\lambda_n(t)) = p \times q$, $p_2(\lambda_n(t)) = ae^{i2\pi nt}$ (here p_i is the projection of $(\mathbb{P}^1 \times \mathbb{P}^1 - \Delta) \times \mathbb{A}^1_*$ onto the ith summand, $i = 1, 2$). The loops $\lambda_n(t)$ and $\lambda_m(t)$ are homotopic if and only if $n = m$. The class of the loop $\lambda_1(t)$ is a generator of the group $\pi_1((\mathbb{P}^1 \times \mathbb{P}^1 - \Delta) \times \mathbb{A}^1_*)$. Now consider a path $\gamma(t)$ in $(\mathbb{P}^1 \times \mathbb{P}^1 - \Delta) \times \mathbb{A}^1_*$ such that $\gamma(0) = x_1$, $\gamma(1) = x_2$ and $p_2(\gamma(t)) = ae^{i\pi t}$ for

all $t \in [0, 1]$. Then the class of the loop $\tau(\gamma(t))$ in $\pi_1(\mathbb{Y}_1^*)$ is not contained in the subgroup $\pi_1((\mathbb{P}^1 \times \mathbb{P}^1 - \Delta) \times \mathbb{A}_*^1) \subset \pi_1(\mathbb{Y}_1^*)$.

On the other hand, $\tau(\gamma^\zeta(t)) = \tau(\gamma(t))$ for all $t \in [0, 1]$, $\gamma^\zeta(0) = x_2$, $\gamma^\zeta(1) = x_1$, and $p_2(\gamma^\zeta(t)) = -ae^{i\pi t}$. Therefore the product of $\gamma(t)$ and $\gamma^\zeta(t)$ is a loop $\varphi(t)$ in $(\mathbb{P}^1 \times \mathbb{P}^1 - \Delta) \times \mathbb{A}_*^1$ such that $\varphi(0) = \varphi(1) = x_1$, $p_2(\varphi(t)) = ae^{i2\pi t}$. Hence $\varphi(t)$ can be deformed into $\lambda_1(t)$. Thus the square of the class of $\tau(\gamma(t))$ is a generator of the subgroup $\pi_1((\mathbb{P}^1 \times \mathbb{P}^1 - \Delta) \times \mathbb{A}_*^1)$. Therefore $\pi_1(\mathbb{Y}_1^*) = \mathbb{Z}$.

4) Note that $\pi_2(\mathrm{SL}_2) = 0$ (because SL_2 can be contracted to SU_2 and SU_2 is homeomorphic to the three-dimensional sphere). It is known that if $M \to N$ is a covering then $\pi_n(M) \cong \pi_n(N)$ for all $n \geq 2$. Combining these two observations we obtain $\pi_2(X)$ for the solids X nos. 2, 4, 5, 9, 10, 11. Because $\pi_2(\mathbb{A}^1) = \pi_2(\mathbb{A}_*^1) = 0$, we also have $\pi_2(X)$ for solids X nos. 12, 13, 14, 15. From the homotopy sequence of the fibration $\mathrm{SL}_2 \xrightarrow{T} \mathrm{SL}_2/T$ we have $\pi_2(\mathbb{P}^1 \times \mathbb{P}^1 - \Delta) = \mathbb{Z}$. Because $\mathbb{P}^1 \times \mathbb{P}^1 - \Delta$ is a two-fold covering of $\mathbb{P}^2 - C$, we have $\pi_2(\mathbb{P}^2 - C) = \mathbb{Z}$. This yields $\pi_2(X)$ for the solids X nos. 3, 6, 7. Because $\mathrm{SL}_2(n)$ contracts to SL_2, and H_n contracts to T, it follows from Theorem 3.2 and the homotopy sequence of the fibration $\mathrm{SL}_2(n) \xrightarrow{H_n} \mathrm{SL}_2(n)/H_n$ that $\pi_2(\mathbb{X}_n) = \mathbb{Z}$. Finally, because $\mathbb{X}_0^*$ is a two-fold covering of $\mathbb{Y}_1^*$, we have $\pi_2(\mathbb{Y}^*) = \mathbb{Z}$.

Thus we have determined the invariants $\operatorname{Pic} X$, $\pi_i(X)$, $i = 1, 2$, and $\operatorname{rk}(X)$ for the solids X mentioned in the theorem. Note that $\mathbb{X}_1^* \cong \mathrm{SL}_2 \not\cong \mathbb{A}^3$ (because $\mathbb{A}^3$ is contractible and SL_2 is homotopically equivalent to the three-dimensional sphere). Now, as is directly seen from the table, to complete the proof of the theorem we have to show that $\mathbb{X}_n \not\cong \mathbb{X}_m$ when $|n| \neq |m|$.

It follows from the proof of Theorem 3.2 that $\mathbb{X}_d \cong \mathrm{SL}_2(d)/H_d$ for any $d \in \mathbb{N}$. Because $\mathrm{SL}_2(d)$ is connected and simply connected, it follows from [33] that any one-dimensional vector bundle on X_d can be endowed with a unique $\mathrm{SL}_2(d)$-homogeneous structure and, moreover, assigning to this bundle the character of the group H_d by which H_d acts on the fiber over the point of X_d corresponding to the coset eH_d we obtain an isomorphism $\operatorname{Pic} X_d \to \operatorname{Hom}(H_d, k^*) \cong \mathbb{Z}$. It is easy to see that the weights of T for the isotropy representation of the homogeneous space $\mathrm{SL}_2(d)/H_d$ are as follows: χ^3, χ^{-2}, and χ^d. Because H_d is a semidirect product of T and an n-dimensional unipotent group, we may identify $\operatorname{Hom}(H_d, k^*)$ with $\operatorname{Hom}(T, k^*)$. It now follows from the previous remarks that the character $\chi^{-d} \in \operatorname{Hom}(H_d, k^*)$ corresponds (under the aforementioned isomorphism) to the one-dimensional bundle on $\mathrm{SL}_2(d)/H_d$, which is the third exterior power of the cotangent bundle. Therefore the absolute value of the canonical class of the solid X_d (as an element of the group $\mathbb{Z} = \operatorname{Pic} X_d$) is equal to d. This (along with Proposition 3.4) finishes the proof of the theorem. □

Appendix

By the fall of 1990, after the papers included in this dissertation had been published, various authors obtained new and important results related to its subject matter. To acquaint the reader with these developments I have included in this appendix the formulations of the new results along with brief comments. The details can be found in the original papers.

At the end of the Bibliography there is a complementary list of references. Its entries are indicated by adding the letter A, e.g. [A1].

Each part of each section of this appendix has its own numbering for the results and formulas contained in it.

The ground field in each section of the appendix is subject to the same assumptions as in the corresponding chapter of the main text.

§1. Appendix to Chapter 1

Condition b) of part 1, §2, Chapter 1 is equivalent to the fact that the algebra $k[G/H]$ is affine (see [77]).

The observable subgroups H of G for which the algebra $k[G/H]$ is affine are called Grosshans subgroups [A41–A43, A45, A3, A34] (Grosshans clarified their role in [77]).

The group G can always be assumed to be a closed subgroup of $\mathrm{GL}_n(k)$ for some n. Under this assumption, a subgroup H is Grosshans in G if and only if it is Grosshans in $\mathrm{GL}_n(k)$ (see [77]).

Because of that, the original Hilbert's 14th problem is now considered as the classification problem for Grosshans subgroups of reductive subgroups (we should remark that Sukhanov [A10] obtained a classification of observable subgroups). Taking into account that $k[G/H]$ is the algebra of invariants of the natural action of the reductive group $H/\operatorname{Rad}_u H$ on $k[G/\operatorname{Rad}_u H]$ and that a unipotent subgroup is always observable, we have that if $\operatorname{Rad}_u H$ is Grosshans in G then H is Grosshans in G as well. Therefore the case of a unipotent H is of primary importance for the original Hilbert's 14th problem.

Any one-dimensional unipotent subgroup is Grosshans (this is a classical result of Weitzenböck [A49] (see also [77, A3, A9]). On the other hand,

Nagata's counterexample to Hilbert's 14th problem [118] shows that the commutative unipotent subgroup H of $G = \mathrm{GL}_{32}(\mathbb{C})$ consisting of all the matrices of the form

$$\begin{pmatrix} 1 & \lambda_1 & & & & 0 \\ 0 & 1 & & & & \\ & & \ddots & & & \\ & & & & 1 & \lambda_{16} \\ & 0 & & & 0 & 1 \end{pmatrix},$$

where $(\lambda_1, \dots, \lambda_{16})$ runs through the solution set of the system $\sum_{j=1}^{16} \alpha_{ij} x_j = 0$; $i = 1, 2, 3$, with the coefficients α_{ij} being algebraically independent over $\mathbb{Q}$, is not a Grosshans subgroup.

In this counterexample the subgroup H is in "general position" with respect to the root onc-parameter unipotent subgroups of G. It is natural to consider the other extreme: the unipotent subgroups H that are in the most "special position" or, more precisely, are generated by some of those one-parameter subgroups. These are exactly the unipotent subgroups H of G for which $N_G(H)$ contains a maximal torus of G (they are also called regular subgroups). As is known, [9, 10, 15], they can be described by the closed (under addition) subsystems S of the system R_+ of positive roots of G. Some of them were shown to be indeed Grosshans: according to [84] and [A20], this is true for unipotent radicals of parabolic subgroups of G. There is also the following general conjecture (sometimes called the Popov-Pommerening conjecture [A41–A43, A21–A23, A31, A34, A35]):

CONJECTURE. *Any regular unipotent subgroup H of a reductive group G is Grosshans in G.*

At the present moment it is supported by the following evidence.

THEOREM 1 [A41]. *Let $G = \mathrm{GL}_n(k)$, B and P, respectively, be a Borel and a parabolic subgroup of G, $P \supset B$, and H be a regular subgroup such that* $\mathrm{Rad}_u P \supseteq H \supseteq (\mathrm{Rad}_u P) \cap (\mathrm{Rad}_u B)'$. *Then H is Grosshans in G.*

THEOREM 2 [A21]. *Let S be a closed (under addition) subsystem of the system R_+ of positive roots of a reductive group G and H be the unipotent subgroup of G generated by the root one-dimensional subgroups corresponding to the elements of S. If the set $R_+ \backslash S$ is linearly independent over $\mathbb{Q}$ then H is Grosshans in G.*

This theorem implies that whenever $P = B$ the assertion of Theorem 1 is true for any reductive group G. This result was somewhat extended in [A22, A24]. In [A22, A23] the subgroups mentioned in Theorem 2 were characterized as the regular unipotent subgroups H of G such that $\dim(V^H)_\lambda \leq 1$ for any finite-dimensional G-module V and any character $\lambda \in \mathrm{Hom}(T, k^*)$, where T is a maximal torus of $N_G(H)$ that acts naturally on V^H.

THEOREM 3 [A32–A35]. *If each connected simple normal subgroup of a reductive group* G *is either of rank* ≤ 2 *or a group of type* A_n *with* $n \leq 4$ *then the above conjecture holds for the group* G.

§2. Appendix to Chapter 2

We keep the notation of §1, Chapter 2.

1. In [A26] Kempf found the number M in the case when G is a torus:

THEOREM 1. *Let* $V = k^n$ *and* $G = T \subset \mathrm{GL}_n(k)$ *be a torus. Let* $s = \dim T$ *and the number* t *be defined exactly as in part* 5 *of* §2, *Chapter* 2. *Then the degrees of the elements of a minimal homogeneous system of generators of the algebra* $k[V]^G$ *do not exceed the number*

$$n \cdot \begin{pmatrix} \mathrm{LCM}\; a, \\ a \in \mathbb{Z} \\ 0 < a \leq n \cdot s! \cdot t^s \end{pmatrix}.$$

The estimates for M for finite groups [121], tori (Theorem 1), and connected semisimple groups (the theorem from part 5 of §2, Chapter 2), combined together, allow us to obtain an estimate for M for any reductive group G (one should use the facts that G/G^0 is finite, Z_{G^0} is a torus, and G^0/Z_{G^0} is connected and semisimple; for details see [A26]).

The problem of determining M for a semisimple but not necessarily connected group G was considered by Veles′ko in [A1]. Using the just-mentioned reduction to the cases when G is connected or finite, along with Noether's bound [121] and the bound from Chapter 2, he showed that the number $[G\colon G^0] \cdot M^0$, where M^0 is the estimate for the G^0-module V obtained in Chapter 2, can be taken as a number of M for G.

2. As was mentioned in part 4 of §1, Chapter 2, if there exists a homogeneous system of parameters $\theta_1, \dots, \theta_m$ of $k[V]^G$ with $\deg \theta_i = d_i$ then the Poincaré series of $k[V]^G$ can be written as

$$F(k[V]^G, x) = \frac{x^{e_1} + \dots + x^{e_l}}{(1 - x^{d_1}) \cdots (1 - x^{d_m})}, \qquad e_1 \leq \dots \leq e_l. \tag{1}$$

The assumption that the converse is also true is known as the Mallows–Sloane conjecture [A39]. It is not true in general: Stanley [148] gave a counterexample in which G was a finite group. Dixmier [A16] gives a counterexample where G is connected, namely, $G = \mathrm{SL}_2$ and $V = S_5 \oplus S_1$ (in the notation of §3, Chapter 2). More precisely, written in the irreducible form, the rational function $F(k[V]^G, x)$ is

$$F(k[V]^G, x) = \frac{1 + x^6 + 6x^8 + 3x^{10} + 6x^{12} + x^{14} + x^{20}}{(1 - x^4)^2 (1 - x^6)^2 (1 - x^8)}. \tag{2}$$

It was shown in [A16] that the minimum of the numbers l in (1) (taken over all homogeneous systems of parameters) equals 38 (whereas the sum of the coefficients in the numerator of (2) equals $19 < 38$).

In the counterexamples of Stanley and Dixmier the representation of G on V is reducible. Dixmier [A17] put forth the following

CONJECTURE. *The Mallows–Sloane conjecture is true for irreducible representations of reductive groups.*

According to [A16], the Mallows–Sloane conjecture is true for SL_2-modules S_d for $d \leq 9$ and SL_3-module of ternary quartics.

3. It was conjectured in [42] that for any SL_2-module S_d the ideal I (see parts 2 and 3 of §3, Chapter 2) is not radical if $d \geq 5$. This was proved by Dixmier in [A15].

4. By Theorem 1 from part 1 of §1, Chapter 3, the number q equals $-\deg F(k[V]^G, x)$. In this form it can be defined for any reductive group G and not necessarily for a connected semisimple group (in the general case however the algebra $k[V]^G$ is not Gorenstein so that there is no functional equation (4) from part 4 of §1, Chapter 2). When we determined an estimate for M in Chapter 2 we used the nonnegativity of q. Knop [A28] obtained much more precise information about q.

THEOREM 1. *Let G be a reductive, not necessarily connected group and V be a finite-dimensional G-module. Then the number $q = -\deg F(k[V]^G, x)$ is not smaller than the transcendence degree m of the algebra $k[V]^G$. The equality $q = m$ occurs exactly when $k[V/V^{G^0}]^G = k$ and all linear transformations of V^{G^0} defined by the elements of G have determinant equal to 1. If the action of G on V is stable then q is not smaller than the transcendence degree of the algebra $k[V]^{(G^0)'}$.*

From this it now follows that the method used in part 4 of §1, Chapter 2, to reduce the computation of M to the determination of an upper bound D for the degrees $d_1, \dots, d_m$ of the elements $\theta_1, \dots, \theta_m$ of a homogeneous system of parameters of $k[V]^G$ can be carried over without change to the case of any reductive group G. More precisely, by definition, $e_l = \sum_{j=1}^m d_j - q$ and, since $q \geq m \geq 0$, we can, as in part 4 of §1, Chapter 2, take $M = mD$. We should remark however that the assumption that the group G be connected and semisimple is essential for the argument in Chapter 2 where a bound D was found.

§3. Appendix to Chapter 3

We keep the notation and conventions of Chapter 3.

1. With regard to the proof of the finiteness theorem (Theorem 6 of Chapter 3), geometric condition (14) is important only because it guarantees the equality $q = n$. This condition, in turn, is, in that context, just a means to an upper bound on the dimensions of representations with free algebras of invariants, depending only on the group (like $n \leq 2\dim G$). The existence of such a bound is in fact equivalent to Theorem 1. But it follows easily from Proposition 4 that it is not necessary to have the equality $n = q$ in order to find a bound; it suffices to have a suitable inequality between n and q. In this respect I conjectured in [43] that the inequality $q \leq n$ is always true.

Knop [A27] proved that conjecture:

THEOREM 1. *For any representation* $G \to \mathrm{GL}(V)$ *of the connected semisimple group* G *the number* q *does not exceed* $n = \dim V$.

As was mentioned in [A27], the ideas of Panyushev [A6, A7] played a key role in the proof.

In fact a more general statement was proved in [A27]. Namely, let an algebraic group G act regularly on an irreducible affine algebraic variety X. Suppose that the algebra $k[X]$ is positively graded: $k[X] = \bigoplus_{d \geq 0} k[X]_d$, $\dim k[X]_d < \infty$, $k[X]_0 = k$, and that the action of G on $k[X]$ respects the grading (i.e., all spaces $k[X]_d$ are invariant). Then $k[X]^G$ is a graded subalgebra of $k[X]$. Let A be an affine graded subalgebra of $k[X]^G$, $Y = \operatorname{Specm} A$, and $\pi\colon X \to Y$ be the morphism induced by the inclusion of A in $k[X]$.

Since the algebras $k[X]$ and $k[Y]$ are affine, we have that their Poincaré series $F(k[X], x)$ and $F(k[Y], x)$ are the power series expansions of certain rational functions of x. Let

$$q_X = -\deg(F(k[X], x) \qquad \text{and} \qquad q_Y = -\deg F(k[Y], x).$$

THEOREM 2 [A27]. *Suppose that*

a) *both* X *and* Y *are normal and Gorenstein,*

b) $\operatorname{codim}_X(X \setminus U) \geq 2$, *where*

$$U = \{x \in X \mid x \notin \operatorname{sing} X,\ \pi(x) \notin \operatorname{sing} Y,\ \pi \text{ is smooth at } x\},$$

c) $R := \{x \in X \mid \dim Gx = \dim X - \dim Y\} \neq \varnothing$.

Then $q_X \geq q_Y$.

THEOREM 3 [A27]. *The conditions of Theorem* 2 *are satisfied if the group* G *is connected and semisimple,* X *has rational singularities, the algebra* $k[X]$ *is factorial, and* $A = k[X]^G$ (*i.e.,* $Y = X/G$).

It is clear that Theorem 1 follows from Theorems 2 and 3.

According to Theorem 4 of Chapter 3, the equality $q = n$ is in a sense typical. Knop [A27] characterized the actions for which this equality holds:

THEOREM 4. *Suppose that the conditions of Theorem* 2 *are satisfied. Then the following are equivalent*:

(1) $q_X = q_Y$,

(2) $\operatorname{codim}_X\{x \in X \mid \dim G_x > 0\} \geq 2$.

If X *is a vector space and the action of* G *on* X *is linear, then Conditions* (1) *and* (2) *are equivalent to*

(3) *there exists a two-dimensional linear subspace* L *of* X *such that the group* G_x *is finite for any* $x \in L$, $x \neq 0$.

Using Theorem 4, Knop and Littelmann [A30] found all representations of connected simple groups and all irreducible representations of connected semisimple groups for which $q < n$. Their classification is as follows (we use the numbering of the fundamental weights from Bourbaki [9, 10]):

THEOREM 5. *Let* G *be a connected simply connected simple group. Then Table* 1 *below consists of all* (*up to an outer automorphism of* G) G-*modules* V *with*

$$V^G = 0 \quad \textit{and} \quad q < n = \dim V.$$

In order to give the corresponding result for irreducible representations of connected semisimple groups we recall the castling construction (see [133]).

Any irreducible representation ρ of dimension bigger than one of any connected semisimple group can be written (nonuniquely) in the form $(H \times \mathrm{SL}_d, W \otimes k^d)$, where (H, W) is an irreducible representation of a connected semisimple group H, and $1 \leq d \leq n-1$, $n = \dim W$. The representation $(H \times \mathrm{SL}_{n-d}, V^* \otimes k^{n-s})$ is then said to be obtained from ρ by the castling transform.

Chains of castling transforms (including those of length zero) connecting pairs of representations give rise to an equivalence relation on the set of irreducible representations of dimension bigger than one of all possible connected semisimple groups. In each equivalence class the representations of minimal dimension give rise to isomorphic linear groups. Those representations of minimal dimension are exactly the representations whose dimensions cannot be lowered by one castling transform. They are also called castling reduced. The algebras of invariants of castling equivalent representations are isomorphic (see [A36]).

THEOREM 6. *The number* $q/\dim V$ *is the same for all castling equivalent representations. The set of the irreducible castling reduced* G-*modules* V, *where* G *is a connected simply-connected semisimple group, for which this number is less than one is given* (*up to an outer automorphism of* G) *by Table* 2 (*in this table* F *denotes the type of any nontrivial semisimple group*, ω' *denotes any of its nontrivial dominant weights, and* $d' = \dim R(\omega')$; *the prime and double prime mark the fundamental weights of the second and, respectively, third simple factor*).

Table 1: Series

Type of G	V	Condition
A_n	$rR(\omega_1) + sR(\omega_n)$	$r, s \leq n$, or
		$r = n+1$, $s \leq n-1$
	$R(\omega_2) + rR(\omega_1) + sR(\omega_n)$	$r = 0$, $s \leq n$, or
		$r = 0$, $s = n+1$, n is odd, or
		$r = 1$, $s \leq n$, or
		$r \geq 2$, $r+s \leq n+1$
	$pR(\omega_2) + qR(\omega_{n-1})$ $+rR(\omega_1) + sR(\omega_n)$	$p+q=2$, $r+s \leq 2$, n is odd
	$2R(\omega_2)$	n is odd
	$R(\omega_2) + R(\omega_{n-1})$	n is even
	$2R(\omega_2) + R(\omega_1)$	n is even
	$2R(\omega_2) + R(\omega_n)$	n is even
	$R(\omega_2) + R(\omega_{n-1}) + R(\omega_1)$	n is even
	$2R(\omega_2) + 2R(\omega_1)$	n is even
	$2R(\omega_2) + R(\omega_1) + R(\omega_n)$	n is even
	$2R(\omega_2) + 2R(\omega_n)$	n is even
	$R(\omega_2) + R(\omega_{n-1}) + 2R(\omega_1)$	n is even
	$R(\omega_2) + R(\omega_{n-1})$ $+R(\omega_1) + R(\omega_n)$	n is even
	$R(2\omega_1) + rR(\omega_1) + sR(\omega_n)$	$r+s \leq n-1$
	$R(2\omega_1) + R(\omega_2)$	
	$R(2\omega_1) + R(\omega_{n-1})$	
	$R(\omega_1 + \omega_n)$	
B_n	$rR(\omega_1)$	$r \leq 2n-1$
	$R(\omega_2)$	$n \geq 3$
C_n	$rR(\omega_1)$	$r \leq 2n$
	$R(\omega_2)$	
	$R(\omega_1 + \omega_2)$	
	$R(2\omega_1 + \omega_2)$	
	$R(2\omega_1)$	
D_n	$rR(\omega_1)$	$r \leq 2n-2$
	$R(\omega_2)$	

Table 1: Sporadic cases

Type of G	V
A_3	$3R(\omega_2)$
	$3R(\omega_2)+R(\omega_1)$
	$4R(\omega_2)$
A_4	$2R(\omega_2)+R(\omega_3)$
A_5	$2R(\omega_2)+R(\omega_4)$
	$3R(\omega_2)$
	$R(\omega_3)$
	$R(\omega_3)+R(\omega_1)$
	$R(\omega_3)+2R(\omega_1)$
	$R(\omega_3)+R(\omega_1)+R(\omega_5)$
	$R(\omega_3)+3R(\omega_1)$
	$R(\omega_3)+2R(\omega_1)+R(\omega_5)$
	$R(\omega_3)+3R(\omega_1)+R(\omega_5)$
	$R(\omega_3)+2R(\omega_1)+2R(\omega_5)$
	$R(\omega_3)+R(\omega_2)$
	$R(\omega_3)+R(\omega_2)+R(\omega_1)$
	$R(\omega_3)+R(\omega_2)+R(\omega_5)$
	$2R(\omega_3)$
A_6	$R(\omega_3)$
	$R(\omega_3)+R(\omega_1)$
	$R(\omega_3)+R(\omega_6)$
	$R(\omega_3)+2R(\omega_1)$
	$R(\omega_3)+R(\omega_1)+R(\omega_6)$
	$R(\omega_3)+2R(\omega_6)$
	$R(\omega_3)+3R(\omega_1)$
	$R(\omega_3)+2R(\omega_1)+R(\omega_6)$
	$R(\omega_3)+R(\omega_1)+2R(\omega_6)$
	$R(\omega_3)+3R(\omega_6)$
A_7	$R(\omega_3)$
	$R(\omega_3)+R(\omega_1)$
	$R(\omega_3)+R(\omega_7)$

Type of G	V
B_3	$R(\omega_3)$
	$R(\omega_3)+R(\omega_1)$
	$R(\omega_3)+2R(\omega_1)$
	$R(\omega_3)+3R(\omega_1)$
	$2R(\omega_3)$
	$2R(\omega_3)+R(\omega_1)$
	$2R(\omega_3)+2R(\omega_1)$
	$3R(\omega_3)$
	$3R(\omega_3)+R(\omega_1)$
	$4R(\omega_3)$
B_4	$R(\omega_4)$
	$R(\omega_4)+R(\omega_1)$
	$R(\omega_4)+2R(\omega_1)$
	$R(\omega_4)+3R(\omega_1)$
	$R(\omega_4)+4R(\omega_1)$
	$2R(\omega_4)$
	$2R(\omega_4)+R(\omega_1)$
	$2R(\omega_4)+2R(\omega_1)$
	$3R(\omega_4)$
B_5	$R(\omega_5)$
	$R(\omega_5)+R(\omega_1)$
	$R(\omega_5)+2R(\omega_1)$
	$R(\omega_5)+3R(\omega_1)$
	$R(\omega_5)+4R(\omega_1)$
	$2R(\omega_5)$
B_6	$R(\omega_6)$
	$R(\omega_6)+R(\omega_1)$
	$R(\omega_6)+2R(\omega_1)$
C_2	$R(\omega_1)+2R(\omega_2)$
C_3	$R(\omega_3)$
	$R(\omega_3)+R(\omega_1)$
	$R(\omega_3)+2R(\omega_1)$
	$2R(\omega_2)$

TABLE 1: SPORADIC CASES (continued)

Type of G	V
D_4	$R(\omega_1) + R(\omega_3)$
	$R(\omega_1) + R(\omega_3) + R(\omega_4)$
	$2R(\omega_1) + R(\omega_3)$
	$2R(\omega_1) + R(\omega_3) + R(\omega_4)$
	$2R(\omega_1) + 2R(\omega_3)$
	$3R(\omega_1) + R(\omega_3)$
	$2R(\omega_1) + 2R(\omega_3) + R(\omega_4)$
	$3R(\omega_1) + R(\omega_3) + R(\omega_4)$
	$3R(\omega_1) + 2R(\omega_3)$
	$4R(\omega_1) + R(\omega_3)$
D_5	$R(\omega_5)$
	$R(\omega_5) + R(\omega_1)$
	$R(\omega_5) + 2R(\omega_1)$
	$R(\omega_5) + 3R(\omega_1)$
	$R(\omega_5) + 4R(\omega_1)$
	$R(\omega_5) + 5R(\omega_1)$
	$2R(\omega_5)$
	$2R(\omega_5) + R(\omega_1)$
	$2R(\omega_5) + 2R(\omega_1)$
	$2R(\omega_5) + 3R(\omega_1)$
	$R(\omega_4) + R(\omega_5)$
	$R(\omega_4) + R(\omega_5) + R(\omega_1)$
	$R(\omega_4) + R(\omega_5) + 2R(\omega_1)$
	$R(\omega_4) + R(\omega_5) + 3R(\omega_1)$
	$3R(\omega_5)$
	$3R(\omega_5) + R(\omega_1)$
	$R(\omega_4) + 2R(\omega_5)$
	$R(\omega_4) + 2R(\omega_5) + R(\omega_1)$

Type of G	V
D_6	$R(\omega_6)$
	$R(\omega_6) + R(\omega_1)$
	$R(\omega_6) + 2R(\omega_1)$
	$R(\omega_6) + 3R(\omega_1)$
	$R(\omega_6) + 4R(\omega_1)$
	$R(\omega_6) + 5R(\omega_1)$
	$2R(\omega_6)$
	$2R(\omega_6) + R(\omega_1)$
	$R(\omega_5) + R(\omega_6)$
	$R(\omega_5) + R(\omega_6) + F(\omega_1)$
D_7	$R(\omega_7)$
	$R(\omega_7) + R(\omega_1)$
	$R(\omega_7) + 2R(\omega_1)$
	$R(\omega_7) + 3R(\omega_1)$
E_6	$R(\omega_1)$
	$2R(\omega_1)$
	$R(\omega_1) + R(\omega_6)$
	$3R(\omega_1)$
	$2R(\omega_1) + R(\omega_6)$
	$R(\omega_2)$
E_7	$R(\omega_1)$
	$R(\omega_7)$
	$2R(\omega_7)$
E_8	$R(\omega_8)$
F_4	$R(\omega_1)$
	$R(\omega_4)$
	$2R(\omega_4)$
G_2	$R(\omega_1)$
	$2R(\omega_1)$
	$3R(\omega_1)$
	$R(\omega_2)$

TABLE 2: SERIES

Type of G	V	Condition
$A_n + F$	$R(\omega_1) \otimes R(\omega')$	$d' \leq n+1$
$B_n + F$	$R(\omega_1) \otimes R(\omega')$	$d' \leq 2n-1$; $(F, \omega') \not\simeq \mathrm{SL}_{d'}, \mathrm{Sp}_{d'}$
$C_n + F$	$R(\omega_1) \otimes R(\omega')$	$d' \leq 2n$; $(F, \omega') \not\simeq \mathrm{SL}_{d'}, \mathrm{Sp}_{d'}, \mathrm{SO}_{d'}$
$D_n + F$	$R(\omega_1) \otimes R(\omega')$	$d' \leq 2n-2$; $(F, \omega') \not\simeq \mathrm{SL}_{d'}, \mathrm{Sp}_{d'}$
$A_m + B_n$	$R(\omega_1) \otimes R(\omega_1')$	$m \leq n-1$
$A_m + C_n$	$R(\omega_1) \otimes R(\omega_1')$	$m \leq n-1$
$A_m + D_n$	$R(\omega_1) \otimes R(\omega_1')$	$m \leq n-1$
$B_m + C_n$	$R(\omega_1) \otimes R(\omega_1')$	
$C_m + C_n$	$R(\omega_1) \otimes R(\omega_1')$	
$C_m + D_n$	$R(\omega_1) \otimes R(\omega_1')$	
$C_m + A_1$	$R(\omega_1) \otimes 2R(\omega_1')$	
$C_m + A_1 + A_1$	$R(\omega_1) \otimes R(\omega_1') \otimes R(\omega_1'')$	
$C_m + A_3$	$R(\omega_1) \otimes R(\omega_2')$	
$A_m + A_m + A_1$	$R(\omega_1) \otimes R(\omega_1') \otimes R(\omega_1'')$	
$A_m + A_1$	$R(\omega_2) \otimes R(\omega_1')$	$m \geq 3$

TABLE 2: SPORADIC CASES

Type of G	V
$A_3 + A_1$	$R(\omega_2) \otimes R(2\omega_1')$
	$R(\omega_2) \otimes R(3\omega_1')$
$A_3 + A_2$	$R(\omega_2) \otimes R(\omega_1')$
$A_3 + A_1 + A_1$	$R(\omega_2) \otimes R(\omega_1') \otimes R(\omega_1'')$
$A_4 + A_2$	$R(\omega_2) \otimes R(\omega_1')$
$A_5 + A_2$	$R(\omega_2) \otimes R(\omega_1')$
$A_5 + A_1$	$R(\omega_2) \otimes R(2\omega_1')$
	$R(\omega_3) \otimes R(\omega_1')$
$B_3 + A_1$	$R(\omega_3) \otimes R(\omega_1')$
	$R(\omega_3) \otimes R(2\omega_1')$
	$R(\omega_3) \otimes R(3\omega_1')$
$B_3 + A_2$	$R(\omega_3) \otimes R(\omega_1')$
$B_3 + C_2$	$R(\omega_3) \otimes R(\omega_1')$
$B_3 + A_1 + A_1$	$R(\omega_3) \otimes R(\omega_1') \otimes R(\omega_1'')$
$B_3 + A_3$	$R(\omega_3) \otimes R(\omega_1')$
$B_4 + A_1$	$R(\omega_4) \otimes R(\omega_1')$
	$R(\omega_4) \otimes R(2\omega_1')$
$B_4 + A_2$	$R(\omega_4) \otimes R(\omega_1')$
$B_5 + A_1$	$R(\omega_5) \otimes R(\omega_1')$
$C_3 + A_1$	$R(\omega_2) \otimes R(\omega_1')$
$D_5 + A_1$	$R(\omega_5) \otimes R(\omega_1')$
	$R(\omega_5) \otimes R(2\omega_1')$
$D_5 + A_2$	$R(\omega_5) \otimes R(\omega_1')$
$D_6 + A_1$	$R(\omega_6) \otimes R(\omega_1')$
$E_6 + A_1$	$R(\omega_1) \otimes R(\omega_1')$
	$R(\omega_1) \otimes R(2\omega_1')$
$E_6 + A_2$	$R(\omega_1) \otimes R(\omega_1')$
$E_7 + A_1$	$R(\omega_7) \otimes R(\omega_1')$
$F_4 + A_1$	$R(\omega_4) \otimes R(\omega_1')$
$G_2 + A_1$	$R(\omega_1) \otimes R(\omega_1')$
	$R(\omega_1) \otimes R(2\omega_1')$
$G_2 + A_2$	$R(\omega_1) \otimes R(\omega_1')$

2. The numbers q and m for all of the representations mentioned in Theorems 5 and 6 above were found in [A30]. Thus at the present moment the value of q is known for all G-modules V where either V is simple or G is simple (and connected). Theorem 4 of part 1 together with Theorem 2 from part 3 of §1, Chapter 3, and its corollary allow us to compute the value of $2\tau/\gamma$, where γ and τ are the coefficients of the Laurent expansion 2 from part 3 of §1, Chapter 3, for the aforementioned modules. We should remark that, because of Theorem 2 from part 3 of §1, Chapter 3, one can interpret Theorem 1 from part 4 of the preceding chapter of this appendix as a statement about the nonnegativity of τ for connected semisimple groups.

3. One of the important facts used in Chapters 2 and 3 is that algebra $k[V]^G$ is Gorenstein (which implied the existence of functional equation (1) from part 2 of §1, Chapter 3 for $F(k[V]^G, x)$).

For an arbitrary reductive group G the algebra $k[V]^G$ is not necessarily Gorenstein. It was conjectured in [150] that if G is unimodular, i.e., $G \subset \mathrm{SL}(V)$, then $k[V]^G$ is Gorenstein. It is indeed the case when G is finite. In general however it turned out not to be true as is seen from the following result of Knop [A23].

THEOREM 1. *Let $G \subset \mathrm{GL}(V)$ be a reductive group. Then there exists a reductive group $H \subset \mathrm{SL}(W)$ such that the algebras $k[V]^G$ and $k[W]^H$ are isomorphic.*

Knop [A28] also obtained the following generalizations of the results mentioned in part 1 of this chapter and part 4 of the previous chapter of this appendix.

Let G be a reductive, not necessarily connected group and V be a finite-dimensional G-module. Let $\lambda_V \in \mathrm{Hom}(G, k^*)$ be the character given by the formula $\lambda_V(g) = \det g$ where $g \in G$ in the right-hand side is considered as a linear transformation of V. Let q denote the number $-\deg F(k[V]^G, x)$.

THEOREM 2. *Suppose that*

(1) *the action of G on V is stable,*

(2) $\mathrm{codim}_V\{v \in V \mid \dim G_v > 0\} \geq 2$,

(3) $\lambda_V = \lambda_{\mathrm{ad}}$.

Then the algebra $k[V]^G$ is Gorenstein.

THEOREM 3. (1) *Suppose that the action of G on V is stable and the generic stabilizer is finite. Let e be the minimal degree of a semi-invariant polynomial on V of weight $\lambda_V \cdot \lambda_{\mathrm{ad}}^{-1}$. Then $q \leq \dim V + e$.*

(2) *If $\mathrm{Hom}(G, k^*) = \{1\}$, then $q \leq \dim V$.*

If $\mathrm{codim}_V\{v \in V \mid \dim G_v > 0\} \geq 2$, then the equality holds in both cases.

The same argument as in part 6 of §3, Chapter 3, allows us to use the inequality from the first part of this theorem for the classification problem of G-modules V with free algebra of invariants. More precisely, we can restrict ourselves to the case $(V^*)^G = 0$ and then it follows from Proposition 4 in part 6 of §3, Chapter 3 (the proof of which is valid, without changes, for any G), and the inequality above that if $k[V]^G$ is free, the action of G on V is stable, and the generic stabilizer is finite then $\dim V \leq 2\dim G + e$ and if, in addition, the action is irreducible then $\dim V \leq (3\dim G + 1 + e)/2$. In particular, if G has no nontrivial characters and the generic stabilizer is finite then the corollary of Proposition 4 from part 4 of §6, Chapter 3, is still true (in this case the action is automatically stable [29]).

§4. Appendix to Chapter 4

1. The method from Chapter 4 carries over, without major changes, to nonconnected semisimple groups G. Indeed, Veles′ko [A1] showed that the assumption about the connectivity of G in the main theorem of Chapter 4 can be removed. He argues along the lines of Chapter 4, obtaining however intermediate estimates in somewhat more general situations (e.g., instead of the case $G = K \times \mathrm{SL}_2$, where K is a finite abelian group, considered in part 5 of §2, Chapter 4, he obtains estimates for $\mathrm{hd}\, k[V]^G$ and $\mathrm{ed}\, k[V]^G$ in the case $G = H \times \mathrm{SL}_2$, where H is any finite group; similarly the case when G is a one-dimensional diagonalizable group considered in part 2 of §2, Chapter 4, is replaced with the case when G^0 is a one-dimensional torus).

2. Gordeev [A4] suggested a new approach to finding upper bounds for $\mathrm{hd}\, k[V]^G$ for semisimple (not necessarily connected) groups $G \subset \mathrm{GL}(V)$ based on the investigation of the so-called class of the linear group G. This is the number

$$\mathrm{cl}\, G = \mathrm{cl}_V\, G = \min_{g\in G,\, g\neq 1} \mathrm{codim}\{v \in V \mid gv = v\}.$$

The rough idea is that the smaller $\mathrm{cl}\, G$ is then the more "special" is the group and the "better" is its algebra of invariants (i.e., the smaller $\mathrm{hd}\, k[V]^G$ is). This point of view is supported by the following examples:

(1) If H is a transitive primitive permutation group of a finite set S and the number

$$\min_{h\in H,\, h\neq 1} |\{s \in S \mid hs \neq s\}|$$

is sufficiently small with respect to $|S|$ then, by the Jordan theorem, H is either symmetric or alternating group of the S.

(2) If $H \subset \mathrm{GL}(V)$ is a finite linear group and the algebra $k[V]^H$ is free, then, by the Chevalley-Shephard-Todd theorem, H is generated by reflections and therefore $\mathrm{cl}\, H = 1$.

(3) If $H \subset \mathrm{GL}(V)$ is a finite linear group and algebra $k[V]^H$ is a complete intersection then, by [95, 19], H is generated by elements having exactly two eigenvalues different from 1, and therefore $\mathrm{cl}\, H = 2$.

(4) If $\mathfrak{G} \in \mathrm{End}(V)$ is a simple Lie algebra acting irreducibly on V and containing an operator of a sufficiently small with respect to $\dim V$ rank then $\mathfrak{G}$ is either $\mathrm{sl}(V)$, or $\mathrm{so}(V)$, or $\mathrm{sp}(V)$.

For any semisimple group G of positive rank let $d(G)$ denote the LCM of the numbers $d(H)$, where H runs over simple normal subgroups of G and

$$d(H) = \begin{cases} r+1, & \text{if } G \text{ is of type } A_r, \\ 2(r-1), & \text{if } G \text{ is of type } D_r, \\ 3, & \text{if } G \text{ is of type } E_6, \\ 2 & \text{in other cases.} \end{cases}$$

The next result was proved in [A4].

THEOREM 1. *Let* $G \subset \mathrm{GL}(V)$ *be a semisimple (not necessarily connected) group of positive rank. Then*

$$\operatorname{hd} k[V]^G \geq ((\operatorname{cl} G)/d(G)) - 2\dim G - 1.$$

The key idea in the proof of this inequality is as follows. Let $H \subset \mathrm{GL}(W)$ be a reductive group with $W^H = 0$, $f_1, \dots, f_m$ a minimal homogeneous system of generators of $k[W]^H$, $r = \dim W/H$, $x_1, \dots, x_n$ coordinates on W, $\mathfrak{V}(J)$ the variety of zeros of the ideal J of $k[W]$ generated by the minors of order r of $(\partial f_i/\partial x_j)$. Since the canonical morphism $\pi_{W,H}$ is smooth at the points of the principal (i.e., open in W) Luna stratum [108], we have that $\mathfrak{V}(J)$ is contained in the complement of that stratum. Suppose now that there are only two Luna strata in W. Then automatically those strata are $\mathfrak{N} = \pi_{W,H}^{-1}(\pi_{W,H}(0))$ and $W\backslash\mathfrak{N}$. Thus $\mathfrak{V}(J) \subset \mathfrak{N}$. Therefore $\operatorname{codim}_W \mathfrak{V}(J) = \operatorname{ht}(J) \geq \operatorname{codim}_W \mathfrak{N}$ where $\operatorname{ht} J$ denotes the height of J. But, according to the general result of Bruns [A14] about heights of determinantal ideals, we must have $\operatorname{ht} J \leq m + \dim W - 2r + 1$. Since $m - r = \operatorname{hd} k[W]^H$, and $\dim W - r$ is the dimension of the generic fiber of $\pi_{W,H}$, we now have the key inequality

$$\operatorname{hd} k[W]^H \geq \operatorname{codim}_W \mathfrak{N} - \begin{pmatrix} \text{the dimension of the generic} \\ \text{fiber of } \pi_{W,H} \end{pmatrix} - 1. \qquad (1)$$

In the general case, when the number of Luna strata is more than two, the use of inequality (1) can be justified by the monotonicity theorem for homological dimension (see Theorem 1.2 of §1, Chapter 4). More precisely, let $G \subset \mathrm{GL}(V)$ be a semisimple group. Suppose that the generic stabilizer for the action of G on V is trivial. If $v \in V$ is a point from a subprincipal Luna stratum with closed orbit, $H = G_v$ and W is a submodule complementary to N_v^H in the slice-module (N, H_v) (see part 5 of §0, Chapter 4) then the H-module W has exactly two Luna strata, and therefore inequality (1) holds. Since $\operatorname{hd} k[V]^G \geq \operatorname{hd} k[W]^H$, we now have a lower bound for $\operatorname{hd} k[V]^G$.

This result is immediately applicable when G is finite, since in this case $\mathfrak{N} = 0$, the fibers of $\pi_{W,H}$ are finite, $N_v = V$, and $\operatorname{cl}_V G \leq \operatorname{cl}_V H \leq \dim W$. We thus obtain

THEOREM 2 [A4]. *If* $G \subset \mathrm{GL}(V)$ *is a finite group then* $\operatorname{hd} k[V]^G \geq \operatorname{cl} G - 1$.

If $\operatorname{rk} G$ is positive, then, in order to apply (1), in [A4], in terms of $\operatorname{cl} G$, the following two results are obtained: first, a sufficient condition for the triviality of the generic stabilizer for the action of G on V (it suffices to have $\operatorname{cl} G > \dim G$) and, second, an estimate for $\operatorname{codim} \mathfrak{N}$ (this is done by the same argument as in the proof of Theorem 5 from part 6 of Chapter 5, i.e., with the aid of the Hilbert–Mumford Theorem). This proves the theorem.

Gordeev [A4] also found lower bounds for $\operatorname{cl} G$ in terms of $\dim V$. This, combined with Theorem 1, yields the following result.

THEOREM 3 [A4]. *Let $G \subset \mathrm{GL}(V)$ be a semisimple, not necessarily connected group of positive rank r. Then*

$$(1) \qquad \operatorname{hd} k[V]^G \geq \frac{\sqrt{\dim V}}{12 \cdot [G\colon G^0] \cdot |Z_{G^0}| \cdot r \cdot d(G)} - 2\dim G - 1, \quad \textit{if } V^G = 0,$$

$$(2) \qquad \operatorname{hd} k[V]^G \geq \frac{\dim V}{12 \cdot [G\colon G^0] \cdot r \cdot d(G)} - 2\dim G - 1,$$

$$\operatorname{hd} k[V]^G \geq \frac{\sqrt{\dim V}}{34 \cdot [G\colon G^0] \cdot d(G)} - 2\dim G - 1,$$

if the action of G on V is irreducible.

3. In [A18, A19] one can find new information about homological dimension and embedding dimension for algebras of invariants of binary forms (see part 2 of §10 and parts 5–9 of §2, Chapter 4).

More precisely, one finds in [A19] the answer to the question about the minimal value of $\operatorname{hd} k[S_n]^{\mathrm{SL}_2}$ for $n \geq 7$, $n \neq 8$, which I raised in [A44]. It turns out that this minimum equals 25 and is attained at $n = 7$.

In [A18] one finds the best up-to-date asymptotic characteristics of the number $\omega_n = \operatorname{ed} k[S_n]^{\mathrm{SL}_2}$ as $n \to \infty$. More precisely, let $p(n)$ be the number of partitions (with repetitions) of n (it is known that $p(n) \underset{n\to\infty}{\sim} e^{\pi\sqrt{2n/3}/4n\sqrt{3}}$). The following theorem is proved in [A18].

THEOREM 1. (1) *For odd $n \to \infty$ we have that*

$$\liminf \omega_n \cdot \left[\frac{\sqrt{n}}{\log n \log\log n} p(n)\right]^{-1} > 0.$$

(2) *For any $\varepsilon > 0$ and $n \equiv 2 \pmod 4$, $n \to \infty$,*

$$\liminf \omega_n \cdot \left[e^{((\pi/\sqrt{3})-\varepsilon)(\log\log n/\log n)\sqrt{n}} \cdot p\left(\frac{n}{2}\right)\right]^{-1} > 0$$

(3) *For $n \equiv 0 \pmod 4$ and $n \to \infty$*

$$\liminf \omega_n \cdot \left[\frac{\sqrt{n}}{\log n \log\log n} p\left(\frac{n}{4}\right)\right]^{-1} > 0.$$

The proof is based on the monotonicity theorem for embedding dimension (see Theorem 1.1 from part 2 of §1, Chapter 4).

§5. Appendix to Chapter 5

1. Since the paper [37] (which constitutes a major part of Chapter 5) was published the classifications of representations with free modules of covariants and equidimensional representations have been extended to reducible representations of connected simple groups (Schwarz [136] and Adamovich [1]), irreducible representations of connected semisimple groups (Littelmann [A36]) and almost all of the representations of connected semisimple groups that are a product of two normal simple subgroups that are not of type A at the same time (Wehlau [A47, A48]). We should remark that the key ideas remained the same as in Chapter 5, namely, Theorems 1, 3, 4, 5 of that chapter and the corresponding monotonicity theorems (according to which the freeness of the module of covariants and equidimensionality are preserved under passage to subrepresentations and slice-representations); of course the technical problems became more complicated and required some extra work.

It follows a posteriori from those classifications that for the classes of representations considered in [37, 136, 1] and [A36, A47, A48] the following general conjecture, which was first formulated in [36] and then gained some popularity (sometimes under the name of the "Russian Conjecture") (see [1, 91, 103, A36, A47, A48]), is true:

CONJECTURE. *The module of covariants of any equidimensional representation of a connected semisimple group is free.*

Thus it is now known that for all representations of connected simple groups and irreducible representations of connected semisimple groups the condition of equidimensionality is equivalent to the freeness of the module of covariants.

It is quite possible that this conjecture foreshadows some general result from commutative algebra (see the corresponding discussion in [A2]).

The results of the aforementioned classification are as follows (the numbering of fundamental weights is taken from Bourbaki [9, 10]).

THEOREM 1. *The set of equidimensional G-modules of any connected simply-connected simple group G coincides with the set of G-modules with free modules of covariants and is exhausted (up to an outer automorphism of G) by all possible submodules of the G-modules from Table 3 below.*

THEOREM 2. *The set of irreducible equidimensional G-modules of any connected simply-connected semisimple group G coincides with the set of irreducible G-modules with free modules of covariants. For nonsimple groups G all irreducible castling reduced equidimensional G-modules are exactly (up to an outer automorphism of G) the modules from Table 4 below (in this table the regular, double, and triple primes indicate the fundamental weights of the first, respectively, second or third simple summands).*

(For castling see part 1, Chapter 3, of this appendix.)

TABLE 3: SERIES

Type of G	V	Condition
A_{n-1}	$nR(\omega_1)$	
	$pR(\omega_1) + qR(\omega_{n-1})$	$q \le p < n,\ p + q = n + 1$
	$R(\omega_2) + 2R(\omega_1) + mR(\omega_{n-1})$	$m = [(n-1)/2]$
	$R(\omega_2) + R(\omega_1) + mR(\omega_{n-1})$	$n = 2m$
	$R(\omega_2) + mR(\omega_{n-1})$	$n = 2m - 1$
	$2R(\omega_2) + 2R(\omega_1)$	
	$2R(\omega_2) + R(\omega_1) + R(\omega_{n-1})$	n is odd
	$2R(\omega_2) + 2R(\omega_{n-1})$	n is even
	$2R(\omega_2) + R(\omega_{n-1})$	$n \ge 5$ is odd
	$R(\omega_2) + R(\omega_{n-2}) + R(\omega_1)$	n is even
	$R(\omega_2) + R(\omega_{n-2}) + R(\omega_1) + R(\omega_{n-1})$	
	$R(2\omega_1) + R(\omega_1) + mR(\omega_{n-1})$	$m = [(n-1)/2]$
	$R(2\omega_1) + mR(\omega_{n-1})$	$n = 2m$
	$R(2\omega_1) + R(\omega_2) + R(\omega_1)$	
	$R(2\omega_1) + R(\omega_2) + R(\omega_{n-1})$	n is even
	$R(2\omega_1) + R(\omega_{n-2}) + R(\omega_1)$	n is even
	$R(2\omega_1) + R(\omega_{n-2}) + R(\omega_{n-1})$	n is even
	$2R(2\omega_1)$	
	$R(2\omega_1) + R(2\omega_{n-1})$	
	$R(\omega_1 + \omega_{n-1}) + R(\omega_1)$	
B_n	$(n+1)R(\omega_1)$	
	$R(\omega_2) + R(\omega_1)$	
	$R(2\omega_1)$	
C_n	$(n+1)R(\omega_1)$	
	$R(\omega_2) + 2R(\omega_1)$	
	$R(2\omega_1) + R(\omega_1)$	
D_n	$nR(\omega_1)$	
	$R(\omega_2) + R(\omega_1)$	
	$R(2\omega_1)$	

TABLE 3: SPORADIC CASES

Type of G	V
A_1	$R(3\omega_1)$
	$R(4\omega_1)$
A_2	$R(3\omega_1)$
A_3	$3R(\omega_2)$
	$R(2\omega_2)$
	$R(\omega_1+\omega_3)+R(\omega_2)$
A_5	$R(\omega_3)+3R(\omega_1)$
	$R(\omega_3)+2R(\omega_1)+2R(\omega_5)$
	$R(\omega_3)+R(\omega_2)+R(\omega_1)$
	$2R(\omega_3)$
	$R(\omega_3)+R(2\omega_1)$
A_6	$R(\omega_3)+2R(\omega_1)+R(\omega_6)$
	$R(\omega_3)+2R(\omega_6)$
A_7	$R(\omega_3)+R(\omega_1)$
	$R(\omega_3)+R(\omega_7)$
	$R(\omega_4)$
A_8	$R(\omega_3)$
B_3	$3R(\omega_1)+R(\omega_3)$
	$R(\omega_1)+2R(\omega_3)$
	$3R(\omega_3)$
B_4	$3R(\omega_1)+R(\omega_4)$
	$R(\omega_1)+2R(\omega_4)$
B_5	$3R(\omega_1)+R(\omega_5)$
	$2R(\omega_5)$
B_6	$2R(\omega_1)+R(\omega_6)$

Type of G	V
C_2	$3R(\omega_2)$
	$2R(\omega_2)+R(\omega_1)$
	$R(2\omega_1)+R(\omega_2)$
	$R(2\omega_2)$
C_3	$2R(\omega_2)$
	$R(\omega_3)+2R(\omega_1)$
C_4	$R(\omega_4)$
D_4	$3R(\omega_1)+R(\omega_3)$
	$2R(\omega_1)+2R(\omega_3)$
	$2R(\omega_1)+R(\omega_3)+R(\omega_4)$
D_5	$4R(\omega_1)+R(\omega_4)$
	$2R(\omega_1)+2R(\omega_4)$
	$2R(\omega_1)+R(\omega_4)+R(\omega_5)$
	$2R(\omega_4)+R(\omega_5)$
D_6	$3R(\omega_1)+R(\omega_5)$
	$R(\omega_1)+2R(\omega_5)$
	$R(\omega_1)+R(\omega_5)+R(\omega_6)$
D_7	$R(\omega_1)+R(\omega_6)$
D_8	$R(\omega_7)$
G_2	$3R(\omega_1)$
	$R(\omega_2)$
F_4	$2R(\omega_4)$
	$R(\omega_1)$
E_6	$3R(\omega_1)$
	$2R(\omega_1)+R(\omega_6)$
	$R(\omega_2)$
E_7	$2R(\omega_7)$
	$R(\omega_1)$
E_8	$R(\omega_8)$

TABLE 4: SERIES

Type of G	V	Condition
$A_n + F$, where F is the type of any nontrivial semisimple group	$R(\omega_1) \otimes R(\omega')$, where ω' is the highest weight of any locally faithful irreducible representation	$\dim R(\omega') \leq n + 1$
$A_m + B_n$	$R(\omega_1) \otimes R(\omega_1')$	$n \geq m + 1$
$A + C_n$	$R(\omega_1) \otimes R(\omega_1')$	$n \geq m + 1$
$A_m + D_n$	$R(\omega_1) \otimes R(\omega_1')$	$n \geq m + 1$
$B_m + B_n$	$R(\omega_1) \otimes R(\omega_1')$	
$B_m + C_n$	$R(\omega_1) \otimes R(\omega_1')$	
$B_m + D_n$	$R(\omega_1) \otimes R(\omega_1')$	
$C_m + C_n$	$R(\omega_1) \otimes R(\omega_1')$	
$C_m + D_n$	$R(\omega_1) \otimes R(\omega_1')$	
$D_m + D_n$	$R(\omega_1) \otimes R(\omega_1')$	
$B_n + A_1$	$R(\omega_1) \otimes R(2\omega_1')$	
$C_n + A_1$	$R(\omega_1) \otimes R(2\omega_1')$	
$D_n + A_1$	$R(\omega_1) \otimes R(2\omega_1')$	
$B_n + A_1 + A_1$	$R(\omega_1) \otimes R(\omega_1') \otimes R(\omega_1'')$	
$C_n + A_1 + A_1$	$R(\omega_1) \otimes R(\omega_1') \otimes R(\omega_1'')$	
$D_n + A_1 + A_1$	$R(\omega_1) \otimes R(\omega_1') \otimes R(\omega_1'')$	
$B_n + A_3$	$R(\omega_1) \otimes R(\omega_2')$	
$C_n + A_3$	$R(\omega_1) \otimes R(\omega_2')$	
$D_n + A_3$	$R(\omega_1) \otimes R(\omega_2')$	
$A_n + A_1$	$R(\omega_2) \otimes R(\omega_1')$	$n = 3, 5, 7$ and $n \geq 4$ is even
$C_n + B_3$	$R(\omega_1) \otimes R(\omega_3')$	
$C_n + A_2 + A_1$	$R(\omega_1) \otimes R(\omega_1') \otimes R(\omega_1'')$	
$C_n + A_1$	$R(\omega_1) \otimes R(3\omega_1')$	

TABLE 4: SPORADIC CASES

Type of G	V
$A_5 + A_1$	$R(\omega_3) \otimes R(\omega_1')$
$A_5 + A_2$	$R(\omega_2) \otimes R(\omega_1')$
$A_4 + A_4$	$R(\omega_2) \otimes R(\omega_1')$
$A_4 + A_3$	$R(\omega_2) \otimes R(\omega_1')$
$A_4 + A_2$	$R(\omega_2) \otimes R(\omega_1')$
$A_4 + A_1$	$R(\omega_2) \otimes R(2\omega_1')$
$A_3 + A_3$	$R(\omega_2) \otimes R(\omega_2')$
$A_3 + A_2$	$R(\omega_2) \otimes R(\omega_1')$
$A_3 + A_1$	$R(\omega_2) \otimes R(2\omega_1')$
	$R(2\omega_1) \otimes R(\omega_1')$
$A_2 + A_2$	$R(2\omega_1) \otimes R(\omega_1')$
$A_2 + A_1$	$R(2\omega_1) \otimes R(\omega_1')$
$A_1 + A_1$	$R(3\omega_1) \otimes R(\omega_1')$
	$R(2\omega_1) \otimes R(2\omega_1')$
$B_5 + A_1$	$R(\omega_5) \otimes R(\omega_1')$
$B_4 + A_2$	$R(\omega_4) \otimes R(\omega_1')$
$B_4 + A_1$	$R(\omega_4) \otimes R(\omega_1')$
$B_3 + A_3$	$R(\omega_3) \otimes R(\omega_1')$
$B_3 + A_2$	$R(\omega_3) \otimes R(\omega_1')$
$B_3 + A_1$	$R(\omega_3) \otimes R(2\omega_1')$
	$R(\omega_3) \otimes R(\omega_1')$

Type of G	V
$C_3 + A_1$	$R(\omega_3) \otimes R(\omega_1')$
$D_6 + A_1$	$R(\omega_5) \otimes R(\omega_1')$
$D_5 + A_3$	$R(\omega_4) \otimes R(\omega_1')$
$D_5 + A_2$	$R(\omega_4) \otimes R(\omega_1')$
$D_5 + A_1$	$R(\omega_4) \otimes R(2\omega_1')$
	$R(\omega_4) \otimes R(\omega_1')$
$E_6 + A_2$	$R(\omega_1) \otimes R(\omega_1')$
$E_6 + A_1$	$R(\omega_1) \otimes R(\omega_1')$
$E_7 + A_1$	$R(\omega_7) \otimes R(\omega_1')$
$G_2 + C_2$	$R(\omega_1) \otimes R(\omega_1')$
$G_2 + A_2$	$R(\omega_1) \otimes R(\omega_1')$
$G_2 + A_1$	$R(\omega_1) \otimes R(2\omega_1')$
	$R(\omega_1) \otimes R(\omega_1')$
$B_3 + A_1 + A_1$	$R(\omega_3) \otimes R(\omega_1') \otimes R(\omega_1'')$
$A_3 + A_1 + A_1$	$R(\omega_2) \otimes R(\omega_1') \otimes R(\omega_1'')$
$A_3 + A_3 + A_1$	$R(\omega_1) \otimes R(\omega_1') \otimes R(\omega_1'')$
$A_3 + C_2 + A_1$	$R(\omega_1) \otimes R(\omega_1') \otimes R(\omega_1'')$
$A_2 + A_2 + A_2$	$R(\omega_1) \otimes R(\omega_1') \otimes R(\omega_1'')$
$A_2 + A_2 + A_1$	$R(\omega_1) \otimes R(\omega_1') \otimes R(\omega_1'')$
$A_1 + A_1 + A_1$	$R(2\omega_1) \otimes R(\omega_1') \otimes R(\omega_1'')$
	$R(\omega_1) \otimes R(\omega_1') \otimes R(\omega_1'')$
$A_1 + A_1 + A_1 + A_1$	$R(\omega_1) \otimes R(\omega_1') \otimes R_1(\omega_1'') \otimes R(\omega_1)$

The behavior of equidimensionality under castling is as follows (see [A36]). If $(G, V) = (H \times \mathrm{SL}_p, W \otimes k^p)$, $(\widetilde{G}, \widetilde{V}) = (H \times \mathrm{SL}_q, W^* \otimes k^q)$, $p + q = \dim W$, $1 \leq p \leq q$, then:

(1) if $(\widetilde{G}, \widetilde{V})$ is equidimensional, then (G, V) is also equidimensional,

(2) if (G, V) is equidimensional, then $(\widetilde{G}, \widetilde{V})$ is equidimensional if and only if $p + 1 \geq \dim V/G$.

From this the following description of equidimensional representations that are not castling reduced was deduced in [A36]: first, representations from the class of castling reduced representations (V, G) with $\dim V/G \leq 2$; second, representations (G, V) from the classes of representations $(A_2 + A_2 + A_2, \omega_1 + \omega_1' + \omega_1'')$, $(C_2 + A_2 + A_1, \omega_1 + \omega_1' + \omega_1'')$, and $(A_1 + A_1 + A_1, 2\omega_1 + \omega_1' + \omega_1'')$ for which G is the product of exactly three factors (there are infinitely many such representatives in each class); third, representations $(A_{4n-4} + C_n + A_1, \omega_1 + \omega_1' + \omega_1'')$, $(A_5 + A_{11}, \omega_2 + \omega_1')$, $(D_5 + A_{11}, \omega_4 + \omega_1')$, $(B_4 + A_{12}, \omega_4 + \omega_1')$, $(E_6 + A_{23}, \omega_1 + \omega_1')$.

2. In Lemma 3 from part 11 of §1, Chapter 5 the condition that $(df_1)_v, \dots, (df_m)_v$ are linearly independent means, in invariant terms, that morphism $\pi_{V,G}$ is smooth at v. The orbit of this point has maximal dimension. Panyushev [A6, A7] discovered that for a rather large class of connected semisimple linear groups $G \subset \mathrm{GL}(V)$ with free algebra $k[V]^G$ of invariants morphism $\pi_{V,G}$ is automatically smooth at any point whose orbit is of maximal dimension. Developing his ideas, Knop [A27] strengthened the results of Panyushev.

THEOREM 1 [A27]. *Let G be an algebraic group acting regularly on an algebraic variety X and let $\pi: X \to Y$ be a dominant morphism that is constant on orbits. Suppose that:*

(a) *X and Y are both normal and Gorenstein,*

(b) $\mathrm{codim}_X(X \setminus U) \geq 2$, *where* $U = \{x \in X \mid x \notin \mathrm{sing}\, X, \pi(x) \notin \mathrm{sing}\, Y$, *$\pi$ is smooth at x*$\}$,

(c) $R := \{x \in X \mid \dim Gx = \dim X - \dim Y\} \neq \varnothing$.

Then

$$U \supseteq R \cap (X \setminus \mathrm{sing}\, X) \cap \pi^{-1}(Y \setminus \mathrm{sing}\, Y).$$

THEOREM 2 [A27]. *The conditions of Theorem 1 are satisfied if G is connected and semisimple, X has only rational singularities, the algebra $k[X]$ is factorial, $Y = X/G$ and $\pi = \pi_{X,G}$.*

(Compare with Theorems 1 and 2 in part 1, §3 of this appendix.)

THEOREM 3 [A27]. *Let G be a connected semisimple group, V be a G-module and $v \in V$ be a point such that $\pi_{V,G}(v) \notin \mathrm{sing}\, V/G$. Suppose that* $\mathrm{codim}_V\{u \in V \mid \dim G_u > 0\} \geq 2$. *Then $\pi_{V,G}$ is smooth at v if and only if Gv is of maximal dimension.*

Notice that, because of Theorem 4 from part 1, §3 of this appendix, the hypothesis of this theorem is equivalent to the equality $q_V = \dim V$.

From Lemma 3 in part 11 of §1, Chapter 5 and Theorems 1 and 2 we now have

THEOREM 4 [A27]. *Let G be a connected semisimple group, V be a G-module with free algebra of invariants, and $\mathfrak{N} = \pi_{V,G}^{-1}(\pi_{V,G}(0))$. Then:*

(a) *if the orbit of $v \in \mathfrak{N}$ is open in $\mathfrak{N}$ then there is a section S in V (in the sense of part 11 of §1, Chapter 5) containing v,*

(b) *if each irreducible component of the variety $\mathfrak{N}$ contains an open orbit then the ideal I generated by homogeneous invariants of positive degree is radical.*

Notice that, by Theorem 3, if $\operatorname{codim}_V\{u \in V \mid \dim G_u > 0\} \geq 2$, then Condition (a) of Theorem 4 is not only sufficient but also necessary for the existence of a section.

In particular, it now follows from the above results that for semisimple θ-groups, introduced in [13], there is always a section and that the ideal I is radical (this result was first obtained by Panyushev in [A7]).

3. As was mentioned (and utilized) in [37], Kostant's method used in the proofs of existence for sections of the adjoint representation [101] and isotropy groups of symmetric spaces [102], can be extended to a more general situation via a certain "arithmetic equality." The precise statement is as follows (see [A3]).

Let G be a connected reductive group and V be a G-module with free algebra of invariants. Let $d_1 \leq \cdots \leq d_r$ be the degrees of the elements of a minimal homogeneous system of generators of this algebra. Let v be a nonsingular point of the variety $\mathfrak{N}$ such that it is contained in an irreducible component of $\mathfrak{N}$ of codimension r and its orbit is conic (for example, this is the case if Gv is open in $\mathfrak{N}$). It follows from the assumption on the orbit Gv that in the Lie algebra of G there exists a semisimple element h such that its eigenvalues are rational and $hv = 2v$. Let L be an r-dimensional subspace of V such that

$$hL \subset L \qquad \text{and} \qquad V = \Theta_{v,\mathfrak{N}} \oplus L.$$

Let $c_1 \geq \cdots \geq c_r$ be the eigenvalues of h on L. Then for all i we have that $c_i \leq 0$.

THEOREM 1. (1) $2d_i \geq 2 - c_i$ *for all* i.

(2) *The morphism $\pi_{V,G}$ is smooth at v if and only if equality holds in* (1) *for all* i.

The argument identical to that of Lemma 3 from part 11 of §1, Chapter 5, shows that $\pi_{V,G}$ is smooth at v if and only if there exists a section S containing v (for such a section one can take $v + L$). From this and Theorem 1 we deduce

THEOREM 2. *There exists a section containing* v *if and only if*

$$\sum_{i=1}^{r} d_i = r - \frac{1}{2}\sum_{i=1}^{r} c_i . \tag{1}$$

From this and Theorem 4 of the previous part we now have

THEOREM 3. *Let* G *be a connected semisimple group,* V *be a* G*-module with free algebra of invariants, and* $\mathfrak{N} = \pi_{V,G}^{-1}(\pi_{V,G}(0))$. *Then equality* (1) *holds for any point* $v \in \mathfrak{N}$ *whose orbit is open in* $\mathfrak{N}$.

It was conjectured in [A37] that the arithmetic equality (1) holds for all θ-groups (for some point v whose orbit is open in $\mathfrak{N}$). Thus Theorem 3 shows that this conjecture is true for semisimple θ-groups.

4. Let G be a connected reductive group and T be a maximal torus. When speaking about weights, roots, etc., of G we shall always mean that they are considered relative to T.

Let $\rho\colon G \to \mathrm{GL}(V)$ be a representation. The module of covariants of the G-module V is a direct sum of its submodules—isotypic components of the G-module $k[V]$, [103, A3]. For any dominant weight Λ denote by $k[V]_\Lambda$ the isotypic component of type Λ, i.e., the sum of the submodules of $k[V]$ of type $R(\Lambda)$. The module of covariants is free if and only if all of its submodules $k[V]_\Lambda$ (the "local" modules of covariants, [A3]) are free. In general, as we see, they are not free. Proceeding along the lines of part 4 of §1, Chapter 2, we can replace this condition with another one. Namely let $\theta_1, \dots, \theta_m$ be a homogeneous system of parameters of $k[V]^G$. Then $k[V]_\Lambda$ is a module of finite type over $P = k[\theta_1, \dots, \theta_m]$, and one can ask whether this module is free. As is known [139], if this condition is satisfied for one system of parameters, it is also satisfied for any other. In this case $k[V]_\Lambda$ is called a Cohen-Macaulay module and because of that we call Λ a Cohen-Macaulay weight with respect to the representation ρ. By the Hochster-Roberts Theorem [85], the trivial character 0 is always Cohen-Macaulay. However, as the examples from [150] show, for a given ρ there may exist non-Cohen-Macaulay weights.

Stanley [150] investigated the problem of finding Cohen-Macaulay weights for a given representation ρ. He formulated the following conjecture.

Let $\lambda_1, \dots, \lambda_n$ be all of the weights of T on V (taken with their multiplicities) and $\chi \in \mathrm{Hom}(T, k^*)$. Suppose that the equation

$$x_1\lambda_1 + \cdots + x_n\lambda_n = \chi \tag{1}$$

has a real solution $x_1 = u_1, \dots, x_n = u_n$ with the following properties

(a) $u_1 \le 0, \dots, u_n \le 0$,

(b) if $x_1 = b_1, \dots, x_n = b_n$ is an integral solution of (1) and $b_i \ge u_i$ for all i, then $b_i \ge 0$ for all i. In this case χ is called a *critical character for* (T, V) (compare with parts 4, 5 of §1, Chapter 3).

It is clear that if $-\chi$ is contained in the interior of the zonohedron Z_ρ (see part 6 of §1, Chapter 3) then χ is critical for ρ. In this case χ is called *strongly critical for* ρ.

A dominant weight $\Lambda \in \mathrm{Hom}(T, k^*)$ is called *critical* (resp., *strongly critical*) for ρ if all of the characters $\lambda_i - \sum_{\alpha\in S}\alpha$, where S runs through the finite subsets of the root system of G, are critical (resp., strongly critical) for ρ. In the language of zonohedra, Λ is strongly critical for ρ means that all zonohedra $Z_{\mathrm{Ad}} - \lambda_i$ are contained in the interior of the zonohedron Z_ρ. It now follows that for a given ρ there exist only finitely many strongly critical weights Λ.

STANLEY'S CONJECTURE. *Any critical with respect to ρ dominant weight is Cohen-Macaulay with respect to ρ.*

Van den Bergh "almost" proved this conjecture in [A46]:

THEOREM 1. *Let G be a connected semisimple group and suppose that the generic stabilizer for V is finite. Suppose also that $k[V]_\Lambda \neq 0$. Then Λ is critical with respect to (G, V) if and only if Λ is strongly critical with respect to (G, V). In this case Λ is Cohen-Macaulay with respect to (G, V).*

The next result shows that the property of local modules $k[V]_\Lambda$ to be Cohen-Macaulay is related to the equidimensionality (and therefore is of interest with respect to the "Russian Conjecture" from part 1 of this chapter).

THEOREM 2. *Let $G \subset \mathrm{GL}(V)$ be a connected semisimple group. Then the G-module V is equidimensional if and only if the $k[V]$-module $k[V]_\Lambda$ is Cohen-Macaulay for any Λ.*

The proof does not seem to have been published but is known to experts (the statement and the proof were communicated to me by Bram Broer who also mentioned that this result is known to Schwarz and van den Bergh).

Because the number of different dominant weights Λ for which $k[V]_\Lambda \neq 0$ is, in general, infinite, this theorem shows that Stanley's conjecture gives only a sufficient but by far not necessary condition for the Cohen-Macaulayness of a dominant weight. The problem of determining the Cohen-Macaulay weights with respect to a given ρ seems rather natural.

Similar to $k[V]^G$, the naturally defined Poincaré series $F(k[V]_\Lambda, x)$ for the modules $k[V]_\Lambda$ are the power series expansions with respect to x of certain rational functions of x. For those series Stanley [150] proved the following theorem providing an analogue for the functional equation (4) from part 4 of §1, Chapter 2.

THEOREM 3. *Let G be a connected semisimple group and Λ be a critical dominant weight with respect to a representation (G, V). Then $F(k[V]_\Lambda, x^{-1}) = (-1)^{\dim V/G} x^{\dim V} F(k[V]_{\Lambda^*}, x)$. (Here, by definition, $R(\Lambda^*) = R(\Lambda)^*$.)*

§6. Appendix to Chapter 6

Below we use the terminology and notation from part 8 of §1, Chapter 6 and parts 3–6 of § 2, Chapter 6.

1. The following additional facts about normal three-dimensional affine quasihomogeneous varieties X of SL_2 containing a fixed point can be found in Kraft's book [103].

Let $m_n(X)$ denote the multiplicity of the SL_2-module S_n in the SL_2-module $k[X]$. Since $k[X]^G = k$ we have that this multiplicity is finite.

THEOREM 1 (Interpretation of height in terms of multiplicity). *Let h_X and n_X denote the height and, respectively, stabilizer of the variety X. Then $h_X = n_X \cdot \lim_{n\to\infty} m_n(X)/n$.*

Let $\mathrm{Aut}_{SL_2} X$ denote the group of SL_2-equivariant automorphisms of the algebraic variety X.

THEOREM 2 (On factorization). *There exists a normal three-dimensional affine quasihomogeneous variety $\widetilde{X}$ of SL_2 with a fixed point such that $n_{\widetilde{X}} = 1$, and X is the quotient of $\widetilde{X}$ by a finite subgroup $F \subset \mathrm{Aut}_{SL_2} \widetilde{X}$. Such an $\widetilde{X}$ is defined uniquely up to an isomorphism, F is isomorphic to the generic stabilizer on X, and $h_{\widetilde{X}} = h_X$.*

It is clear that the converse is also true: for any X with $n_X = 1$ and any finite subgroup $H \subset \mathrm{Aut}_{SL_2} X$ the quotient X/H is a normal three-dimensional affine quasihomogeneous variety of SL_2 with a fixed point.

THEOREM 3 (On automorphisms). *Let $n_X = 1$. Consider the embedding $SL_2 \hookrightarrow X$, $g \mapsto gwv$, where $w = \left(\begin{smallmatrix} 0 & 1 \\ -1 & 0 \end{smallmatrix}\right)$ and $v \in X$ is a canonical point. Then the action of the group B on SL_2 by right translations can be extended to X and thus gives rise to the natural embedding of B in $\mathrm{Aut}_{SL_2} X$. This embedding is in fact an equality: $B = \mathrm{Aut}_{SL_2} X$.*

An explicit description of *normal* three-dimensional affine quasihomogeneous varieties of SL_2 with a fixed point and with given height h and stabilizer n (compare with Theorem 6 from part 7 of §2, Chapter 6) is as follows (in [103] the proofs are only given for $n = 1$ but the argument carries over to the general case).

Let $(r_1, s_1), \dots, (r_t, s_t)$ denote the set of indecomposable elements (i.e., a minimal system of generators) of the additive monoid

$$M_{h,n} = \{(r, s) \in \mathbb{N} \times \mathbb{N} \mid (r, s) \neq (0, 0), (1-h)s \leq hr, r - s \equiv 0 (\mathrm{mod}\, n)\}:$$

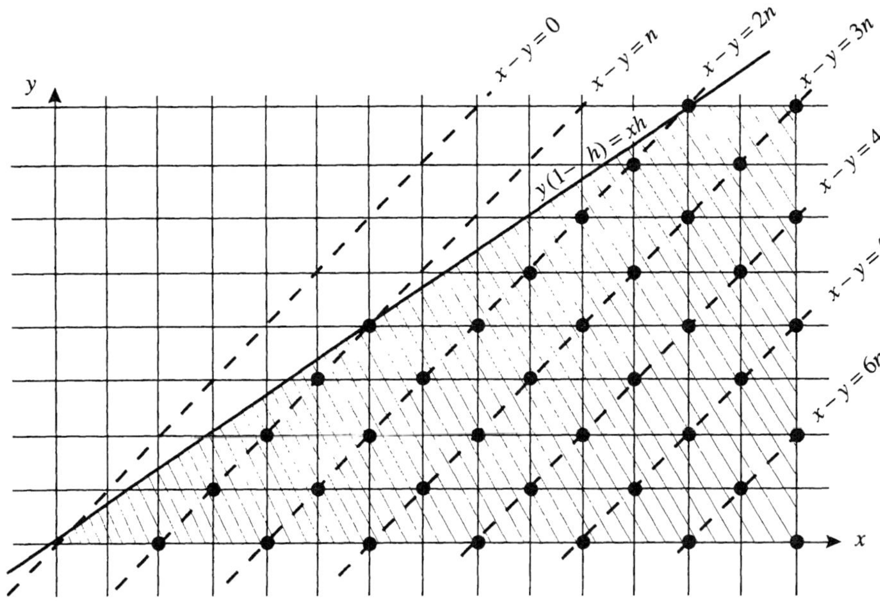

Consider the point $v = x^{r_1}y^{s_1} \oplus \cdots \oplus x^{r_t}y^{s_t} \in S_{r_1+s_1} \oplus \cdots \oplus S_{r_t+s_t}$ and variety $X = \overline{\mathrm{SL}_2 v}$. It is three-dimensional and contains the fixed point $o = 0 \oplus \cdots \oplus 0$.

THEOREM 4 (The explicit construction). *The variety X constructed above is normal, $h_X = h$ and $n_X = n$.*

Panyushev [A8] gave, in terms of continuous fractions, the following arithmetic algorithm for finding the generators (r_i, s_i) of the monoid $M_{h,n}$.

For the rational number $h' = nh/(1-2h)$ consider its continuous fraction expansion:

$$h' = a_0 + \cfrac{1}{a_1 + \cfrac{1}{a_2 + \cfrac{}{\ddots \, \cfrac{1}{a_s}}}} = [a_0; a_1, \ldots, a_s]. \tag{1}$$

The number s will be called the *length* of the fraction.

Beginning with h', we perform the following procedure, obtaining new continuous fractions, until we come to zero:

if the length of the continuous fraction is even we subtract 1 from the last term, and if it is odd we just discard the last term (of course, the length is considered to become smaller either when the last term is discarded or when it reaches the value of 1): $[b_0; b_1, \ldots, b_{l-1}, 1] = [b_0; b_1, \ldots, b_{l-1} + 1])$.

We obtain a sequence of rational numbers $h' = d_1 > d_2 > \cdots > d_t = 0 (= 0/1))$.

For example, $d_1 = [1; 3, 2, 2] = 22/17$, $d_2 = [1; 3, 2] = 9/7$, $d_3 = [1; 3, 1] = [1; 4] = 5/4$, $d_4 = [1; \ ] = 1$, $d_5 = [0; \ ] = 0$.

Consider the irreducible fraction $d_i = p_i/q_i$; $p_i, q_i \in \mathbb{Z}$. Then $\{(r_i, s_i) = (nq_i + p_i, p_i), i = 1, \ldots, t\}$ is a minimal system of generators of the monoid $M_{h,n}$.

One can now describe, in terms of those generators, the SL_2-module structure on the tangent space of X at the fixed point o.

THEOREM 5 (On tangent space).

$$\Theta_{o,X} = S_{r_1+s_1} \oplus \cdots \oplus S_{r_t+s_t}.$$

(In [103] a proof is given for $n_X = 1$, but it can be carried over to the general case (see [A12]).)

In particular, it follows that $X \hookrightarrow S_{r_1+s_1} \oplus \cdots \oplus S_{r_t+s_t}$ is a minimal possible equivariant linear embedding of X. Moreover it also follows that if $v \in S_n$ is a form such that $0 \in X = \overline{\mathrm{SL}_2 v}$ and $\dim X = 3$, then X is not normal (this result is due to Bartels [A11]).

2. As was mentioned in part 9 of §2, Chapter 6, the stabilizer and height do not form a complete set of invariants for nonnormal varieties. Bartels [A12] investigated nonnormal three-dimensional affine quasihomogeneous varieties of SL_2 with a fixed point. She showed that in this case there are continuous systems of invariants and for such varieties one can find moduli spaces of arbitrarily high dimension.

3. Let X be a normal three-dimensional affine quasihomogeneous variety of SL_2 with a fixed point. Panyushev [A8] gave a description of the minimal equivariant resolution of singularities $\pi: \widetilde{X} \to X$ (of course, $\widetilde{X}$ is also a quasihomogeneous variety of SL_2, although no longer affine).

As it turns out, in the open (in X) orbit one can find a point v such that $Y = \overline{Bv}$ is an (abstract) toroidal embedding. The technique of toroidal embeddings allows us to construct a minimal B-equivariant resolution of singularities $\widetilde{Y} \to Y$. This morphism gives rise to an SL_2-equivariant morphism $\mathrm{SL}_2 \times_B Y \to X$. Their composition $\mathrm{SL}_2 \times_B \widetilde{Y} \to X$ is the desired resolution.

A description of the exceptional fiber $\pi^{-1}(o)$ of this resolution in terms of the height $h = h_X$ and stabilizer $n = n_X$ was given in [A8]. More precisely, consider again the number $h' = (nh)/(1-2h)$. Let $h' = f_0/g_0$, $f_0, g_0 \in \mathbb{Z}$, be an irreducible fraction.

If $f_0 > 1$ consider continuous fraction (1) from part 1 of this chapter and, beginning with h', proceed as described below (this procedure is dual to the one described above) until we reach either the fraction $[0; l]$ or the fraction [;], which is, by definition, equal to $1/0 = \infty$ (the latter case is only possible when $a_0 > 0$):

If the length of the continuous fraction is even we discard the last term, and if the length is odd we subtract 1 from it.

At the end we obtain a sequence of rational numbers $h' = h_0 < h_1 < \cdots < h_m \leq \infty$.

For example, (1) $h_0 = [0; 3, 2, 2] = 5/17$, $h_1 = [0; 3, 2, 1] = [0; 3, 3] = 3/10$, $h_2 = [0; 3] = 1/3$,

(2) $h_0 = [1; 3, 2, 2] = 22/17$, $h_1 = [1; 3, 2, 1] = [1; 3, 3] = 13/10$, $h_2 = [1; 3] = 4/3$, $h_3 = [1; 2] = 3/2$, $h_4 = [1; 1] = [2; \ ] = 2$, $h_5 = [\ ; \ ] = 1/0 = \infty$.

Let $h_i = f_i/g_i$; $f_i, g_i \in \mathbb{Z}$ be an irreducible fraction.

THEOREM 1. (1) *If* $f_0 = 1$ *then*

$$\pi^{-1}(o) = \mathbb{P}^1.$$

(2) *If* $f_0 > 1$, *then* $\pi^{-1}(o)$ *has exactly* m *irreducible components, each being the standard rational ruled surface of type* F_{e_i}, *where*

$$e_i = \begin{cases} 2f_i + ng_i & \text{if } h_i \neq \infty, \\ 0 & \text{if } h_i = \infty \end{cases}$$

with

$$F_{e_i} \cap F_{e_j} = \begin{cases} \varnothing & \text{if } |i-j| > 1, \\ \mathbb{P}^1 & \text{if } |i-j| = 1, \end{cases}$$

and in the latter case $F_{e_i} \cap F_{e_j}$ *is the fiber in each of* F_{e_i}, F_{e_j} *(considered as standard bundles over* $\mathbb{P}^1$*).*

4. By the authors' own account, the paper [31], whose content constitutes Chapter 6, along with the theory of toroidal embeddings [99] became the starting point for the general theory of embeddings of homogeneous spaces of algebraic groups developed by Luna and Vust in [A38] (an embedding of a homogeneous space G/H is a quasihomogeneous variety of the group G with a base point v from the open orbit such that $G_v = H$). I cannot discuss their results here because it would require too many auxiliary concepts and much explanation. I should only remark that the main tool is the analysis of equivariant discrete valuations of the field $k(G/H)$. As it turned out [A38], the most thorough classification results for embeddings of G/H can only be expected when the number $c(G/H) = \max_{x \in G/H} \operatorname{codim} Bx$, where B is a Borel subgroup of G, is small (this number is called the *complexity* of the homogeneous space G/H). The condition $c(G/H) = 0$ means that B has an open orbit in G/H. Subgroups H of G such that $c(G/H) = 0$ are called *spherical.* In this case G/H is called a *spherical homogeneous space* and an embedding of it is called a *spherical variety.* The theory of embeddings of spherical homogeneous spaces of connected reductive groups contains the theory of toroidal embeddings as a special case; similar to the latter, it offers a dictionary between the properties of such embeddings and the properties of certain fans (i.e., polyhedral cones in $\mathbb{Q}^n$) and integral points contained in those fans; in comparison with the toroidal case, some of those points could, in addition, be "colored". In this language one can obtain a complete classification of normal embeddings of G/H as well as a description of equivariant morphisms between them for certain types of spherical subgroups. Geometry and topology of spherical embeddings are being actively investigated. They turned out to be related to the classical problems of enumerative geometry (Schubert calculus and Hilbert's 15th problem). See [A40, A13, A29] and the literature therein.

Homogeneous spaces of the form SL_2/F, where F is a finite group, have complexity 1. The results of Chapter 6 yield a classification of normal affine embeddings of such spaces. Applying their theory, Luna and Vust [A38] extended this classification in the case $F = \{1\}$ to all (not necessarily affine) normal embeddings. Jauslin-Moser did the same for any finite group F [A25] (she also considered some geometric properties of such embeddings, like the question of smoothness and the structure of the Chow ring). Again it turned out that the classification is given by a set of discrete data. This set has the convenient geometric interpretation as a colored diagram ("hedgehog"). For a more precise description, using those terms, I will give in conclusion a classification of *full* normal embeddings of SL_2 (see [A37, A38, A25]).

To each point of the projective line $\mathbb{P}^1$ we associate a copy of the interval $[0,1]$ and consider the "hedgehog" H obtained by glueing together all left endpoints at point 0:

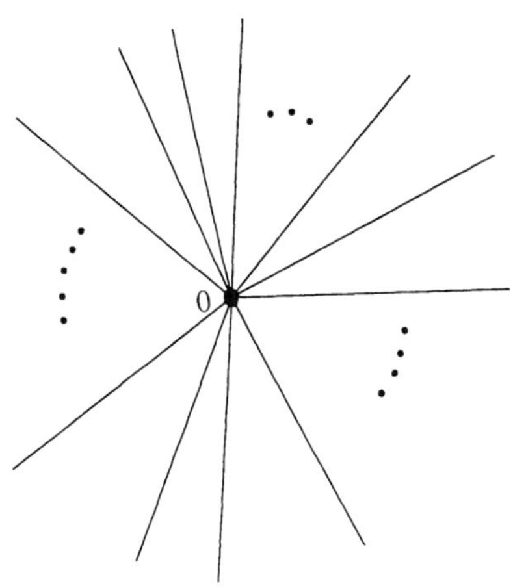

Any point $a \in H$ is contained in one of the chosen intervals (the "spines" of H) and therefore is a real number. We denote it $|a|$.

A *fan* is, by definition, a finite set F of points from H with the following properties:

(a) all points of F are rational,

(b) if all points of F are chosen from the same spine then among them there is a point a such that $|a| < 1/2$,

(c) F cannot consist of two points a and b with $|a| = |b| = 1$.

The connected components of $H\backslash F$ are called *segments*. If a segment consists of all points a such that $|a| > \alpha$ for some $\alpha > 1/2$ then it is called *special*.

A *colored fan* is, by definition, a fan some of whose special segments are marked (for example, colored in red).

THEOREM 1. *Full normal embeddings of* SL_2 *are classified by colored fans.*

Given a fan, one can graphically recover the combinatorics of the orbital decomposition: red segments correspond to the fixed points, the remaining segments correspond to the orbits of the form $\mathbb{P}^1 \simeq \mathrm{SL}_2/B$, and the points from $F\backslash\{0\}$ correspond to the two-dimensional orbits (with the incidence relation being preserved). For example,

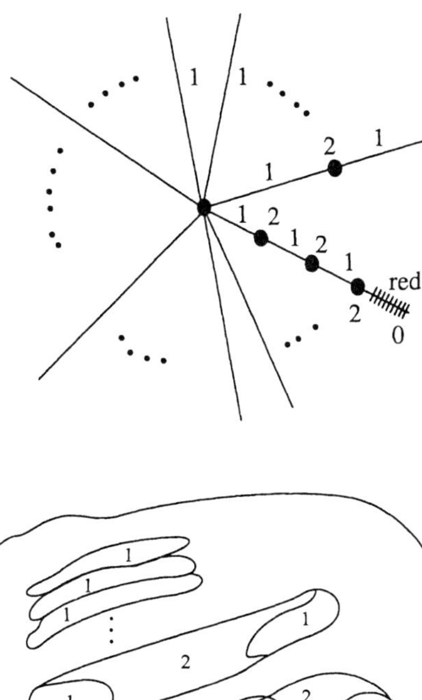

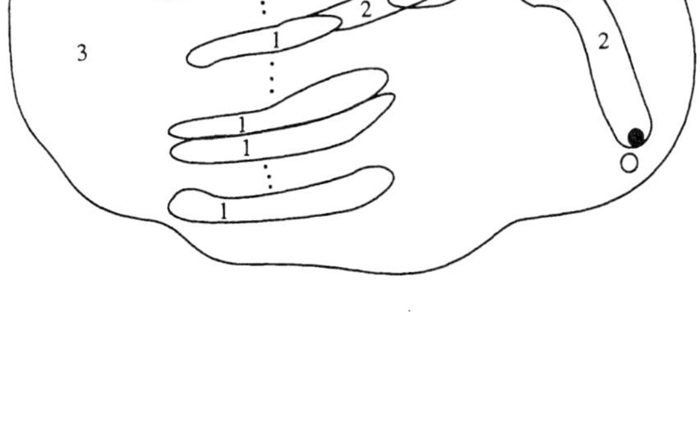

Bibliography

1. O. M. Adamovich, *Equidimensional representations of simple algebraic groups*, Geometric Methods in Problems of Algebra and Analysis, vyp. 2 (A. V. Chernavskiĭ, ed.), Yaroslav. Gos. Univ., Yaroslavl, 1980, pp. 120–125; English transl. in Amer. Math. Soc. Transl. (2) **128** (1986), 25–29.

2. O. M. Adamovich and E. O. Golovina, *Simple linear groups having a free algebra of invariants*, Questions of Group Theory and Homological Algebra, vyp. 2, Yaroslav. Gos. Univ., Yaroslavl, 1979, pp. 3–41; English transl. in Selecta Math. Sovietica **3** (1983/84), 183–220.

3. E. M. Andreev, E. B. Vinberg, and A. G. Elashvili, *The largest dimension orbits of semisimple Lie groups*, Funktsional. Anal. i Prilozhen. **1** (1967), no. 4, 3–7; English transl. in Functional Anal. Appl. **1** (1967).

4. E. M. Andreev and V. L. Popov, *On stationary subgroups of points in general position in the representation space of a semisimple Lie group*, Funktsional. Anal. i Prilozhen. **5** (1971), no. 4, 1–8; English transl. in Functional Anal. Appl. **5** (1971).

5. M. F. Atiyah and I. G. MacDonald, *Introduction to commutative algebra*, Addison-Wesley, Reading, MA, 1969.

6. D. N. Akhiezer, *Dense orbits with two ends*, Izv. Akad. Nauk SSSR Ser. Mat. **41** (1977), 308–344; English transl. in Math. USSR-Izv. **11** (1977).

7. ——, *Algebraic groups acting transitively on the complement of a homogeneous hypersurface*, Dokl. Akad. Nauk SSSR **245** (1979), 281–284; English transl. in Soviet Math. Dokl. **20** (1979).

8. F. A. Bogomolov, *Holomorphic tensors and vector bundles on projective varieties*, Izv. Akad. Nauk SSSR Ser. Mat. **42** (1978), 1227–1287; English transl. in Math. USSR-Izv. **13** (1979).

9. N. Bourbaki, *Groupes et algèbres de Lie*, Chapitres IV-VI, Hermann, Paris, 1968.

10. ——, *Groupes et algèbres de Lie*, Chapitres VII, VIII, Hermann, Paris, 1975.

11. H. Weyl, *The classical groups. Their invariants and representations*, 2nd ed., Princeton University Press, Princeton, NJ, 1946.

12. E. B. Vinberg, *Algebraic transformation groups of maximal rank*, Mat. Sb. **88** (1972), 493–503; English transl. in Math. USSR-Sb. **17** (1972).

13. ———, *The Weyl groups of a graded Lie algebra*, Izv. Akad. Nauk SSSR Ser. Mat. **40** (1976), 488–526; English transl. in Math. USSR-Izv. **10** (1976).

14. B. N. Kimelfeld and E. B. Vinberg, *Homogeneous domains on flag varieties and spherical subgroups of semisimple Lie groups*, Funktsional. Anal. i Prilozhen. **12** (1978), no. 3, 12–19; English transl. in Functional Anal. Appl. **12** (1978).

15. A. L. Onishchik and E. B. Vinberg, *Seminar on algebraic groups and Lie groups*, 1967/68, Izdat. Moskov. Gos. Univ., Moscow, 1969; English transl., Springer-Verlag, Berlin and New York, 1990.

16. V. L. Popov and E. B. Vinberg, *On a class of quasihomogeneous affine varieties*, Izv. Akad. Nauk SSSR Ser. Mat. **36** (1972), 749–763; English transl. in Math. USSR-Izv. **6** (1972).

17. A. G. Elashvili and E. B. Vinberg, *A classification of three-vectors of a nine-dimensional space*, Trudy Sem. Vektor. Tenzor. Anal. **18** (1978), 197–233; English transl. in Selecta Math. Sovietica **7** (1988), 63–98.

18. M. Kh. Gizatullin, *Affine surfaces which are quasihomogeneous with respect to an algebraic group*, Izv. Akad. Nauk SSSR Ser. Mat. **35** (1971), 738–753; English transl. in Math USSR-Izv. **5** (1971).

19. N. L. Gordeev, *Invariants of linear groups generated by matrices with two eigenvalues different from one*, Zap. Nauchn. Sem. Leningrad. Otdel. Mat. Inst. Steklov. (LOMI) **114** (1972), 120–130; English transl. in J. Soviet Math. **27** (1984).

20. G. B. Gurevich, *Foundations of the theory of algebraic invariants*, OGIZ, Moscow, 1948; English transl., Noordhoff, Groningen, 1964.

21. J. Dieudonné, J. Carrell, and D. Mumford, *Geometric invariant theory*, "Mir", Moscow, 1974. (Russian) *

22. E. B. Dynkin, *Semisimple subalgebras of semisimple Lie algebras*, Mat. Sb. **30** (1952), 349–402; English transl. in Amer. Math. Soc. Transl. Ser. (2) **6** (1957), 111–244.

23. V. G. Kač, *Concerning the question of description of the orbit space of a linear algebraic group*, Uspekhi Mat. Nauk **30** (1975), no. 6, 173–174. (Russian)

24. S. Mac Lane, *Homology*, Academic Press, New York; Springer-Verlag, Berlin, 1963.

25. D. I. Panyushev, *On orbit spaces of finite and linear groups*, Izv. Akad. Nauk SSSR Ser. Mat. **46** (1982), 95–99; English transl. in Math. USSR-Izv. **20** (1983).

26. A. M. Popov, *Irreducible simple linear groups with finite stationary subgroups of general position*, Funktsional. Anal. i Prilozhen. **9** (1975), no. 1, 81–82; English transl. in Functional Anal. Appl. **9** (1975).

* *Editor's Note*: This book is a translation of J. A. Dieudonné and J. B. Carrell, *Invariant theory, old and new*, Academic Press, New York, 1971 and D. B. Mumford, *Geometric invariant theory*, Academic Press, New York, 1965.

27. ——, *Stationary subgroups of general position for some actions of simple Lie groups*, Funktsional. Anal. i Prilozhen. **10** (1976), no. 3, 88–90; English transl. in Functional Anal. Appl. **10** (1976).

28. ——, *Irreducible semisimple linear Lie groups with finite stationary subgroups of general position*, Funktsional. Anal. i Prilozhen. **12** (1978), no. 2, 91–92; English transl. in Functional Anal. Appl. **12** (1978).

29. V. L. Popov, *Stability criteria for the action of a semisimple group on a factorial manifold*, Izv. Akad. Nauk SSSR Ser. Mat. **34** (1970), 523–531; English transl. in Math. USSR-Izv. **4** (1970).

30. ——, *On the stability of the action of an algebraic group on an algebraic variety*, Izv. Akad. Nauk SSSR Ser. Mat. **36** (1972), 371–385; English transl. in Math. USSR-Izv. **6** (1972).

31. ——, *Quasihomogeneous affine algebraic varieties of* SL_2, Izv. Akad. Nauk SSSR Ser. Mat. **37** (1973), 792–832; English transl. in Math. USSR-Izv. **7** (1973).

32. ——, *A classification of affine algebraic surfaces quasihomogeneous with respect to an algebraic group*, Izv. Akad. Nauk SSSR Ser. Mat. **37** (1973), 1038–1055; English transl. in Math. USSR-Izv. **7** (1973).

33. ——, *The Picard groups of homogeneous spaces of linear algebraic groups and one-dimensional homogeneous vector bundles*, Izv. Akad. Nauk SSSR Ser. Mat. **38** (1974), 294–322; English transl. in Math. USSR-Izv. **8** (1974).

34. ——, *Structure of the closures of orbits in spaces of finite-dimensional linear representations of the group* SL *(2)*, Mat. Zametki **16** (1974), 943–950; English transl. in Math. Notes **16** (1974).

35. ——, *A classification of three-dimensional affine algebraic varieties quasihomogeneous with respect to an algebraic group*, Izv. Akad. Nauk SSSR Ser. Mat. **39** (1975), 566–609; English transl. in Math. USSR-Izv. **9** (1975).

36. ——, *On a classification of the representations exceptional in the sense of Igusa*, Funktsional. Anal. i Prilozhen. **9** (1975), no. 4, 83–84; English transl. in Functional Anal. Appl. **9** (1975).

37. ——, *Representations with free module of covariants*, Funktsional. Anal. i Prilozhen. **10** (1976), no. 3, 91–92; English transl. in Functional Anal. Appl. **10** (1977).

38. ——, *A classification of spinors of dimension* 14, Trudy Moskov. Mat. Obshch. **37** (1978), 173–217; English transl. in Trans. Moscow Math. Soc. **1** (1980).

39. ——, *Algebraic curves with infinite automorphism groups*, Mat. Zametki **23** (1978), 183–195; English transl. in Math. Notes **23** (1978).

40. ——, *On Hilbert's theorem on invariants*, Dokl. Akad. Nauk SSSR **249** (1979), 551–555; English transl. in Soviet Math. Dokl. **20** (1979).

41. ——, *On Hilbert's* 14*th problem*, XV All-Union Algebraic Conference (Krasnoyarsk, 1979), Abstracts, part 1, 123. (Russian)

42. ——, *The constructive theory of invariants*, Izv. Akad. Nauk SSSR Ser. Mat. **45** (1981), 1100–1120; English transl. in Math. USSR-Izv. **19** (1982).

43. ——, *A finiteness theorem for representations with free algebra of invariants*, Izv. Akad. Nauk SSSR Ser. Mat. **46** (1982), 347–370; English transl. in Math. USSR-Izv. **20** (1983).

44. ——, *Syzygies in invariant theory*, Izv. Akad. Nauk SSSR Ser. Mat. **47** (1983), 544–622; English transl. in Math. USSR-Izv. **22** (1984).

45. M. Rosenlicht, *Equivalence relations on algebraic curves*, Ann. of Math. (2) **56** (1952), 169–191.

46. J.-P. Serre, *Groupes algébriques et corps de classes*, Publ. Inst. Math. Univ. Nancago, Hermann, Paris, 1959; English transl., Springer-Verlag, New York and Berlin, 1988.

47. K. S. Sibirskiĭ, *Introduction to the algebraic theory of invariants of differential equations*, "Shtiintsa", Kishinev, 1982; English transl., Manchester Univ. Press, Manchester, 1988.

48. T. Springer, *Invariant theory*, Lecture Notes in Math., vol. 585, Springer-Verlag, Berlin and New York, 1977.

49. D. Hadzhiev, *A description of closed orbits and the closures of nonclosed orbits of irreducible representations of the Lorenz group*, Dokl. Akad. Nauk UzSSR **12** (1966), 3–6. (Russian)

50. J. E. Humphreys, *Linear algebraic groups*, Springer-Verlag, New York, 1975.

51. R. Hartshorne, *Algebraic geometry*, Springer-Verlag, New York, 1977.

52. A. G. Elashvili, *A canonical form and stationary subalgebras of points in general position for simple linear Lie groups*, Funktsional. Anal. i Prilozhen. **6** (1972), no. 1, 51–62; English transl. in Functional Anal. Appl. **6** (1972).

53. ——, *Stationary subalgebras of points in general position for irreducible linear Lie groups*, Funktsional. Anal. i Prilozhen. **6** (1972), no. 2, 65–78; English transl. in Functional Anal. Appl. **6** (1972).

54. S. Aronhold, *Ueber eine fundamentale Begründung der Invariantentheorie*, J. Reine Angew. Math. **63** (1862), 281–345.

55. A. Ash, D. Mumford, M. Rapaport, and Y. Tai, *Smooth compactification of locally symmetric varieties*, Lie groups: History, Frontiers and Applications, vol. IV, Math. Sci. Press, Brookline, MA, 1975.

56. D. Bartels, *On nonnormality of affine quasi-homogeneous* SL$(2, C)$*-varieties*, Lecture Notes in Math., vol. 924, Springer-Verlag, Berlin and New York, 1962, pp. 384–399.

57. A. Bialynicki-Birula, *On algebraic actions of* SL(2), Bull. Acad. Polon. Sci. Sér. Sci. Math. Astronom. Phys. **26** (1978), 293–294.

58. A. Bialynicki-Birula, G. Hochschild, and G. D. Mostow, *Extensions of representations of algebraic linear groups*, Amer. J. Math. **85** (1963), 131–144.

59. G. Boole, *Exposition of a general theory of linear transformations*, parts I, II, Cambridge Math. J. **III** (1843), 1–20; 106–119.

60. A. Borel, *Sur la cohomologie des espaces fibrés principaux et des espaces homogènes des groupes de Lie compacts*, Ann. of Math. **57** (1953), 115–207.

61. W. Borho and H. Kraft, *Uber Bahnen und deren Deformationen bei linearen Aktionen reduktiver Gruppen*, Comment. Math. Helv. **54** (1979), 61–104.

62. J. F. Boutot, *Rational singularities and quotients by reductive groups*, Tagung über Transformationsgruppen und Invariantentheorie (Bonn, 22–28 Sept. 1988).

63. E. Brieskorn, *Singular elements of semisimple algebraic groups*, Actes Congrès Int. Math. (Nice, 1970), vol. 2, Gauthier-Villars, Paris, 1971, pp. 279–284.

64. M. Brion, *Sur la théorie des invariants*, Publ. Math. de l'Univ. Pierre et Marie Curie **45** (1982), 1–192.

65. A. G. Brower and A. M. Cohen, *The Poincaré series of the polynomials invariant under* SU_2 *in its irreducible representations of degree* ≤ 17, Math. Centrum, ZW, 134/79, Amsterdam, 1979, pp. 1–20.

66. C. Chevalley, *Invariants of finite groups generated by reflections*, Amer. J. Math. **67** (1955), 778–782.

67. R. Cicurel, *Sur les orbites affines des G-variétés admettant une orbite ouverte*, C. R. Acad. Sci. Paris Ser I. Math. **274** (1972), 1316–1318.

68. C. De Concini, D. Eisenbud, and C. Procesi, *Young diagrams and determinantal varieties*, Invent. Math. **56** (1980), 129–165.

69. C. De Concini and C. Procesi, *Symmetric functions, conjugacy classes and the flag variety*, Invent. Math. **64** (1981), 203–220.

70. M. Demazure, *Sous-groupes algébriques de rang maximum du groupe de Cremona*, Ann. Sci. École Norm. Sup. **4** (1970), 507–588.

71. J. Dixmier, *Sur les invariants des formes binaires*, C. R. Acad. Sci. Paris Ser. I Math. **292** (1981), 987–990.

72. V. Gatti and E. Viniberghi, *Spinors of* 13-*dimensional space*, Adv. Math. **30** (1978), 137–155.

73. D. Gieseker, *Some applications of geometric invariant theory to moduli problems*, Proc. Internat. Congr. Math. (Helsinki, 1978), Acad. Sci. Fennica, Helsinki, 1980, pp. 525–528.

74. P. Gordan, *Beweis, dass jede Covariante und Invariante einer binären Form eine ganze Function mit numerischen Coefficienten einer endlichen Anzahl solcher Formen ist*, J. Reine Angew. Math. **69** (1868), 323–354.

75. ——, *Vorlesungen über Invariantentheorie* (*herausgegeben von G. Kerschensteiner*), II. Teubner, Leipzig, 1887.

76. J. H. Grace and A. Young, *The algebra of invariants*, Cambridge Univ. Press, Cambridge, 1903.

77. F. Grosshans, *Observable groups and Hilbert's fourteenth problem*, Amer. J. Math. **95** (1973), 229–253.

78. W. Haboush, *Reductive groups are geometrically reductive*, Ann. of Math. **102** (1975), 67–83.

79. W. Hesselink, *Desingularisation of varieties of nullforms*, Invent. Math. **55** (1979), 141–163.

80. D. Hilbert, *Über die Theorie der algebraischen Formen*, Gesammelte Abhandlungen, vol. II, Springer-Verlag, Berlin and New York, 1970, pp. 199–257.

81. ——, *Über die vollen Invariantensysteme*, Gesammelte Abhandlungen, vol. II, Springer-Verlag, Berlin and New York, 1970, pp. 287–344.

82. ——, *Mathematische Probleme*, Gesammelte Abhandlungen, vol. III, Springer-Verlag, Berlin and New York, 1970, pp. 290–329.

83. H. Hiller, *Geometry of Coxeter groups*, Pitman, Boston, MA, and London, 1982.

84. G. Hochschild and G. D. Mostow, *Unipotent groups in invariant theory*, Proc. Nat. Acad. Sci. U.S.A. **70** (1973), 646–648.

85. M. Hochster and J. Roberts, *Rings of invariants of reductive groups acting on regular rings are Cohen-Macaulay*, Adv. Math. **13** (1974), 125–175.

86. J.-i. Igusa, *Modular forms and projective invariants*, Amer. J. Math. **89** (1967), 816–855.

87. ——, *On certain representations of semisimple algebraic groups*, Lecture Notes in Math., vol. 185, 1971, Springer-Verlag, Berlin, pp. 76–86.

88. ——, *On certain representations of semisimple algebraic groups and the arithmetic of the corresponding invariants*, Invent. Math. **12** (1971), 62–94.

89. ——, *Geometry of absolutely admissible representations*, Number Theory, Algebraic Geometry and Commutative Algebra, in honor of Yasuo Akizuki, Kinokuniya, Tokyo, 1973, pp. 373–452.

90. T. Jósefiak, P. Pragacz, and J.Weyman, *Resolutions of determinantal varieties*, Astérisque **87-88** (1981), 109–189.

91. V. G. Kač, *Some remarks on nilpotent orbits*, J. Algebra **64** (1980), 190–213.

92. ——, *Infinite root systems, representations of graphs and invariant theory*, Invent. Math. **56** (1980), 57–92.

93. ——, *Infinite root systems, representations of graphs and invariant theory* II, J. Algebra **77** (1982), 141–162.

94. V. G. Kač, V. L. Popov, and E. B. Vinberg, *Sur les groupes algébriques dont l'algèbre des invariants est libre*, C. R. Acad. Sci. Paris Ser. I Math. **283** (1976), 857–878.

95. V. G. Kač and K.-i. Watanabe, *Finite linear groups whose ring of invariants is a complete intersection*, Bull. Amer. Math. Soc. **6** (1982), 221–223.

96. T. Kaga and T. Watabe, *On 4-dimensional quasihomogeneous affine algebraic varieties of reductive algebraic groups*, Sci. Rep. Niigata Univ. Ser. A **15** (1978), 51–69.

97. G. Kempf, *The Hochster-Roberts theorem in invariant theory*, Michigan Math. J. **26** (1979), 19–32.

98. ——, *Some quotient surfaces are smooth*, Michigan Math. J. **27** (1980), 295–299.

99. G. Kempf, F. Knudsen, D. Mumford, and B. Saint-Donant, *Toroidal embeddings*, Lecture Notes in Math., vol. 339, Springer-Verlag, Berlin and New York, 1973.

100. F. Klein, *Vergleichende Betrachtungen über neuere geometrische Forschungen*, Erlangen, 1872.

101. B. Kostant, *Lie group representations on polynomial rings*, Amer. J. Math. **85** (1963), 327–404.

102. B. Kostant and S. Rallis, *Orbits and representations associated with symmetric spaces*, Amer. J. Math. **93** (1971), 753–809.

103. H. Kraft, *Geometrische Methoden in der Invariantentheorie*, Friedr. Vieweg & Sohn, Braunschweig, 1984.

104. H. Kraft and C. Procesi, *Closures of conjugacy classes of matrices are normal*, Invent. Math. **53** (1979), 227–247.

105. ——, *Minimal singularities in* GL_n, Invent. Math. **62** (1981), 503–515.

106. M. Krämer, *Eine Klassifikation bestimmter Untergruppen kompakter zusammenhängender Liegruppen*, Comm. Algebra **3** (1975), 691–737.

107. A. Lascoux, *Syzygies des variétés déterminantales*, Adv. Math. **30** (1978), 202–237.

108. D. Luna, *Slices étales*, Bull. Soc. Math. France **33** (1973), 81–105.

109. ——, *Fonctions différentiables invariantes sous l'opération d'un groupe réductif*, Ann. Inst. Fourier **26** (1976), 33–49.

110. ——, *Adhérences d'orbite et invariants*, Invent. Math. **29** (1975), 231–238.

111. D. Luna and R. W. Richardson, *A generalization of the Chevalley restriction theorem*, Duke Math. J. **46** (1979), 487–496.

112. H. Matsumura, *Commutative algebra*, Benjamin Cummings Publ. Co., Inc., Reading, MA, 1970.

113. R. V. Moody and J. Patera, *Fast recursion formula for weight multiplicities*, Bull. Amer. Math. Soc. **7** (1982), 237–242.

114. D. Mumford, *Geometric invariant theory*, Academic Press, New York, 1965.

115. D. Mumford and K. Suominen, *Introduction to the theory of moduli*, Algebraic Geometry, Proc. Fifth Nordic Summer School in Math. (Oslo, 1970), Wolters-Noordhoff, Groningen, 1972, pp. 171–222.

116. D. Mumford, *Stability of projective varieties*, Lectures given at the Inst. Hautes Études Sci. (Bures-sur-Yvette, March-April 1976), L'Enseignement Mathématique, Geneva, 1977.

117. D. Mumford and J. Fogarty, *Geometric invariant theory*, 2nd ed., Ergeb. Math. Grenzgeb. vol. 34, Springer-Verlag, Berlin and New York, 1982.

118. M. Nagata, *On the* 14*th problem of Hilbert*, Amer. J. Math. **81** (1959), 766–772.

119. ——, *Invariants of a group in an affine ring*, J. Math. Kyoto Univ. **3** (1964), 369–377.

120. ——, *Lectures on the fourteenth problem of Hilbert*, Tata Inst. of Fund. Research, Bombay, 1965.

121. E. Noëther, *Der Endlichkeitssats der Invarianten endlicher Gruppen*, Math. Ann. **77** (1915), 89–92.

122. T. Oda, *Torus embeddings and applications*, Tata Inst. of Fund. Research, Bombay, Springer-Verlag, Berlin and New York, 1978.

123. F. Pauer, *Normale Einbettungen von* G/U, Math. Ann. **257** (1981), 371–396.

124. V. L. Popov, *Constructive invariant theory*, Young Tableaux and Schur Functions in Algebra and Geometry, (Toruń, 1980), Astérisque, 87–88, Soc. Math. France, Paris, 1981, pp. 303–334.

125. ——, *Constructive invariant theory*, Astérisque, 87-88, Soc. Math. France, Paris, 1981, pp. 303–334.

126. R. W. Richardson, *Affine coset spaces of reductive algebraic groups*, Bull. London Math. Soc. **9** (1977), 38–41.

127. M. Rosenlicht, *Some basic theorems on algebraic groups*, Amer. J. Math. **78** (1956), 401–443.

128. ——, *Some rational questions on algebraic groups*, Ann. of Math. **43** (1957), 25–50.

129. ——, *Toroidal algebraic groups*, Proc. Amer. Math. Soc. **12** (1961), 984–988.

130. ——, *On quotient varieties and the affine embedding of certain homogeneous spaces*, Trans. Amer. Math. Soc. **101** (1961), 211–223.

131. ——, *A remark on quotient spaces*, An. Acad. Brasil. Ciênc. **35** (1963), 487–489.

132. G. Salmon, *Leçons d'algèbre supérieure*, Gauthier-Villars, Paris, 1890.

133. M. Sato and T. Kimura, *A classification of irreducible prehomogeneous vector spaces and their relative invariants*, Nagoya Math. J. **65** (1977), 1–155.

134. I. Schur, *Vorleungen über Invariantentheorie*, Springer-Verlag, Berlin and New York, 1968.

135. G. W. Schwarz, *Representations of simple Lie groups with regular rings of invariants*, Invent. Math. **49** (1978), 167–191.

136. ——, *Representations of simple Lie groups with a free module of covariants*, Invent. Math. **50** (1978), 1–12.

137. J. Sekiguchi and Y. Shimizu, *Simple singularities and infinitesimally symmetric spaces*, Proc. Japan Acad. Ser. A Math. Sci. **57** (1981), 42–46.

138. Séminaire Chevalley: Anneaux de Chow, Secr. Math., E. N. S., Paris, 1958.

139. J.-P. Serre, *Algèbre locale, Multiplicités*, Lecture Notes in Math., vol. 11, Springer-Verlag, Berlin, 1965.

140. F. J. Servedio, *Prehomogeneous vector spaces and varieties*, Trans. Amer. Math. Soc. **176** (1973), 421–444.

141. J. Shah, *A complete moduli space for* K3 *surfaces of degree* 2, Ann. of Math. (2) **112** (1980), 485–510.

142. G. C. Shephard and J. A. Todd, *Finite unitary reflection groups*, Canad. J. Math. **6** (1954), 274–304.

143. T. Shioda, *On the graded ring of invariants of binary octavicts*, Amer. J. Math. **89** (1967), 1022–1046.

144. N. J. A. Sloane, *Error-correcting codes and invariant theory: New applications of the nineteenth-century technique*, Amer. Math. Monthly **84** (1977), 82–107.

145. P. Slodowy, *Simple singularities and simple algebraic groups*, Lecture Notes in Math., vol. 815, Springer-Verlag, Berlin, 1980.

146. T. Springer, *Representations with a free algebra of invariants*, Math. Forschungsinst. Oberwolfach, Tagungsbericht 26/1982, Algebraische Gruppen 20.6 bis 26.6, 1982, p. 7.

147. R. P. Stanley, *Combinatorial reciprocity theorems*, Adv. Math. **14** (1974), 194–253.

148. ——, *Hilbert functions of graded algebras*, Adv. Math. **28** (1978), 57–83.

149. ——, *Invariants of finite groups and their applications to combinatorics*, Bull. Amer. Math. Soc. **1** (1979), 475–511.

150. ——, *Combinatorics and invariant theory*, Proc. Sympos. Pure Math. **34** (1979), 345–355.

151. R. Steinberg, *Conjugacy classes in algebraic groups*, Lecture Notes in Math., vol. 366, Springer-Verlag, Berlin, 1974.

152. H. Sumihiro, *Equivariant completion* I, J. Math. Kyoto Univ. **14** (1974), 1–28.

153. J. J. Sylvester, *Tables of the generating functions and ground forms of the binary quantics of the first ten orders*, Collected Math. Papers, vol. III, Chelsea, 1973, pp. 283–311.

154. Th. Vust, *Sur le type principal d'orbites d'un module rationnel*, Comment. Math. Helv. **49** (1974), 408–416.

155. ——, *Opération de groupes réductifs dans un type de cônes presque homogènes*, Bull. Soc. Math. France **102** (1974), 317–334.

156. K.-i. Watanabe, *Certain invariant subrings are Gorenstein* I, Osaka J. Math. **11** (1974), 1–8.

157. *Young tableaux and Schur functors in algebra and geometry*, Astérisque, 87–88, Soc. Math. France, Paris, 1981.

References for the Appendix

A1. A. E. Veles′ko, *Finiteness theorems in invariant theory of nonconnected semisimple algebraic groups*, preprint, Inst. Mat. Akad. Nauk. BSSR, no. 29(339), 1988, pp. 3–22. (Russian)

A2. E. B. Vinberg, *Effective invariant theory*, Algebra, Izdat. Moskov. Univ., Moscow, 1982, pp. 27–33. (Russian)

A3. E. B. Vinberg and V. L. Popov, *Invariant theory*, Itogi Nauki i Tekhniki: Sovrem. Probl. Mat. Fund. Naprav., vol. 55, VINITI, Moscow, 1989, pp. 137–314; English transl., Encyclopaedia of Math. Sci., vol. 55, Springer-Verlag (to appear).

A4. N. L. Gordeev, *Coranks of elements of linear groups, and complexity of algebras of invariants*, Algebra i Analiz **2** (1990), no. 2, 39–64; English transl. in Leningrad Math. J. **2** (1991), 245–267.

A5. Yu. G. Zarkhin, *Simple linear Lie algebras containing an operator of a small rank*, preprint, Computer Center of NCBI, Acad. Sci. USSR, Pushcin., 1986, pp. 3–15.

A6. D. I. Panyushev, *Regular elements in spaces of linear representations of reductive algebraic groups*, Izv. Akad. Nauk SSSR Ser. Mat. **48** (1984), 411–419; English transl. in Math. USSR-Izv. **24** (1985), 383–390.

A7. ——, *Regular elements in spaces of linear representations* II, Izv. Akad. Nauk SSSR Ser. Mat. **49** (1985), 979–985; English transl. in Math. USSR-Izv. **27** (1986), 279–284.

A8. ——, *Resolution of singularities of affine normal quasihomogeneous* SL_2*-varieties*, Funktsional. Anal. i Prilozhen. **22** (1988), no. 4, 94–95; English transl. in Functional Anal. Appl. **22** (1988) 338–339.

A9. V. L. Popov, *Contractions of actions of reductive algebraic groups*, Mat. Sb. **130** (1986), 310–334; English transl. in Math. USSR-Sb. **58** (1987), 311–336.

A10. A. A. Sukhanov, *Description of observable subgroups of linear algebraic groups*, Mat. Sb. **137** (1988), 90–102; English transl. in Math. USSR-Sb. **65** (1990), 97–108.

A11. D. Bartels, *On nonnormality of affine quasihomogeneous* $\mathrm{SL}(2, \mathbb{C})$*-varieties*, Lecture Notes in Math., vol. 924, Springer-Verlag, Berlin and New York, 1981, pp. 384–399.

A12. ——, *Quasihomogene affine Varietäten für* $\mathrm{SL}(2, \mathbb{C})$, Lecture Notes in Math., vol. 1146, Springer-Verlag, Berlin and New York, 1985, pp. 1–105.

A13. M. Brion, *Spherical varieties: An introduction*, Progr. Math. **80** (1989), 11–26.

A14. W. Bruns, *The Eisenbud-Evans generalized principal theorem and determinantal ideals*, Proc. Amer. Math. Soc. **81** (1981), 19–24.

A15. J. Dixmier, *Sur les invariants des formes binaires*, C. R. Acad. Sci. Paris Ser. I Math. **292** (1981), 987–990.

A16. ——, *Série de Poincaré et systèmes de paramètres pour les invariants des formes binaires*, Acta Sci. Math. **45** (1983), 151–160.

A17. ——, *Sur les invariants du groupe symétrique dans certaines représentations*, II, preprint IHES/M, no. 33 (1989), 1–35.

A18. J. Dixmier, P. Erdös, and J. L. Nicolas, *Sur le nombre d'invariants fondamentaux des formes binaires*, C. R. Acad. Sci. Paris Sér. I Math. **305** (1987), 319–322.

A19. J. Dixmier and D. Lazard, *Le nombre minimum d'invariants fondamentaux pour les formes binaires de degreé* 7, Portugal. Math. **43** (1985–86), 377–392.

A20. F. Grosshans, *The invariants of unipotent radicals of parabolic subgroups*, Invent. Math. **73** (1983), 1–9.

A21. ——, *Hilbert's fourteenth problem for non-reductive groups*, Math. Z. **193** (1986), 95–103.

A22. J. Horvath, *Invariants of certain unipotent groups*, Ph. D. thesis, Yale University, 1987.

A23. ——, *Weight spaces of invariants of certain unipotent group actions*, J. Algebra **126** (1989), 293–299.

A24. ——, *On a problem of Pommerening in invariant theory*, J. Algebra **126** (1989), 300–309.

A25. L. Jauslin-Moser, *Normal embeddings of* SL(2)/Γ, Thèse 2259 Univ. de Genève. Sect. Math., 1987, pp. 1–103.

A26. G. Kempf, *Computing invariants*, Lecture Notes in Math., vol. 1278, Springer-Verlag, Berlin and New York, 1987, pp. 81–94.

A27. F. Knop, *Über die Glattheit von Quotientenabbildungen*, Manuscripta Math. **56** (1986), 419–427.

A28. ——, *Der kanonische Modul eines Invariantenrings*, J. Algebra **127** (1989), 40–54.

A29. ——, *The Luna-Vust theory of spherical embeddings*, Proc. Hyderabad Conference on Algebraic Groups (1989) preprint, 1990.

A30. F. Knop and P. Littelmann, *Der Grad erzeugender Funktionen von Invariantenringen*, Math. Z. **196** (1987), 211–229.

A31. Tan Lin, *The invariant theory of unipotent subgroups of reductive algebraic groups*, Diss. Univ. California at Los Angeles, 1986.

A32. ——, *On the Popov-Pommerening conjecture for groups of type* G_2, Algebras, Groups, and Geometries (to appear).

A33. ——, *On the Popov-Pommerening conjecture for groups of type* G_2, II: *the quasi-closed case* (to appear).

A34. ——, *Some recent developments in the Popov-Pommerening conjecture*, Group Actions, and Invariant Theory (Montreal, PQ, 1988), CMS Conf. Proc., 10, Amer. Math. Soc., Providence, RI, 1989, pp. 207–220.

A35. ——, *On the Popov-Pommerening conjecture for groups of type* A_n, Proc. Amer. Math. Soc. **106** (1989), 611–617.

A36. P. Littelmann, *Koreguläre und äquidimensionale Darstellungen*, J. Algebra **123** (1989), 193–222.

A37. D. Luna, SL(2)*-embeddings*, Tagung über Transformationsgruppen und Invariantentheorie (Bonn, 22–28 Sept. 1988).

A38. D. Luna and T. Vust, *Plongements d'espaces homogènes*, Comment. Math. Helv. **58** (1983), 186–245.

A39. C. L. Mallows and N. J. A. Sloane, *On the invariants of a linear group of order* 336, Proc. Cambridge Philos. Soc. **74** (1973), 435–440.

A40. F. Pauer, *Normale Eibettungen von sphärischen homogenen Räumen*, DMV Sem. **13** (1989), 145–155.

A41. K. Pommerening, *Invarianten unipotenter Gruppen*, Math. Z. **176** (1981), 359–374.

A42. ——, *Invariants of unipotent groups (a survey)*, Lecture Notes in Math., vol. 1278, Springer-Verlag, Berlin and New York, 1987, pp. 8–17.

A43. ——, *Ordered sets with the standardizing property and straightening laws for algebras of invariants*, Adv. Math. **63** (1987), 271–290.

A44. V. L. Popov, *Homological dimension of algebras of invariants*, J. Reine Angew. Math. **341** (1983), 157–173.

A45. ——, *Modern developments in invariant theory*, Proc. Internat. Congr. Math. (Berkeley, California, 1986), vol. 1, Amer. Math. Soc., Providence, RI, 1987, pp. 394–406.

A46. M. Van den Bergh, *Cohen–Macaulayness of modules of covariants*, Univ. instelling, Antwerpen, 1990, preprint 90–06, pp. 1–33.

A47. D. L. Wehlau, *Some recent results on the Popov conjecture*, Group Actions and Invariant Theory (Montreal, PQ, 1988), CMS Conf. Proc., 10, Amer. Math. Soc., Providence, RI, 1989, pp. 221–228.

A48. ——, *Equidimensionality and the Popov conjecture*, Ph.D. Thesis, Brandeis Univ., 1989.

A49. R. Weitzenböck, *Über die Invarianten von linearen Gruppen*, Acta Math. **58** (1932), 231–293.

Subject Index

Recent Titles in This Series

(Continued from the front of this publication)

97 **Itiro Tamura,** Topology of foliations: An introduction, 1992
96 **A. I. Markushevich,** Introduction to the classical theory of Abelian functions, 1992
95 **Guangchang Dong,** Nonlinear partial differential equations of second order, 1991
94 **Yu. S. Il'yashenko,** Finiteness theorems for limit cycles, 1991
93 **A. T. Fomenko and A. A. Tuzhilin,** Elements of the geometry and topology of minimal surfaces in three-dimensional space, 1991
92 **E. M. Nikishin and V. N. Sorokin,** Rational approximations and orthogonality, 1991
91 **Mamoru Mimura and Hirosi Toda,** Topology of Lie groups, I and II, 1991
90 **S. L. Sobolev,** Some applications of functional analysis in mathematical physics, third edition, 1991
89 **Valeriĭ V. Kozlov and Dmitriĭ V. Treshchëv,** Billiards: A genetic introduction to the dynamics of systems with impacts, 1991
88 **A. G. Khovanskiĭ,** Fewnomials, 1991
87 **Aleksandr Robertovich Kemer,** Ideals of identities of associative algebras, 1991
86 **V. M. Kadets and M. I. Kadets,** Rearrangements of series in Banach spaces, 1991
85 **Mikio Ise and Masaru Takeuchi,** Lie groups I, II, 1991
84 **Đáo Trọng Thi and A. T. Fomenko,** Minimal surfaces, stratified multivarifolds, and the Plateau problem, 1991
83 **N. I. Portenko,** Generalized diffusion processes, 1990
82 **Yasutaka Sibuya,** Linear differential equations in the complex domain: Problems of analytic continuation, 1990
81 **I. M. Gelfand and S. G. Gindikin, Editors,** Mathematical problems of tomography, 1990
80 **Junjiro Noguchi and Takushiro Ochiai,** Geometric function theory in several complex variables, 1990
79 **N. I. Akhiezer,** Elements of the theory of elliptic functions, 1990
78 **A. V. Skorokhod,** Asymptotic methods of the theory of stochastic differential equations, 1989
77 **V. M. Filippov,** Variational principles for nonpotential operators, 1989
76 **Phillip A. Griffiths,** Introduction to algebraic curves, 1989
75 **B. S. Kashin and A. A. Saakyan,** Orthogonal series, 1989
74 **V. I. Yudovich,** The linearization method in hydrodynamical stability theory, 1989
73 **Yu. G. Reshetnyak,** Space mappings with bounded distortion, 1989
72 **A. V. Pogorelev,** Bendings of surfaces and stability of shells, 1988
71 **A. S. Markus,** Introduction to the spectral theory of polynomial operator pencils, 1988
70 **N. I. Akhiezer,** Lectures on integral transforms, 1988
69 **V. N. Saliĭ,** Lattices with unique complements, 1988
68 **A. G. Postnikov,** Introduction to analytic number theory, 1988
67 **A. G. Dragalin,** Mathematical intuitionism: Introduction to proof theory, 1988
66 **Ye Yan-Qian,** Theory of limit cycles, 1986
65 **V. M. Zolotarev,** One-dimensional stable distributions, 1986
64 **M. M. Lavrent'ev, V. G. Romanov, and S. P. Shishat·skiĭ,** Ill-posed problems of mathematical physics and analysis, 1986
63 **Yu. M. Berezanskiĭ,** Selfadjoint operators in spaces of functions of infinitely many variables, 1986
62 **S. L. Krushkal', B. N. Apanasov, and N. A. Gusevskiĭ,** Kleinian groups and uniformization in examples and problems, 1986

(See the AMS catalog for earlier titles)